COMPUTATIONAL MATHEMATICS

Models, Methods, and Analysis
with MATLAB and MPI

COMPUTATIONAL MATHEMATICS

MATHEMATICS

Models, Methods, and Analysis
with MATLAB and MPI

ROBERT E. WHITE

CHAPMAN & HALL/CRC

A CRC Press Company
Boca Raton   London   New York   Washington, D.C.

Library of Congress Cataloging-in-Publication Data

White, R. E. (Robert E.)
    Computational mathematics : models, methods, and analysis with MATLAB and MPI /
Robert E. White.
        p. cm.
    Includes bibliographical references and index.
    ISBN 1-58488-364-2 (alk. paper)
        1. Numerical analysis. 2. MATLAB. 3. Computer interfaces. 4. Parallel programming
(Computer science) I. Title.

QA297.W495 2003
519.4—dc21   2003055207

Visit the CRC Press Web site at www.crcpress.com

Contents

List of Figures

List of Tables

Preface

This book evolved from the need to migrate computational science into undergraduate education. It is intended for students who have had basic physics, programming, matrices and multivariable calculus.

The choice of topics in the book has been influenced by the Undergraduate Computational Engineering and Science Project (a United States Department of Energy funded effort), which was a series of meetings during the 1990s. These meetings focused on the nature and content for computational science undergraduate education. They were attended by a diverse group of science and engineering teachers and professionals, and the continuation of some of these activities can be found at the Krell Institute, http://www.krellinst.org. Variations of Chapters 1-4 and 6 have been taught at North Carolina State University in fall semesters since 1992. The other four chapters were developed in 2002 and taught in the 2002-03 academic year.

The department of mathematics at North Carolina State University has given me the time to focus on the challenge of introducing computational science materials into the undergraduate curriculum. The North Carolina Supercomputing Center, http://www.ncsc.org, has provided the students with valuable tutorials and computer time on supercomputers. Many students have made important suggestions, and Carol Cox Benzi contributed some course materials with the initial use of MATLAB®. MATLAB is a registered trademark of The MathWorks, Inc. For product information, please contact:

The MathWorks, Inc.
3 Apple Hill Drive
Natick, MA 01760-2098 USA
Tel: 508-647-7000
Fax: 508-647-7001
E-mail: info@mathworks.com
Web: www.mathworks.com <http://www.mathworks.com/>.

I thank my close friends who have listened to me talk about this effort, and especially Liz White who has endured the whole process with me.

Bob White, July 1, 2003

Introduction

Computational science is a blend of applications, computations and mathematics. It is a mode of scientific investigation that supplements the traditional laboratory and theoretical methods of acquiring knowledge. This is done by formulating mathematical models whose solutions are approximated by computer simulations. By making a sequence of adjustments to the model and subsequent computations one can gain some insights into the application area under consideration. This text attempts to illustrate this process as a method for scientific investigation. Each section of the first six chapters is motivated by a particular application, discrete or continuous model, numerical method, computer implementation and an assessment of what has been done.

Applications include heat diffusion to cooling fins and solar energy storage, pollutant transfer in streams and lakes, models of vector and multiprocessing computers, ideal and porous fluid flows, deformed membranes, epidemic models with dispersion, image restoration and value of American put option contracts. The models are initially introduced as discrete in time and space, and this allows for an early introduction to partial differential equations. The discrete models have the form of matrix products or linear and nonlinear systems. Methods include sparse matrix iteration with stability constraints, sparse matrix solutions via variation on Gauss elimination, successive over-relaxation, conjugate gradient, and minimum residual methods. Picard and Newton methods are used to approximate the solution to nonlinear systems.

Most sections in the first five chapters have MATLAB$^{\circledR}$ codes; see [14] for the very affordable current student version of MATLAB. They are intended to be studied and not used as a "black box." The MATLAB codes should be used as a first step towards more sophisticated numerical modeling. These codes do provide a learning by doing environment. The exercises at the end of each section have three categories: routine computations, variation of models, and mathematical analysis. The last four chapters focus on multiprocessing algorithms, which are implemented using message passing interface, MPI; see [17] for information about building your own multiprocessor via free "NPACI Rocks" cluster software. These chapters have elementary Fortran 9x codes to illustrate the basic MPI subroutines, and the applications of the previous chapters are revisited from a parallel implementation perspective.

At North Carolina State University Chapters 1-4 are covered in 26 75-minute lectures. Routine homework problems are assigned, and two projects are required, which can be chosen from topics in Chapters 1-5, related courses or work experiences. This forms a semester course on numerical modeling using partial differential equations. Chapter 6 on high performance computing can be studied after Chapter 1 so as to enable the student, early in the semester, to become familiar with a high performance computing environment. Other course possibilities include: a semester course with an emphasis on mathematical analysis using Chapters 1-3, 8 and 9, a semester course with a focus on parallel computation using Chapters 1 and 6-9 or a year course using Chapters 1-9.

This text is not meant to replace traditional texts on numerical analysis, matrix algebra and partial differential equations. It does develop topics in these areas as is needed and also includes modeling and computation, and so there is more breadth and less depth in these topics. One important component of computational science is parameter identification and model validation, and this requires a physical laboratory to take data from experiments. In this text model assessments have been restricted to the variation of model parameters, model evolution and mathematical analysis. More penetrating expertise in various aspects of computational science should be acquired in subsequent courses and work experiences.

Related computational mathematics education material at the first and second year undergraduate level can be found at the Shodor Education Foundation, whose founder is Robert M. Panoff, web site [22] and in Zachary's book on programming [29]. Two general references for modeling are the undergraduate mathematics journal [25] and Beltrami's book on modeling for society and biology [2]. Both of these have a variety of models, but often there are no computer implemenations. So they are a good source of potential computing projects. The book by Landau and Paez [13] has number of computational physics models, which are at about the same level as this book. Slightly more advanced numerical analysis references are by Fosdick, Jessup, Schauble and Domik [7] and Heath [10].

The computer codes and updates for this book can be found at the web site:

$$\text{http://www4.ncsu.edu/~white.}$$

The computer codes are mostly in MATLAB for Chapters 1-5, and in Fortran 9x for most of the MPI codes in Chapters 6-9. The choice of Fortran 9x is the author's personal preference as the array operations are similar to those in MATLAB. However, the above web site and the web site associated with Pacheco's book [21] do have C versions of these and related MPI codes. The web site for this book is expected to evolve and also has links to sequences of heat and pollution transfer images, book updates and new reference materials.

Chapter 1

Discrete Time-Space Models

The first three sections introduce diffusion of heat in one direction. This is an example of model evolution with the simplest model being for the temperature of a well-stirred liquid where the temperature does not vary with space. The model is then enhanced by allowing the mass to have different temperatures in different locations. Because heat flows from hot to cold regions, the subsequent model will be more complicated. In Section 1.4 a similar model is considered, and the application will be to the prediction of the pollutant concentration in a stream resulting from a source of pollution up stream. Both of these models are discrete versions of the continuous model that are partial differential equations. Section 1.5 indicates how these models can be extended to heat and mass transfer in two directions, which is discussed in more detail in Chapters 3 and 4. In the last section variations of the mean value theorem are used to estimate the errors made by replacing the continuous model by a discrete model. Additional introductory materials can be found in G. D. Smith [23], and in R. L. Burden and J. D. Faires [4].

1.1 Newton Cooling Models

1.1.1 Introduction

Many quantities change as time progresses such as money in a savings account or the temperature of a refreshing drink or any cooling mass. Here we will be interested in making predictions about such changing quantities. A simple mathematical model has the form $u^+ = au + b$ where a and b are given real numbers, u is the present amount and u^+ is the next amount. This calculation is usually repeated a number of times and is a simple example of an of algorithm. A computer is used to do a large number calculations.

Computers use a finite subset of the rational numbers (a ratio of two integers) to approximate any real number. This set of numbers may depend on the computer being used. However, they do have the same general form and are called floating point numbers. Any real number x can be represented by an infinite decimal expansion $x = \pm(.x_1 \cdots x_d \cdots)10^e$, and by truncating this we can define the chopped floating point numbers.

Let x be any real number and denote a *floating point number* by

$$\begin{aligned} fl(x) &= \pm.x_1 \cdots x_d 10^e \\ &= \pm(x_1/10 + \cdots + x_d/10^d)10^e. \end{aligned}$$

This is a floating point number with base equal to 10 where x_1 is not equal to zero, x_i are integers between 0 and 9, the exponent e is also a bounded integer and d is an integer called the precision of the floating point system. Associated with each real number, x, and its floating point approximate number, $fl(x)$, is the *floating point error*, $fl(x) - x$. In general, this error decreases as the *precision*, d, increases. Each computer calculation has some floating point or roundoff error. Moreover, as additional calculations are done, there is an accumulation of these roundoff errors.

Example. Let $x = -1.5378$ and $fl(x) = -0.154 \ 10^1$ where $d = 3$. The roundoff error is

$$fl(x) - x = -.0022.$$

The error will accumulate with any further operations containing $fl(x)$, for example, $fl(x)^2 = .237 \ 10^{-1}$ and

$$fl(x)^2 - x^2 = 2.37 - 2.36482884 = .00517116.$$

Repeated calculations using floating point numbers can accumulate significant roundoff errors.

1.1.2   Applied Area

Consider the cooling of a well stirred liquid so that the temperature does not depend on space. Here we want to predict the temperature of the liquid based on some initial observations. Newton's law of cooling is based on the observation that for small changes of time, h, the change in the temperature is nearly equal to the product of the constant c, the h and the difference in the room temperature and the present temperature of the coffee. Consider the following quantities: u_k equals the temperature of a well stirred cup of coffee at time t_k, u_{sur} equals the surrounding room temperature, and c measures the insulation ability of the cup and is a positive constant. The *discrete form of Newton's law of cooling* is

$$\begin{aligned} u_{k+1} - u_k &= ch(u_{sur} - u_k) \\ u_{k+1} &= (1 - ch)u_k + ch \ u_{sur} \\ &= au_k + b \text{ where } a = 1 - ch \text{ and } b = ch \ u_{sur}. \end{aligned}$$

The long run solution should be the room temperature, that is, u_k should converge to u_{sur} as k increases. Moreover, when the room temperature is constant, then u_k should converge monotonically to the room temperature. This does happen if we impose the constraint

$$0 < a = 1 - ch,$$

called a *stability condition*, on the time step h. Since both c and h are positive, $a < 1$.

1.1.3 Model

The model in this case appears to be very simple. It consists of three constants u_0, a, b and the formula

$$u_{k+1} = au_k + b \qquad (1.1.1)$$

The formula must be used repeatedly, but with different u_k being put into the right side. Often a and b are derived from formulating how u_k changes as k increases (k reflects the time step). The change in the amount u_k is often modeled by $du_k + b$

$$u_{k+1} - u_k = du_k + b$$

where $d = a - 1$. The model given in (1.1.1) is called a *first order finite difference* model for the sequence of numbers u_{k+1}. Later we will generalize this to a sequence of column vectors where a will be replaced by a square matrix.

1.1.4 Method

The "iterative" calculation of (1.1.1) is the most common approach to solving (1.1.1). For example, if $a = \frac{1}{2}, b = 2$ and $u_0 = 10$, then

$$u_1 = \frac{1}{2} \, 10 + 2 = 7.0$$

$$u_2 = \frac{1}{2} \, 7 + 2 = 5.5$$

$$u_3 = \frac{1}{2} \, 5.5 + 2 = 4.75$$

$$u_4 = \frac{1}{2} \, 4.75 + 2 = 4.375$$

If one needs to compute u_{k+1} for large k, this can get a little tiresome. On the other hand, if the calculations are being done with a computer, then the floating point errors may generate significant accumulation errors.

An alternative method is to use the following "telescoping" calculation and the geometric summation. Recall the geometric summation

$$1 + r + r^2 + \cdots + r^k \text{ and } (1 + r + r^2 + \cdots + r^k)(1 - r) = 1 - r^{k+1}$$

Or, for r not equal to 1

$$(1 + r + r^2 + \cdots + r^k) = (1 - r^{k+1})/(1 - r).$$

Consequently, if $|r| < 1$, then

$$1 + r + r^2 + \cdots + r^k + \cdots = 1/(1 - r)$$

is a convergent *geometric series*.

In (1.1.1) we can compute u_k by decreasing k by 1 so that $u_k = au_{k-1} + b$. Put this into (1.1.1) and repeat the substitution to get

$$
\begin{aligned}
u_{k+1} &= a(au_{k-1} + b) + b \\
&= a^2 u_{k-1} + ab + b \\
&= a^2(au_{k-2} + b) + ab + b \\
&= a^3 u_{k-2} + a^2 b + ab + b \\
&\vdots \\
&= a^{k+1} u_0 + b(a^k + \cdots + a^2 + a + 1) \\
&= a^{k+1} u_0 + b(1 - a^{k+1})/(1 - a) \\
&= a^{k+1}(u_0 - b/(1 - a)) + b/(1 - a). \quad (1.1.2)
\end{aligned}
$$

The error for the *steady state solution*, $b/(1 - a)$, will be small if $|a|$ is small, or k is large, or the initial guess u_0 is close to the steady state solution. A generalization of this will be studied in Section 2.5.

Theorem 1.1.1 *(Steady State Theorem) If a is not equal to 1, then the solution of (1.1.1) has the form given in (1.1.2). Moreover, if $|a| < 1$, then the solution of (1.1.1) will converge to the steady state solution $u = au + b$, that is, $u = b/(1 - a)$. More precisely, the error is*

$$u_{k+1} - u = a^{k+1}(u_0 - b/(1 - a)).$$

Example. Let $a = 1/2, b = 2, u_0 = 10$ and $k = 3$. Then (1.1.2) gives

$$u_{3+1} = (1/2)^4(10 - 2/(1 - 1/2)) + 2/(1 - 1/2) = 6/16 + 4 = 4.375.$$

The steady state solution is $u = 2/(1 - \frac{1}{2}) = 4$ and the error for $k = 3$ is

$$u_4 - u = 4.375 - 4 = (\frac{1}{2})^4(10 - 4).$$

1.1.5   Implementation

The reader should be familiar with the information in MATLAB's tutorial. The input segment of the MATLAB code fofdh.m is done in lines 1-12, the execution is done in lines 16-19, and the output is done in line 20. In the following m-file

t is the time array whose first entry is the initial time. The array y stores the approximate temperature values whose first entry is the initial temperature. The value of c is based on a second observed temperature, y_obser, at time equal to h_obser. The value of c is calculated in line 10. Once a and b have been computed, the algorithm is executed by the for loop in lines 16-19. Since the time step $h = 1$, $n = 300$ will give an approximation of the temperature over the time interval from 0 to 300. If the time step were to be changed from 1 to 5, then we could change n from 300 to 60 and still have an approximation of the temperature over the same time interval. Within the for loop we could look at the time and temperature arrays by omitting the semicolon at the end of the lines 17 and 18. It is easier to examine the graph of approximate temperature versus time, which is generated by the MATLAB command plot(t,y).

MATLAB Code fofdh.m

```
1.    % This code is for the first order finite difference algorithm.
2.    % It is applied to Newton's law of cooling model.
3.    clear;
4.    t(1) = 0;      % initial time
5.    y(1) = 200.;     % initial temperature
6.    h = 1;      % time step
7.    n = 300;      % number of time steps of length h
8.    y_obser = 190;      % observed temperature at time h_obser
9.    h_obser = 5;
10.   c = ((y_obser - y(1))/h_obser)/(70 - y(1))
11.   a = 1 - c*h
12.   b = c*h*70
13.   %
14.   % Execute the FOFD Algorithm
15.   %
16.   for k = 1:n
17.       y(k+1) = a*y(k) + b;
18.       t(k+1) = t(k) + h;
19.   end
20.   plot(t,y)
```

An application to heat transfer is as follows. Consider a cup of coffee, which is initially at 200 degrees and is in a room with temperature equal to 70, and after 5 minutes it cools to 190 degrees. By using $h = h\_obser = 5$, $u_0 = 200$ and $u_1 = u\_obser = 190$, we compute from (1.1.1) that $c = 1/65$. The first calculation is for this c and $h = 5$ so that $a = 1 - ch = 60/65$ and $b = ch70 = 350/65$. Figure 1.1.1 indicates the expected monotonic decrease to the steady state room temperature, $u_{sur} = 70$.

The next calculation is for a larger $c = 2/13$, which is computed from a new second observed temperature of $u\_obser = 100$ after $h\_obser = 5$ minutes. In this case for larger time step $h = 10$ so that $a = 1 - (2/13)10 = -7/13$ and $b = ch70 = (2/13)10 \, 70 = 1400/13$. In Figure 1.1.2 notice that the

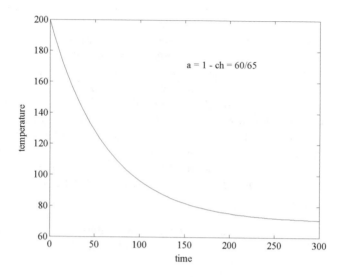

Figure 1.1.1: Temperature versus Time

computed solution no longer is monotonic, but it does converge to the steady state solution.

The model continues to degrade as the magnitude of a increases. In the Figure 1.1.3 the computed solution oscillates and blows up! This is consistent with formula (1.1.2). Here we kept the same c, but let the step size increase to $h = 15$ and in this case $a = 1 - (2/13)15 = -17/13$ and $b = ch70 = (2/13)1050 = 2100/13$. The vertical axis has units multiplied by 10^4.

1.1.6 Assessment

Models of savings plans or loans are discrete in the sense that changes only occur at the end of each month. In the case of the heat transfer problem, the formula for the temperature at the next time step is only an approximation, which gets better as the time step h decreases. The cooling process is continuous because the temperature changes at every instant in time. We have used a discrete model of this, and it seems to give good predictions provided the time step is suitably small. Moreover there are other modes of transferring heat such as diffusion and radiation.

There may be significant accumulation of roundoff error. On a computer (1.1.1) is done with floating point numbers, and at each step there is some new roundoff error $\overline{R}_{k+1}$. Let $U_0 = fl(u_0)$, $A = fl(a)$ and $B = fl(b)$ so that

$$U_{k+1} = AU_k + B + \overline{R}_{k+1}. \tag{1.1.3}$$

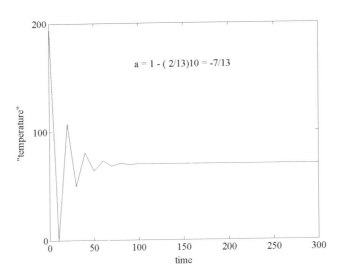

Figure 1.1.2: Steady State Temperature

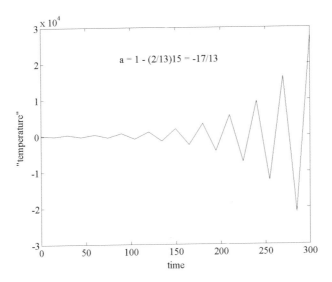

Figure 1.1.3: Unstable Computation

Next, we want to estimate the

$$accumulation\ error = U_{k+1} - u_{k+1}$$

under the assumption that the roundoff errors are uniformly bounded

$$|\overline{R}_{k+1}| \le R < \infty.$$

For ease of notation, we will assume the roundoff errors associated with a and b have been put into the R_{k+1} so that $U_{k+1} = aU_k + b + R_{k+1}$. Subtract (1.1.1) and this variation of (1.1.3) to get

$$
\begin{aligned}
U_{k+1} - u_{k+1} &= a(U_k - u_k) + R_{k+1} && (1.1.4) \\
&= a[a(U_{k-1} - u_{k-1}) + R_k] + R_{k+1} \\
&= a^2(U_{k-1} - u_{k-1}) + aR_k + R_{k+1} \\
&\ \ \vdots \\
&= a^{k+1}(U_0 - u_0) + a^k R_1 + \cdots + R_{k+1}
\end{aligned}
$$

Now let $r = |a|$ and R be the uniform bound on the roundoff errors. Use the geometric summation and the triangle inequality to get

$$|U_{k+1} - u_{k+1}| \le r^{k+1}|U_0 - u_0| + R(r^{k+1} - 1)/(r - 1). \qquad (1.1.5)$$

Either r is less than one, or greater, or equal to one. An analysis of (1.1.4) and (1.1.5) immediately yields the next theorem.

Theorem 1.1.2 (*Accumulation Error Theorem*) *Consider the first order finite difference algorithm. If* $|a| < 1$ *and the roundoff errors are uniformly bounded by* R, *then the accumulation error is uniformly bounded. Moreover, if the roundoff errors decrease uniformly, then the accumulation error decreases.*

1.1.7   Exercises

1. Using fofdh.m duplicate the calculations in Figures 1.1.1-1.1.3.
2. Execute fofdh.m four times for $c = 1/65$, variable $h = 64, 32, 16, 8$ with n = 5, 10, 20 and 40, respectively. Compare the four curves by placing them on the same graph; this can be done by executing the MATLAB command "hold on" after the first execution of fofdh.m
3. Execute fofdh.m five times with $h = 1$, variable $c = 8/65, 4/65, 2/65, 1/65$, and .5/65, and $n = 300$. Compare the five curves by placing them on the same graph; this can be done by executing the MATLAB command "hold on" after the first execution of fofdh.m
4. Consider the application to Newton's discrete law of cooling. Use (1.1.2) to show that if $hc < 1$, then u_{k+1} converges to the room temperature.
5. Modify the model used in Figure 1.1.1 to account for a room temperature that starts at 70 and increases at a constant rate equal to 1 degree every 5

minutes. Use the $c = 1/65$ and $h = 1$. Compare the new curve with Figure 1.1.1.

6. We wish to calculate the amount of a *savings plan* for any month, k, given a fixed interest rate, r, compounded monthly. Denote these quantities as follows: u_k is the amount in an account at month k, r equals the interest rate compounded monthly, and d equals the monthly deposit. The amount at the end of the next month will be the old amount plus the interest on the old amount plus the deposit. In terms of the above variables this is with $a = 1 + r/12$ and $b = d$

$$
\begin{aligned}
u_{k+1} &= u_k + u_k\, r/12 + d \\
&= au_k + b.
\end{aligned}
$$

(a). Use (1.1.2) to determine the amount in the account by depositing $100 each month in an account, which gets 12% compounded monthly, and over time intervals of 30 and 40 years (360 and 480 months).

(b). Use a modified version of fofdh.m to calculate and graph the amounts in the account from 0 to 40 years.

7. Show (1.1.5) follows from (1.1.4).

8. Prove the second part of the accumulation error theorem.

1.2 Heat Diffusion in a Wire

1.2.1 Introduction

In this section we consider heat conduction in a thin electrical wire, which is thermally insulated on its surface. The model of the temperature has the form $u^{k+1} = Au^k + b$ where u^k is a column vector whose components are temperatures for the previous time step, $t = k\Delta t$, at various positions within the wire. The square matrix will determine how heat flows from warm regions to cooler regions within the wire. In general, the matrix A can be extremely large, but it will also have a special structure with many more zeros than nonzero components.

1.2.2 Applied Area

In this section we present a second model of heat transfer. In our first model we considered heat transfer via a discrete version of Newton's law of cooling which involves temperature as only a discrete function of time. That is, we assumed the mass was uniformly heated with respect to space. In this section we allow the temperature to be a function of both discrete time and discrete space.

The model for the diffusion of heat in space is based on empirical observations. The *discrete Fourier heat law* in one direction says that

      (a). heat flows from hot to cold,

      (b). the change in heat is proportional to the
          cross-sectional area,

change in time and

(change in temperature)/(change in space).

The last term is a good approximation provided the change in space is small, and in this case one can use the derivative of the temperature with respect to the single direction. The proportionality constant, K, is called the *thermal conductivity*. The K varies with the particular material and with the temperature. Here we will assume the temperature varies over a smaller range so that K is approximately a constant. If there is more than one direction, then we must replace the approximation of the derivative in one direction by the directional derivative of the temperature normal to the surface.

Fourier Heat Law. Heat flows from hot to cold, and the amount of heat transfer through a small surface area A is proportional to the product of A, the change in time and the directional derivative of the temperature in the direction normal to the surface.

Consider a thin wire so that the most significant diffusion is in one direction, x. The wire will have a current going through it so that there is a source of heat, f, which is from the electrical resistance of the wire. The f has units of (heat)/(volume time). Assume the ends of the wire are kept at zero temperature, and the initial temperature is also zero. The goal is to be able to predict the temperature inside the wire for any future time and space location.

1.2.3   Model

In order to develop a model to do temperature prediction, we will discretize both space and time and let $u(ih, k\Delta t)$ be approximated by u_i^k where $\Delta t = T/maxk, h = L/n$ and L is the length of the wire. The model will have the general form

$$\text{change in heat content} \quad \approx \quad \text{(heat from the source)}$$
$$+\text{(heat diffusion from the right)}$$
$$+\text{(heat diffusion from the left)}.$$

This is depicted in the Figure 1.2.1 where the time step has not been indicated. For time on the right side we can choose either $k\Delta t$ or $(k+1)\Delta t$. Presently, we will choose $k\Delta t$, which will eventually result in the matrix version of the first order finite difference method.

The heat diffusing in the right face (when $(u_{i+1}^k - u_i^k)/h > 0$) is

$$A \, \Delta t \, K(u_{i+1}^k - u_i^k)/h.$$

The heat diffusing out the left face (when $(u_i^k - u_{i-1}^k)/h > 0$) is

$$A \, \Delta t \, K(u_i^k - u_{i-1}^k)/h.$$

Therefore, the heat from diffusion is

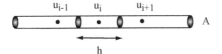

u_{i-1} u_i u_{i+1}

A

h

h is the change in x.
A is the cross-sectional area.
u_i is the approximate temperature.

Figure 1.2.1: Diffusion in a Wire

$$A \, \Delta t \, K (u_{i+1}^k - u_i^k)/h - A \, \Delta t \, K (u_i^k - u_{i-1}^k)/h.$$

The heat from the source is

$$Ah \, \Delta t \, f.$$

The heat content of the volume Ah at time $k\Delta t$ is

$$\rho c u_i^k \, Ah$$

where ρ is the *density* and c is the *specific heat*. By combining these we have the following approximation of the change in the heat content for the small volume Ah:

$$\rho c u_i^{k+1} Ah - \rho c u_i^k \, Ah = Ah \, \Delta t \, f + A \, \Delta t \, K (u_{i+1}^k - u_i^k)/h - A \, \Delta t \, K (u_i^k - u_{i-1}^k)/h.$$

Now, divide by ρcAh, define $\alpha = (K/\rho c)(\Delta t/h^2)$ and explicitly solve for u_i^{k+1}.

Explicit Finite Difference Model for Heat Diffusion.

$$u_i^{k+1} = (\Delta t/\rho c)f + \alpha(u_{i+1}^k + u_{i-1}^k) + (1 - 2\alpha)u_i^k \qquad (1.2.1)$$

$$\text{for } i = 1, ..., n - 1 \text{ and } k = 0, ..., maxk - 1,$$

$$u_i^0 = 0 \text{ for } i = 1, ..., n - 1 \qquad (1.2.2)$$
$$u_0^k = u_n^k = 0 \text{ for } k = 1, ..., maxk. \qquad (1.2.3)$$

Equation (1.2.2) is the initial temperature set equal to zero, and (1.2.3) is the temperature at the left and right ends set equal to zero. Equation (1.2.1) may be put into the matrix version of the first order finite difference method. For example, if the wire is divided into four equal parts, then $n = 4$ and (1.2.1) may be written as three scalar equations for the unknowns u_1^{k+1}, u_2^{k+1} and u_3^{k+1} :

$$u_1^{k+1} = (\Delta t/\rho c)f + \alpha(u_2^k + 0) + (1 - 2\alpha)u_1^k$$
$$u_2^{k+1} = (\Delta t/\rho c)f + \alpha(u_3^k + u_1^k) + (1 - 2\alpha)u_2^k$$
$$u_3^{k+1} = (\Delta t/\rho c)f + \alpha(0 + u_2^k) + (1 - 2\alpha)u_3^k.$$

These three scalar equations can be written as one 3D vector equation

$$u^{k+1} = Au^k + b \text{ where}$$

$$u^k = \begin{bmatrix} u_1^k \\ u_2^k \\ u_3^k \end{bmatrix}, \quad b = (\Delta t/\rho c)f \begin{bmatrix} 1 \\ 1 \\ 1 \end{bmatrix} \text{ and}$$

$$A = \begin{bmatrix} 1-2\alpha & \alpha & 0 \\ \alpha & 1-2\alpha & \alpha \\ 0 & \alpha & 1-2\alpha \end{bmatrix}.$$

An extremely important restriction on the time step Δt is required to make sure the algorithm is stable in the same sense as in Section 1.1 . For example, consider the case $n = 2$ where the above is a single equation, and we have the simplest first order finite difference model. Here $a = 1-2\alpha$ and we must require $a = 1 - 2\alpha < 1$. If $a = 1 - 2\alpha > 0$ and $\alpha > 0$, then this condition will hold. If n is larger than 2, this simple condition will imply that the matrix products A^k will converge to the zero matrix. This will imply there are no blowups provided the source term f is bounded. The illustration of the stability condition and an analysis will be presented in Section 2.5.

Stability Condition for (1.2.1).

$$1 - 2\alpha > 0 \text{ and } \alpha = (K/\rho c)(\Delta t/h^2) > 0.$$

Example. Let $L = c = \rho = 1.0, n = 4$ so that $h = 1/4$, and $K = .001$. Then $\alpha = (K/\rho c)(\Delta t/h^2) = (.001)\Delta t16$ and so that $1 - 2(K/\rho c)(\Delta t/h^2) = 1 - .032\Delta t > 0$. Note if n increases to 20, then the constraint on the time step will significantly change.

1.2.4   Method

The numbers u_i^{k+1} generated by equations (1.2.1)-(1.2.3) are hopefully good approximations for the temperature at $x = i\Delta x$ and $t = (k + 1)\Delta t$. The temperature is often denoted by the function $u(x, t)$. In computer code u_i^{k+1} will be stored in a two dimensional array, which is also denoted by u but with integer indices so that $u_i^{k+1} = u(i, k + 1) \approx u(i\Delta x, (k+1)\Delta t) = $ temperature function. In order to compute all u_i^{k+1}, which we will henceforth denote by $u(i, k + 1)$ with both i and k shifted up by one, we must use a nested loop where the i-loop (space) is the inner loop and the k-loop (time) is the outer loop. This is illustrated in the Figure 1.2.2 by the dependency of $u(i, k + 1)$ on the three previously computed $u(i - 1, k)$, $u(i, k)$ and $u(i + 1, k)$. In Figure 1.2.2 the initial values in (1.2.2) are given on the bottom of the grid, and the boundary conditions in (1.2.3) are on the left and right of the grid.

1.2.5   Implementation

The implementation in the MATLAB code heat.m of the above model for temperature that depends on both space and time has nested loops where the outer

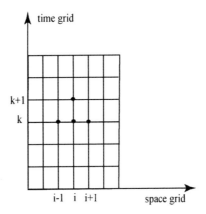

Figure 1.2.2: Time-Space Grid

loop is for discrete time and the inner loop is for discrete space. These loops are given in lines 29-33. Lines 1-25 contain the input data. The initial temperature data is given in the single i-loop in lines 17-20, and the left and right boundary data are given in the single k-loop in lines 21-25. Lines 34-37 contain the output data in the form of a surface plot for the temperature.

MATLAB Code heat.m

```
1.     % This code models heat diffusion in a thin wire.
2.     % It executes the explicit finite difference method.
3.     clear;
4.     L = 1.0;        % length of the wire
5.     T = 150.;       % final time
6.     maxk = 30;      % number of time steps
7.     dt = T/maxk;
8.     n = 10.;        % number of space steps
9.     dx = L/n;
10.    b = dt/(dx*dx);
11.    cond = .001;    % thermal conductivity
12.    spheat = 1.0;   % specific heat
13.    rho = 1.;       % density
14.    a = cond/(spheat*rho);
15.    alpha = a*b;
16.    f = 1.;         % internal heat source
17.    for i = 1:n+1   % initial temperature
18.        x(i) =(i-1)*dx;
19.        u(i,1) =sin(pi*x(i));
20.    end
21.    for k=1:maxk+1  % boundary temperature
```

```
22.          u(1,k) = 0.;
23.          u(n+1,k) = 0.;
24.          time(k) = (k-1)*dt;
25.     end
26.     %
27.     % Execute the explicit method using nested loops.
28.     %
29.     for k=1:maxk       % time loop
30.         for i=2:n;     % space loop
31.             u(i,k+1) = f*dt/(spheat*rho)
                          + (1 - 2*alpha)*u(i,k)
                          + alpha*(u(i-1,k) + u(i+1,k));
32.         end
33.     end
34.     mesh(x,time,u')
35.     xlabel('x')
36.     ylabel('time')
37.     zlabel('temperature')
```

The first calculation given by Figure 1.2.3 is a result of the execution of heat.m with the parameters as listed in the code. The space steps are .1 and go in the right direction, and the time steps are 5 and go in the left direction. The temperature is plotted in the vertical direction, and it increases as time increases. The left and right ends of the wire are kept at zero temperature and serve as heat sinks. The wire has an internal heat source, perhaps from electrical resistance or a chemical reaction, and so, this increases the temperature in the interior of the wire.

The second calculation increases the final time from 150 to 180 so that the time step from increases 5 to 6, and consequently, the stability condition does not hold. Note in Figure 1.2.4 that significant oscillations develop.

The third computation uses a larger final time equal to 600 with 120 time steps. Notice in Figure 1.2.5 as time increases the temperature remains about the same, and for large values of time it is shaped like a parabola with a maximum value near 125.

1.2.6   Assessment

The heat conduction in a thin wire has a number of approximations. Different mesh sizes in either the time or space variable will give different numerical results. However, if the stability conditions hold and the mesh sizes decrease, then the numerical computations will differ by smaller amounts.

The numerical model assumed that the surface of the wire was thermally insulated. This may not be the case, and one may use the discrete version of Newton's law of cooling by inserting a negative source term of $C(u_{sur} - u_i^k)h\ \pi 2r\Delta t$ where r is the radius of the wire. The constant C is a measure of insulation where $C = 0$ corresponds to perfect insulation. The $h\pi 2r$ is

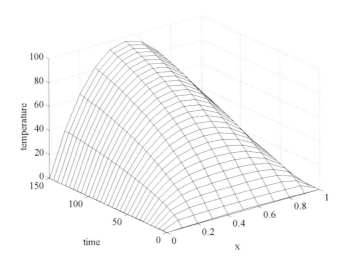

Figure 1.2.3: Temperature versus Time-Space

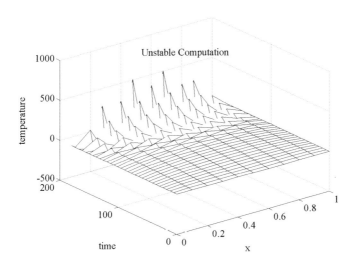

Figure 1.2.4: Unstable Computation

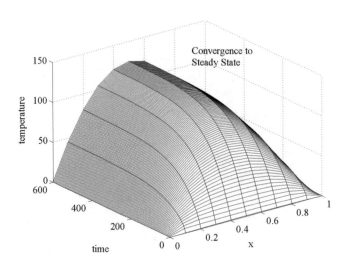

Figure 1.2.5: Steady State Temperature

the lateral surface area of the volume hA with $A = \pi r^2$. Other variations on the model include more complicated boundary conditions, variable thermal properties and diffusion in more than one direction.

In the scalar version of the first order finite difference models the scheme was stable when $|a| < 1$. In this case, u^{k+1} converged to the steady state solution $u = au + b$. This is also true of the matrix version of (1.2.1) provided the stability condition is satisfied. In this case the real number a will be replaced by the matrix A, and A^k will converge to the zero matrix. The following is a more general statement of this.

Theorem 1.2.1 *(Steady State Theorem)   Consider the matrix version of the first order finite difference equation $u^{k+1} = Au^k + b$ where A is a square matrix. If A^k converges to the zero matrix and $u = Au + b$, then, regardless of the initial choice for u^0, u^k converges to u.*

Proof.   Subtract $u^{k+1} = Au^k + b$ and $u = Au + b$ and use the properties of matrix products to get

$$
\begin{aligned}
u^{k+1} - u &= \left(Au^k + b\right) - (Au + b) \\
&= A(u^k - u) \\
&= A(A(u^{k-1} - u)) \\
&= A^2(u^{k-1} - u) \\
&\;\;\vdots \\
&= A^{k+1}(u^0 - u)
\end{aligned}
$$

Since A^k converges to the zero matrix, the column vectors $u^{k+1} - u$ must converge to the zero column vector. ∎

1.2.7 Exercises

1. Using the MATLAB code heat.m duplicate Figures 1.2.3-1.2.5.
2. In heat.m let $maxk = 120$ so that $dt = 150/120 = 1.25$. Experiment with the space step sizes $dx = .2, .1, .05$ and $n = 5, 10, 20$, respectively.
3. In heat.m let $n = 10$ so that $dx = .1$. Experiment with time step sizes $dt = 5, 2.5, 1.25$ and $maxk = 30, 60$ and 120, respectively.
4. In heat.m experiment with different values of the thermal conductivity $cond = .002, .001$ and $.0005$. Be sure to adjust the time step so that the stability condition holds.
5. Consider the variation on the thin wire where heat is lost through the surface of the wire. Modify heat.m and experiment with the C and r parameters. Explain your computed results.
6. Consider the variation on the thin wire where heat is generated by $f = 1 + sin(\pi 10t)$. Modify heat.m and experiment with the parameters.
7. Consider the 3×3 A matrix for (1.2.1). Compute A^k for $k = 10, 100, 1000$ for different values of alpha so that the stability condition either does or does not hold.
8. Suppose $n = 5$ so that there are 4 unknowns. Find the 4×4 matrix version of the finite difference model (1.2.1). Repeat the previous problem for the corresponding 4×4 matrix.
9. Justify the second and third lines in the displayed equations in the proof of the Steady State Theorem.
10. Consider a variation of the Steady State Theorem where the column vector b depends on time, that is, b is replaced by b^k. Formulate and prove a generalization of this theorem.

1.3 Diffusion in a Wire with Little Insulation

1.3.1 Introduction

In this section we consider heat diffusion in a thin electrical wire, which is not thermally insulated on its lateral surface. The model of the temperature will still have the form $u^{k+1} = Au^k + b$, but the matrix A and column vector b will be different than in the insulated lateral surface model in the previous section.

1.3.2 Applied Area

In this section we present a third model of heat transfer. In our first model we considered heat transfer via a discrete version of Newton's law of cooling. That is, we assumed the mass had uniform temperature with respect to space. In the previous section we allowed the temperature to be a function of both

discrete time and discrete space. Heat diffused via the Fourier heat law either
to the left or right direction in the wire. The wire was assumed to be perfectly
insulated in the lateral surface so that no heat was lost or gained through the
lateral sides of the wire. In this section we will allow heat to be lost through
the lateral surface via a Newton-like law of cooling.

1.3.3 Model

Discretize both space and time and let the temperature $u(ih, k\Delta t)$ be approximated by u_i^k where $\Delta t = T/maxk$, $h = L/n$ and L is the length of the wire.
The model will have the general form

$$\text{change in heat in } (hA) \approx \text{(heat from the source)}$$
$$+\text{(diffusion through the left end)}$$
$$+\text{(diffusion through the right end)}$$
$$+\text{(heat loss through the lateral surface)}.$$

This is depicted in the Figure 1.2.1 where the volume is a horizontal cylinder
whose length is h and cross section is $A = \pi r^2$. So the lateral surface area is
$h2\pi r$.

The heat loss through the lateral surface will be assumed to be directly
proportional to the product of change in time, the lateral surface area and to
the difference in the surrounding temperature and the temperature in the wire.
Let c_{sur} be the proportionality constant that measures insulation. If u_{sur} is
the surrounding temperature of the wire, then the heat loss through the small
lateral area is

$$c_{sur}\,\Delta t\,2\pi rh(u_{sur} - u_i^k). \tag{1.3.1}$$

Heat loss or gain from a source such as electrical current and from left and right
diffusion will remain the same as in the previous section. By combining these
we have the following approximation of the change in the heat content for the
small volume Ah:

$$
\begin{aligned}
\rho c u_i^{k+1} Ah - \rho c u_i^k Ah \;=\;& Ah\,\Delta t\,f \\
& +A\,\Delta t\,K(u_{i+1}^k - u_i^k)/h - A\,\Delta t\,K(u_i^k - u_{i-1}^k)/h \\
& +c_{sur}\,\Delta t\,2\pi rh(u_{sur} - u_i^k) \tag{1.3.2}
\end{aligned}
$$

Now, divide by ρcAh, define $\alpha = (K/\rho c)(\Delta t/h^2)$ and explicitly solve for u_i^{k+1}.

Explicit Finite Difference Model for Heat Diffusion in a Wire.

$$
\begin{aligned}
u_i^{k+1} \;=\;& (\Delta t/\rho c)(f + c_{sur}(2/r)u_{sur}) + \alpha(u_{i+1}^k + u_{i-1}^k) \\
& +(1 - 2\alpha - (\Delta t/\rho c)c_{sur}(2/r))u_i^k \tag{1.3.3} \\
\text{for } i \;=\;& 1, ..., n-1 \text{ and } k = 0, ..., maxk - 1, \\
u_i^0 \;=\;& 0 \text{ for } i = 1, ..., n-1 \tag{1.3.4} \\
u_0^k \;=\;& u_n^k = 0 \text{ for } k = 1, ..., maxk. \tag{1.3.5}
\end{aligned}
$$

Equation (1.3.4) is the initial temperature set equal to zero, and (1.3.5) is the temperature at the left and right ends set equal to zero. Equation (1.3.3) may be put into the matrix version of the first order finite difference method. For example, if the wire is divided into four equal parts, then $n = 4$ and (1.3.3) may be written as three scalar equations for the unknowns u_1^{k+1}, u_2^{k+1} and u_3^{k+1} :

$$
\begin{aligned}
u_1^{k+1} &= (\Delta t/\rho c)(f + c_{sur}(2/r)u_{sur}) + \alpha(u_2^k + 0) + \\
&\quad (1 - 2\alpha - (\Delta t/\rho c)c_{sur}(2/r))u_1^k \\
u_2^{k+1} &= (\Delta t/\rho c)(f + c_{sur}(2/r)u_{sur}) + \alpha(u_3^k + u_1^k) + \\
&\quad (1 - 2\alpha - (\Delta t/\rho c)c_{sur}(2/r))u_2^k \\
u_3^{k+1} &= (\Delta t/\rho c)(f + c_{sur}(2/r)u_{sur}) + \alpha(0 + u_2^k) + \\
&\quad (1 - 2\alpha - (\Delta t/\rho c)c_{sur}(2/r))u_3^k.
\end{aligned}
$$

These three scalar equations can be written as one 3D vector equation

$$
u^{k+1} = Au^k + b \text{ where} \tag{1.3.6}
$$

$$
u^k = \begin{bmatrix} u_1^k \\ u_2^k \\ u_3^k \end{bmatrix}, \ b = (\Delta t/\rho c)F \begin{bmatrix} 1 \\ 1 \\ 1 \end{bmatrix},
$$

$$
A = \begin{bmatrix} 1 - 2\alpha - d & \alpha & 0 \\ \alpha & 1 - 2\alpha - d & \alpha \\ 0 & \alpha & 1 - 2\alpha - d \end{bmatrix} \text{ and}
$$

$$
F = f + c_{sur}(2/r)u_{sur} \text{ and } d = (\Delta t/\rho c)c_{sur}(2/r).
$$

An important restriction on the time step Δt is required to make sure the algorithm is stable. For example, consider the case $n = 2$ where equation (1.3.6) is a scalar equation and we have the simplest first order finite difference model. Here $a = 1 - 2\alpha - d$ and we must require $a < 1$. If $a = 1 - 2\alpha - d > 0$ and $\alpha, d > 0$, then this condition will hold. If n is larger than 2, this simple condition will imply that the matrix products A^k will converge to the zero matrix, and this analysis will be presented later in Chapter 2.5.

Stability Condition for (1.3.3).

$$
1 - 2(K/\rho c)(\Delta t/h^2) - (\Delta t/\rho c)c_{sur}(2/r) > 0.
$$

Example.   Let $L = c = \rho = 1.0, r = .05, n = 4$ so that $h = 1/4, K = .001, c_{sur} = .0005, u_{sur} = -10$. Then $\alpha = (K/\rho c)(\Delta t/h^2) = (.001)\Delta t16$ and $d = (\Delta t/\rho c)c_{sur}(2/r) = \Delta t(.0005)(2/.05)$ so that $1 - 2(K/\rho c)(\Delta t/h^2) - (\Delta t/\rho c)c_{sur}(2/r) = 1 - .032\Delta t - \Delta t(.020) = 1 - .052\Delta t > 0$. Note if n increases to 20, then the constraint on the time step will significantly change.

1.3.4 Method

The numbers u_i^{k+1} generated by equations (1.3.3)-(1.3.5) are hopefully good approximations for the temperature at $x = i\Delta x$ and $t = (k + 1)\Delta t$. The temperature is often denoted by the function $u(x, t)$. Again the u_i^{k+1} will be stored in a two dimensional array, which is also denoted by u but with integer indices so that $u_i^{k+1} = u(i, k+1) \approx u(i\Delta x, (k+1)\Delta t) =$ temperature function. In order to compute all u_i^{k+1}, we must use a nested loop where the i-loop (space) is the inner loop and the k-loop (time) is the outer loop. This is illustrated in the Figure 1.2.1 by the dependency of $u(i, k+1)$ on the three previously computed $u(i-1, k)$, $u(i, k)$ and $u(i+1, k)$.

1.3.5   Implementation

A slightly modified version of heat.m is used to illustrated the effect of changing the insulation coefficient, c_{sur}. The implementation of the above model for temperature that depends on both space and time will have nested loops where the outer loop is for discrete time and the inner loop is for discrete space. In the MATLAB code heat1d.m these nested loops are given in lines 33-37. Lines 1-29 contain the input data with additional data in lines 17-20. Here the radius of the wire is $r = .05$, which is small relative to the length of the wire $L = 1.0$. The surrounding temperature is $u_{sur} = -10$. so that heat is lost through the lateral surface when $c_{sur} > 0$. Lines 38-41 contain the output data in the form of a surface plot for the temperature.

MATLAB Code heat1d.m

```
1.      % This code models heat diffusion in a thin wire.
2.      % It executes the explicit finite difference method.
3.      clear;
4.      L = 1.0;       % length of the wire
5.      T = 400.;      % final time
6.      maxk = 100;        % number of time steps
7.      dt = T/maxk;
8.      n = 10.;      % number of space steps
9.      dx = L/n;
10.     b = dt/(dx*dx);
11.     cond = .001;       % thermal conductivity
12.     spheat = 1.0;        % specific heat
13.     rho = 1.;       % density
14.     a = cond/(spheat*rho);
15.     alpha = a*b;
16.     f = 1.;        % internal heat source
17.     dtc = dt/(spheat*rho);
18.     csur = .0005;       % insulation coefficient
19.     usur = -10;        % surrounding temperature
20.     r = .05;        % radius of the wire
21.     for i = 1:n+1       % initial temperature
22.            x(i) =(i-1)*dx;
23.            u(i,1) =sin(pi*x(i));
24.     end
```

```
25.     for k=1:maxk+1        % boundary temperature
26.         u(1,k) = 0.;
27.         u(n+1,k) = 0.;
28.         time(k) = (k-1)*dt;
29.     end
30.     %
31.     % Execute the explicit method using nested loops.
32.     %
33.     for k=1:maxk          % time loop
34.         for i=2:n;         % space loop
35.             u(i,k+1) = (f +csur*(2./r))*dtc
                        + (1-2*alpha - dtc*csur*(2./r))*u(i,k)
                        + alpha*(u(i-1,k)+u(i+1,k));
36.         end
37.     end
38.     mesh(x,time,u')
39.     xlabel('x')
40.     ylabel('time')
41.     zlabel('temperature')
```

Two computations with different insulation coefficients, c_{sur}, are given in Figure 1.3.1. If one tries a calculation with $c_{sur} = .0005$ with a time step size equal to 5, then this violates the stability condition so that the model fails. For $c_{sur} \leq .0005$ the model did not fail with a final time equal to 400 and 100 time steps so that the time step size equaled to 4. Note the maximum temperature decreases from about 125 to about 40 as c_{sur} increases from .0000 to .0005. In order to consider larger c_{sur}, the time step may have to be decreased so that the stability condition will be satisfied.

In the next numerical experiment we vary the number of space steps from $n = 10$ to $n = 5$ and 20. This will change the $h = dx$, and we will have to adjust the time step so that the stability condition holds. Roughly, if we double n, then we should quadruple the number of time steps. So, for $n = 5$ we will let $maxk = 25$, and for $n = 20$ we will let $maxk = 400$. The reader should check the stability condition assuming the other parameters in the numerical model are $u_{sur} = -10$, $c_{sur} = .0005$, $K = .001$, $\rho = 1$ and $c = 1$. Note the second graph in Figure 1.3.1 where $n = 10$ and those in Figure 1.3.2 are similar.

1.3.6   Assessment

The heat conduction in a thin wire has a number of approximations. Different mesh sizes in either the time or space variable will give different numerical results. However, if the stability conditions hold and the mesh sizes decrease, then the numerical computations will differ by smaller amounts. Other variations on the model include more complicated boundary conditions, variable thermal properties and diffusion in more than one direction.

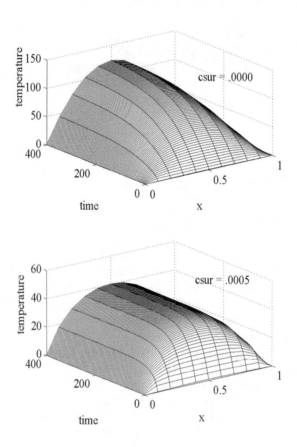

Figure 1.3.1: Diffusion in a Wire with csur = .0000 and .0005

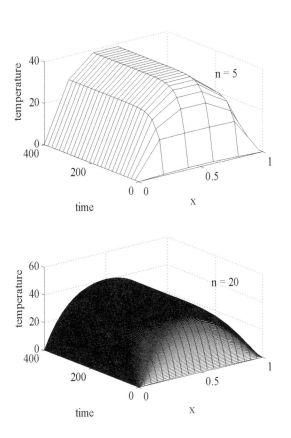

Figure 1.3.2: Diffusion in a Wire with n = 5 and 20

The above discrete model will converge, under suitable conditions, to a *continuum model of heat diffusion*. This is a partial differential equation with initial and boundary conditions similar to those in (1.3.3), (1.3.4) and (1.3.5):

$$\rho c u_t = f + (K u_x)_x + c_{sur}(2/r)(u_{sur} - u) \tag{1.3.7}$$
$$u(x,0) = 0 \text{ and} \tag{1.3.8}$$
$$u(0,t) = 0 = u(L,t) \tag{1.3.9}$$

The partial differential equation in (1.3.6) can be derived from (1.3.2) by replacing u_i^k by $u(ih, k\Delta t)$, dividing by $Ah\,\Delta t$ and letting h and Δt go to 0. Convergence of the discrete model to the continuous model means for all i and k the errors

$$u_i^k - u(ih, k\Delta t)$$

go to zero as h and Δt go to zero. Because partial differential equations are often impossible to solve exactly, the discrete models are often used.

Not all numerical methods have stability constraints on the time step. Consider (1.3.6) and use an implicit time discretization to generate a sequence of ordinary differential equations

$$\rho c(u^{k+1} - u^k)/\Delta t = f + (K u_x^{k+1})_x + c_{sur}(2/r)(u_{sur} - u^{k+1}). \tag{1.3.10}$$

This does not have a stability constraint on the time step, but at each time step one must solve an ordinary differential equation with boundary conditions. The numerical solution of these will be discussed in the following chapters.

1.3.7   Exercises

1.   Duplicate the computations in Figure 1.3.1 with variable insulation coefficient. Furthermore, use $c_{sur} = .0002$ and .0010.

2.   In heat1d.m experiment with different surrounding temperatures $u_{sur} = -5, -10, -20$.

3.   Suppose the surrounding temperature starts at -10 and increases by one degree every ten units of time.

   (a).   Modify the finite difference model (1.3.3) is account for this.

   (b).   Modify the MATLAB code heat1d.m. How does this change the long run solution?

4.   Vary the $r = .01, .02, .05$ and .10. Explain your computed results. Is this model realistic for "large" r?

5.   Verify equation (1.3.3) by using equation (1.3.2).

6.   Consider the 3×3 A matrix version of line (1.3.3) and the example of the stability condition on the time step. Observe A^k for $k = 10, 100$ and 1000 with different values of the time step so that the stability condition either does or does not hold.

7.   Consider the finite difference model with $n = 5$ so that there are four unknowns.

(a).   Find 4×4 matrix version of (1.3.3).

(b).   Repeat problem 6 with this 4×4 matrix

8.   Experiment with variable space steps $h = dx = L/n$ by letting $n = 5, 10, 20$ and 40. See Figures 1.3.1 and 1.3.2 and be sure to adjust the time steps so that the stability condition holds.

9.   Experiment with variable time steps $dt = T/maxk$ by letting $maxk = 100, 200$ and 400 with $n = 10$ and $T = 400$.

10.   Examine the graphical output from the experiments in exercises 8 and 9. What happens to the numerical solutions as the time and space step sizes decrease?

11.   Suppose the thermal conductivity is a linear function of the temperature, say, $K = cond = .001 + .02u$ where u is the temperature.

(a).   Modify the finite difference model in (1.3.3).

(b).   Modify the MATLAB code heat1d.m to accommodate this variation. Compare the numerical solution with those given in Figure 1.3.1.

1.4 Flow and Decay of a Pollutant in a Stream

1.4.1 Introduction

Consider a river that has been polluted upstream. The concentration (amount per volume) will decay and disperse downstream. We would like to predict at any point in time and in space the concentration of the pollutant. The model of the concentration will also have the form $u^{k+1} = Au^k + b$ where the matrix A will be defined by the finite difference model, which will also require a stability constraint on the time step.

1.4.2 Applied Area

Pollution levels in streams, lakes and underground aquifers have become very serious common concern. It is important to be able to understand the consequences of possible pollution and to be able to make accurate predictions about "spills" and future "environmental" policy.

Perhaps, the simplest model for chemical pollution is based on chemical decay, and one model is similar to radioactive decay. A continuous model is $u_t = -du$ where d is a chemical decay rate and $u = u(t)$ is the unknown concentration. One can use Euler's method to obtain a discrete version $u^{k+1} = u^k + \Delta t(-d)u^k$ where u^k is an approximation of $u(t)$ at $t = k\Delta t$, and stability requires the following constraint on the time step $1 - \Delta t d > 0$.

Here we will introduce a second model where the pollutant changes location because it is in a stream. Assume the concentration will depend on both space and time. The space variable will only be in one direction, which corresponds to the direction of flow in the stream. If the pollutant was in a deep lake, then the concentration would depend on time and all three directions in space.

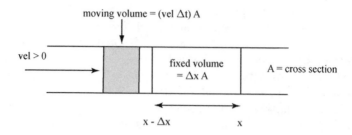

Figure 1.4.1: Polluted Stream

1.4.3   Model

Discretize both space and time, and let the concentration u at $(i\Delta x, k\Delta t)$ be approximated by u_i^k where $\Delta t = T/maxk, \Delta x = L/n$ and L is the length of the stream. The model will have the general form

$$\text{change in amount} \quad \approx \quad (\text{amount entering from upstream})$$
$$-(\text{amount leaving to downstream})$$
$$-(\text{amount decaying in a time interval}).$$

This is depicted in Figure 1.4.1 where the steam is moving from left to right and the stream velocity is positive. For time we can choose either $k\Delta t$ or $(k+1)\Delta t$. Here we will choose $k\Delta t$ and this will eventually result in the matrix version of the first order finite difference method.

Assume the stream is moving from left to right so that the stream velocity is positive, $vel > 0$. Let A be the cross sectional area of the stream. The amount of pollutant entering the left side of the volume $A\Delta x$ $(vel > 0)$ is

$$A(\Delta t\ vel)\ u_{i-1}^k.$$

The amount leaving the right side of the volume $A\Delta x$ $(vel > 0)$is

$$-A(\Delta t\ vel)\ u_i^k.$$

Therefore, the change in the amount from the stream's velocity is

$$A(\Delta t\ vel)\ u_{i-1}^k\ -A(\Delta t\ vel)\ u_i^k.$$

The amount of the pollutant in the volume $A\Delta x$ at time $k\Delta t$ is

$$A\Delta x\ u_i^k.$$

The amount of the pollutant that has decayed, *dec* is decay rate, is

$$-A\Delta x \, \Delta t \, dec \, u_i^k.$$

By combining these we have the following approximation for the change during the time interval Δt in the amount of pollutant in the small volume $A\Delta x$:

$$A\Delta x \, u_i^{k+1} - A\Delta x \, u_i^k = A(\Delta t \, vel)u_{i-1}^k - A(\Delta t \, vel)u_i^k$$
$$-A\Delta x \, \Delta t \, dec \, u_i^k. \tag{1.4.1}$$

Now, divide by $A\Delta x$ and explicitly solve for u_i^{k+1}.

Explicit Finite Difference Model of Flow and Decay.

$$
\begin{aligned}
u_i^{k+1} &= vel(\Delta t/\Delta x)u_{i-1}^k + (1 - vel(\Delta t/\Delta x) - \Delta t \, dec)u_i^k & (1.4.2)\\
i &= 1, ..., n-1 \text{ and } k = 0, ..., maxk - 1,\\
u_i^0 &= \text{given for } i = 1, ..., n-1 \text{ and} & (1.4.3)\\
u_0^k &= \text{given for } k = 1, ..., maxk. & (1.4.4)
\end{aligned}
$$

Equation (1.4.3) is the initial concentration, and (1.4.4) is the concentration far upstream. Equation (1.4.2) may be put into the matrix version of the first order finite difference method. For example, if the stream is divided into three equal parts, then $n = 3$ and (1.4.2) may be written three scalar equations for u_1^{k+1}, u_2^{k+1} and u_3^{k+1} :

$$
\begin{aligned}
u_1^{k+1} &= vel(\Delta t/\Delta x)u_0^k + (1 - vel(\Delta t/\Delta x) - \Delta t \, dec)u_1^k\\
u_2^{k+1} &= vel(\Delta t/\Delta x)u_1^k + (1 - vel(\Delta t/\Delta x) - \Delta t \, dec)u_2^k\\
u_3^{k+1} &= vel(\Delta t/\Delta x)u_2^k + (1 - vel(\Delta t/\Delta x) - \Delta t \, dec)u_3^k.
\end{aligned}
$$

These can be written as one 3D vector equation $u^{k+1} = Au^k + b$

$$
\begin{bmatrix} u_1^{k+1} \\ u_2^{k+1} \\ u_3^{k+1} \end{bmatrix} = \begin{bmatrix} c & 0 & 0 \\ d & c & 0 \\ 0 & d & c \end{bmatrix} \begin{bmatrix} u_1^k \\ u_2^k \\ u_3^k \end{bmatrix} + \begin{bmatrix} du_0^k \\ 0 \\ 0 \end{bmatrix} \tag{1.4.5}
$$

$$\text{where } d = vel \, (\Delta t/\Delta x) \text{ and } c = 1 - d - dec \, \Delta t.$$

An extremely important restriction on the time step Δt is required to make sure the algorithm is stable. For example, consider the case $n = 1$ where the above is a scalar equation, and we have the simplest first order finite difference model. Here $a = 1 - vel(\Delta t/\Delta x) - dec \, \Delta t$ and we must require $a < 1$. If $a = 1 - vel(\Delta t/\Delta x) - dec \, \Delta t > 0$ and $vel, dec > 0$, then this condition will hold. If n is larger than 1, this simple condition will imply that the matrix products A^k converge to the zero matrix, and an analysis of this will be given in Section 2.5.

Stability Condition for (1.4.2).

$$1 - vel(\Delta t/\Delta x) - dec\ \Delta t \text{ and } vel, dec > 0.$$

Example. Let $L = 1.0, vel = .1, dec = .1$, and $n = 4$ so that $\Delta x = 1/4$. Then $1 - vel(\Delta t/\Delta x) - dec\ \Delta t = 1 - .1\Delta t4 - .1\Delta t = 1 - .5\Delta t > 0$. If n increases to 20, then the stability constraint on the time step will change.

In the case where $dec = 0$, then $a = 1 - vel(\Delta t/\Delta x) > 0$ means the entering fluid must must not travel, during a single time step, more than one space step. This is often called the *Courant condition* on the time step.

1.4.4   Method

In order to compute u_i^{k+1} for all values of i and k, which in the MATLAB code is stored in the array $u(i, k + 1)$, we must use a nested loop where the i-loop (space) is inside and the k-loop (time) is the outer loop. In this flow model $u(i, k + 1)$ depends directly on the two previously computed $u(i - 1, k)$ (the upstream concentration) and $u(i, k)$. This is different from the heat diffusion model, which requires an additional value $u(i + 1, k)$ and a boundary condition at the right side. In heat diffusion heat energy may move in either direction; in our model of a pollutant the amount moves in the direction of the stream's flow.

1.4.5   Implementation

The MATLAB code flow1d.m is for the explicit flow and decay model of a polluted stream. Lines 1-19 contain the input data where in lines 12-15 the initial concentration was a trig function upstream and zero downstream. Lines 16-19 contain the farthest upstream location that has concentration equal to .2. The finite difference scheme is executed in lines 23-27, and three possible graphical outputs are indicated in lines 28-30. A similar code is heat1.f90 written in Fortran 9x.

MATLAB Code flow1d.m

```
1.    % This a model for the concentration of a pollutant.
2.    % Assume the stream has constant velocity.
3.    clear;
4.    L = 1.0;      % length of the stream
5.    T = 20.;      % duration of time
6.    K = 200;      % number of time steps
7.    dt = T/K;
8.    n = 10.;      % number of space steps
9.    dx = L/n;
10.   vel = .1;     % velocity of the stream
11.   decay = .1;   % decay rate of the pollutant
12.   for i = 1:n+1    % initial concentration
```

```
13.          x(i) =(i-1)*dx;
14.          u(i,1) =(i<=(n/2+1))*sin(pi*x(i)*2)+(i>(n/2+1))*0;
15.     end
16.     for k=1:K+1      % upstream concentration
17.          time(k) = (k-1)*dt;
18.          u(1,k) = -sin(pi*vel*0)+.2;
19.     end
20.     %
21.     % Execute the finite difference algorithm.
22.     %
23.     for k=1:K      % time loop
24.          for i=2:n+1      % space loop
25.               u(i,k+1) =(1 - vel*dt/dx -decay*dt)*u(i,k)
                                  + vel*dt/dx*u(i-1,k);
26.          end
27.     end
28.     mesh(x,time,u')
29.     % contour(x,time,u')
30.     % plot(x,u(:,1),x,u(:,51),x,u(:,101),x,u(:,151))
```

One expects the location of the maximum concentration to move down-stream and to decay. This is illustrated in Figure 1.4.2 where the top graph was generated by the mesh command and is concentration versus time-space. The middle graph is a contour plot of the concentration. The bottom graph contains four plots for the concentration at four times 0, 5, 10 and 15 versus space, and here one can clearly see the pollutant plume move downstream and decay.

The following MATLAB code mov1d.m will produce a frame by frame "movie" which does not require a great deal of memory. This code will present graphs of the concentration versus space for a sequence of times. Line 1 executes the above MATLAB file flow1d where the arrays x and u are created. The loop in lines 3-7 generates a plot of the concentrations every 5 time steps. The next plot is activated by simply clicking on the graph in the MATLAB figure window. In the pollution model it shows the pollutant moving downstream and decaying.

MATLAB Code mov1d.m

```
1.     flow1d;
2.     lim =[0 1. 0 1];
3.     for k=1:5:150
4.          plot(x,u(:,k))
5.          axis(lim);
6.          k = waitforbuttonpress;
7.     end
```

In Figure 1.4.3 we let the stream's velocity be $vel = 1.3$, and this, with the same other constants, violates the stability condition. For the time step equal

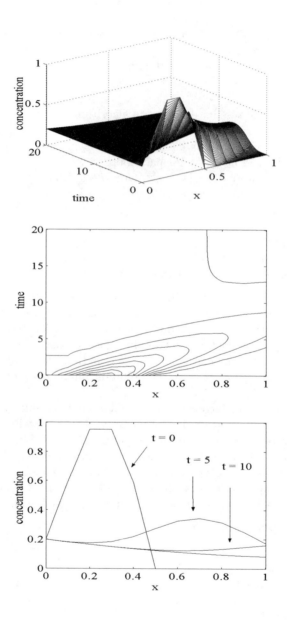

Figure 1.4.2: Concentration of Pollutant

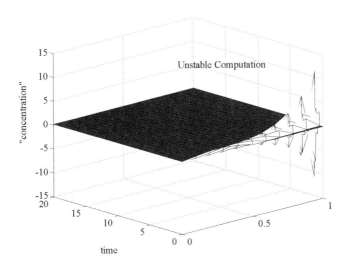

Figure 1.4.3: Unstable Concentration Computation

to .1 and the space step equal to .1, a flow rate equal to 1.3 means that the pollutant will travel .13 units in space, which is more than one space step. In order to accurately model the concentration in a stream with this velocity, we must choose a smaller time step. Most explicit numerical methods for fluid flow problems will not work if the time step is so large that the computed flow for a time step jumps over more than one space step.

1.4.6 Assessment

The discrete model is accurate for suitably small step sizes. The dispersion of the pollutant is a continuous process, which could be modeled by a partial differential equation with initial and boundary conditions:

$$u_t = -vel\ u_x - dec\ u, \tag{1.4.6}$$
$$u(x,0) = \text{given and} \tag{1.4.7}$$
$$u(0,t) = \text{given.} \tag{1.4.8}$$

This is analogous to the discrete model in (1.4.2), (1.4.3) and (1.4.4). The partial differential equation in (1.4.6) can be derived from (1.4.1) by replacing u_i^k by $u(i\Delta x, k\Delta t)$, dividing by $A\Delta x\,\Delta t$ and letting Δx and Δt go to 0. Like the heat models the step sizes should be carefully chosen so that stability holds and the errors

$$u_i^k - u(i\Delta x, k\Delta t)$$

between the discrete and continuous models are small.

Often it is difficult to determine the exact values of the constants *vel* and
dec. Exactly what is the effect of having measurement errors, say of 10%, on
constants *vel*, *dec* or the initial and boundary conditions? What is interaction of
the measurement errors with the numerical errors? The flow rate, *vel*, certainly
is not always constant. Moreover, there may be fluid flow in more than one
direction.

1.4.7   Exercises

1.   Duplicate the computations in Figure 1.4.2.
2.   Vary the decay rate, $dec = .05, .1, 1.$ and 2.0. Explain your computed
results.
3.   Vary the flow rate, $vel = .05, .1, 1.$ and 2.0. Explain your computed
results.
4.   Consider the 3×3 A matrix. Use the parameters in the example of the
stability condition and observe A^k when $k = 10, 100$ and 1000 for different
values of *vel* so that the stability condition either does or does not hold.
5.   Suppose $n = 4$ so that there are four unknowns. Find the 4×4 matrix
description of the finite difference model (1.4.2). Repeat problem 4 with the
corresponding 4×4 matrix.
6.   Verify that equation (1.4.2) follows from equation (1.4.1).
7.   Experiment with different time steps by varying the number of time steps
$K = 100, 200, 400$ and keeping the space steps constant by using $n = 10$.
8.   Experiment with different space steps by varying the number space steps
$n = 5, 10, 20, 40$ and keeping the time steps constant by using $K = 200$.
9.   In exercises 7 and 8 what happens to the solutions as the mesh sizes
decrease, provided the stability condition holds?
10.   Modify the model to include the possibility that the upstream boundary
condition varies with time, that is, the polluting source has a concentration that
depends on time. Suppose the concentration at $x = 0$ is a periodic function
$.1 + .1 \, sin(\pi t/20)$.
   (a).   Change the finite difference model (1.4.2)-(1.4.4) to account for this.
   (b).   Modify the MATLAB code flow1d.m and use it to study this case.
11.   Modify the model to include the possibility that the steam velocity
depends on time. Suppose the velocity of the stream increases linearly over the
time interval from $t = 0$ to $t = 20$ so that $vel = .1 + .01t$.
   (a).   Change the finite difference model (1.4.2)-(1.4.4) to account for this.
   (b).   Modify the MATLAB code flow1d.m and use it to study this case.

1.5   Heat and Mass Transfer in Two Directions

1.5.1   Introduction

The restriction of the previous models to one space dimension is often not very
realistic. For example, if the radius of the cooling wire is large, then one should

expect to have temperature variations in the radial direction as well as in the direction of the wire. Or, in the pollutant model the source may be on a shallow lake and not a stream so that the pollutant may move within the lake in plane, that is, the concentrations of the pollutant will be a function of two space variables and time.

1.5.2 Applied Area

Consider heat diffusion in a thin 2D cooling fin where there is diffusion in both the x and y directions, but any diffusion in the z direction is minimal and can be ignored. The objective is to determine the temperature in the interior of the fin given the initial temperature and the temperature on the boundary. This will allow us to assess the cooling fin's effectiveness. Related problems come from the manufacturing of large metal objects, which must be cooled so as not to damage the interior of the object. A similar 2D pollutant problem is to track the concentration of a pollutant moving across a lake. The source will be upwind so that the pollutant is moving according to the velocity of the wind. We would like to know the concentration of the pollutant given the upwind concentrations along the boundary of the lake, and the initial concentrations in the lake.

1.5.3 Model

The models for both of these applications evolve from partitioning a thin plate or shallow lake into a set of small rectangular volumes, $\Delta x \Delta y T$, where T is the small thickness of the volume. Figure 1.5.1 depicts this volume, and the transfer of heat or pollutant through the right vertical face. In the case of heat diffusion, the heat entering or leaving through each of the four vertical faces must be given by the Fourier heat law applied to the direction perpendicular to the vertical face. For the pollutant model the amount of pollutant, concentration times volume, must be tracked through each of the four vertical faces. This type of analysis leads to the following models in two space directions. Similar models in three space directions are discussed in Sections 4.4-4.6 and 6.2-6.3.

In order to generate a 2D time dependent model for heat transfer diffusion, the Fourier heat law must be applied to both the x and y directions. The continuous and discrete 2D models are very similar to the 1D versions. In the continuous 2D model the temperature u will depend on three variables, $u(x, y, t)$. In (1.5.1) $-(Ku_y)_y$ models the diffusion in the y direction; it models the heat entering and leaving the left and right of the rectangle $h = \Delta x$ by $h = \Delta y$. More details of this derivation will be given in Section 3.2.

Continuous 2D Heat Model for $u = u(x, y, t)$.

$$\rho c u_t - (Ku_x)_x - (Ku_y)_y = f \qquad (1.5.1)$$
$$u(x, y, 0) = \text{given} \qquad (1.5.2)$$
$$u(x, y, t) = \text{given on the boundary} \qquad (1.5.3)$$

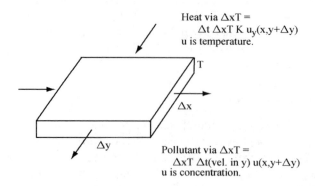

Heat via ΔxT =
    Δt ΔxT K u_y(x,y+Δy)
u is temperature.

Pollutant via ΔxT =
    ΔxT Δt(vel. in y) u(x,y+Δy)
u is concentration.

Figure 1.5.1: Heat or Mass Entering or Leaving

Explicit Finite Difference 2D Heat Model: $u_{i,j}^k \approx u(ih, jh, k\Delta t)$.

$$
\begin{aligned}
u_{i,j}^{k+1} &= (\Delta t/\rho c)f + \alpha(u_{i+1,j}^k + u_{i-1,j}^k + u_{i,j+1}^k + u_{i,j-1}^k) \\
&\quad +(1 - 4\alpha)u_{i,j}^k \qquad\qquad\qquad\qquad\qquad\qquad (1.5.4)\\
\alpha &= (K/\rho c)(\Delta t/h^2), \ i, j = 1, .., n - 1 \text{ and } k = 0, .., maxk - 1,\\
u_{i,j}^0 &= \text{given}, \ i, j = 1, .., n - 1 \qquad\qquad\qquad\qquad (1.5.5)\\
u_{i,j}^k &= \text{given}, \ k = 1, ..., maxk, \text{ and } i, j \text{ on the boundary grid.} \quad (1.5.6)
\end{aligned}
$$

Stability Condition.

$$1 - 4\alpha > 0 \text{ and } \alpha > 0.$$

The model for the dispersion of a pollutant in a shallow lake is similar. Let $u(x, y, t)$ be the concentration of a pollutant. Suppose it is decaying at a rate equal to dec units per time, and it is being dispersed to other parts of the lake by a known wind with constant velocity vector equal to (v_1, v_2). Following the derivations in Section 1.4, but now considering both directions, we obtain the continuous and discrete models. We have assumed both the velocity components are nonnegative so that the concentration levels on the upwind (west and south) sides must be given. In the partial differential equation for the continuous 2D model the term $-v_2 u_y$ models the amount of the pollutant entering and leaving in the y direction for the thin rectangular volume whose base is Δx by Δy.

Continuous 2D Pollutant Model for $u(x, y, t)$.

$$
\begin{aligned}
u_t &= -v_1 u_x - v_2 u_y - dec\, u, & (1.5.7)\\
u(x, y, 0) &= \text{given and} & (1.5.8)\\
u(x, y, t) &= \text{given on the upwind boundary.} & (1.5.9)
\end{aligned}
$$

Explicit Finite Difference 2D Pollutant Model: $u_{i,j}^k \approx u(i\Delta x, j\Delta y, k\Delta t)$.

$$u_{i,j}^{k+1} = v_1(\Delta t/\Delta x)u_{i-1,j}^k + v_2(\Delta t/\Delta y)u_{i,j-1}^k + \qquad (1.5.10)$$
$$(1 - v_1(\Delta t/\Delta x) - v_2(\Delta t/\Delta y) - \Delta t \; dec)u_{i,j}^k$$
$$u_{i,j}^0 = \text{given and} \qquad (1.5.11)$$
$$u_{0,j}^k \text{ and } u_{i,0}^k = \text{given.} \qquad (1.5.12)$$

Stability Condition.

$$1 - v_1(\Delta t/\Delta x) - v_2(\Delta t/\Delta y) - \Delta t \; dec > 0.$$

1.5.4 Method

Consider heat diffusion or pollutant transfer in two directions and let u_{ij}^{k+1} be the approximation of either the temperature or the concentration at (x, y, t) $= (i\Delta x, j\Delta y, (k+1)\Delta t)$. In order to compute all u_{ij}^{k+1}, which will henceforth be stored in the array $u(i, j, k+1)$, one must use nested loops where the j-loop and i-loop (space) are inside and the k-loop (time) is the outer loop. The computations in the inner loops depend only on at most five adjacent values: $u(i, j, k)$, $u(i-1, j, k)$, $u(i+1, j, k)$, $u(i, j-1, k)$, and $u(i, j+1, k)$ all at the previous time step, and therefore, the $u(i, j, k+1)$ and $u(\hat{i}, \hat{j}, k+1)$ computations are independent. The classical order of the nodes is to start with the bottom grid row and move from left to right. This means the outermost loop will be the k-loop (time), the middle will be the j-loop (grid row), and the innermost will be the i-loop (grid column). A notational point of confusion is in the array $u(i, j, k)$. Varying the i corresponds to moving up and down in column j; but this is associated with moving from left to right in the grid row j of the physical domain for the temperature or the concentration of the pollutant.

1.5.5 Implementation

The following MATLAB code heat2d.m is for heat diffusion on a thin plate, which has initial temperature equal to 70 and has temperature at boundary $x = 0$ equal to 370 for the first 120 time steps and then set equal to 70 after 120 time steps. The other temperatures on the boundary are always equal to 70. The code in heat2d.m generates a 3D array whose entries are the temperatures for 2D space and time. The input data is given in lines 1-31, the finite difference method is executed in the three nested loops in lines 35-41, and some of the output is graphed in the 3D plot for the temperature at the final time step in line 43. The 3D plot in Figure 1.5.2 is the temperature for the final time step equal to $Tend = 80$ time units, and here the interior of the fin has cooled down to about 84.

MATLAB Code heat2d.m

```
1.    % This is heat diffusion in 2D space.
```

```
2.      % The explicit finite difference method is used.
3.      clear;
4.      L = 1.0;        % length in the x-direction
5.      W = L;          % length in the y-direction
6.      Tend = 80.;        % final time
7.      maxk = 300;
8.      dt = Tend/maxk;
9.      n = 20.;
10.     % initial condition and part of boundary condition
11.     u(1:n+1,1:n+1,1:maxk+1) = 70.;
12.     dx = L/n;
13.     dy = W/n;       % use dx = dy = h
14.     h = dx;
15.     b = dt/(h*h);
16.     cond = .002;       % thermal conductivity
17.     spheat = 1.0;       % specific heat
18.     rho = 1.;       % density
19.     a = cond/(spheat*rho);
20.     alpha = a*b;
21.     for i = 1:n+1
22.     x(i) =(i-1)*h;        % use dx = dy = h
23.     y(i) =(i-1)*h;
24.     end
25.     % boundary condition
26.     for k=1:maxk+1
27.         time(k) = (k-1)*dt;
28.         for j=1:n+1
29.             u(1,j,k) =300.*(k<120)+ 70.;
30.         end
31.     end
32.     %
33.     % finite difference method computation
34.     %
35.     for k=1:maxk
36.         for j = 2:n
37.             for i = 2:n
39                  u(i,j,k+1) =0.*dt/(spheat*rho)
                            +(1-4*alpha)*u(i,j,k)
                            +alpha*(u(i-1,j,k)+u(i+1,j,k)
                            +u(i,j-1,k)+u(i,j+1,k));
39.             end
40.         end
41.     end
42.     % temperature versus space at the final time
43.     mesh(x,y,u(:,:,maxk)')
```

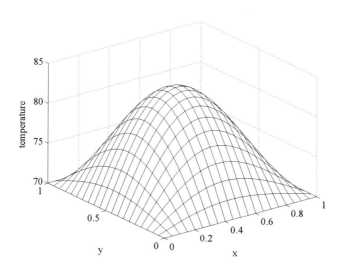

Figure 1.5.2: Temperature at Final Time

The MATLAB code mov2dheat.m generates a sequence of 3D plots of temperature versus space. One can see the heat moving from the hot side into the interior and then out the cooler boundaries. This is depicted for four times in Figure 1.5.3 where the scaling of the vertical axis has changed as time increases. You may find it interesting to vary the parameters and also change the 3D plot to a contour plot by replacing mesh by contour.

MATLAB Code mov2dheat.m

```
1.    % This generates a sequence of 3D plots of temperature.
2.    heat2d;
3.    lim =[0 1 0 1 0 400];
4.    for k=1:5:200
5.        mesh(x,y,u(:,:,k)')
6.        title ('heat versus space at different times' )
7.        axis(lim);
8.        k = waitforbuttonpress;
9.    end
```

The MATLAB code flow2d.m simulates a large spill of a pollutant along the southwest boundary of a shallow lake. The source of the spill is controlled after 25 time steps so that the pollutant plume moves across the lake as depicted by the mesh plots for different times. The MATLAB code flow2d.m generates the 3D array of the concentrations as a function of the x, y and time grid. The input data is given in lines 1-33, the finite difference method is executed in the three nested loops in lines 37-43, and the output is given in lines 44 and 45.

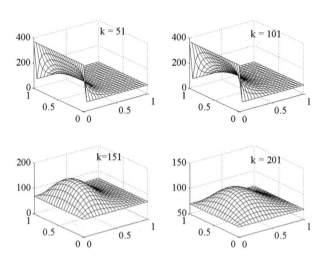

Figure 1.5.3: Heat Diffusing Out a Fin

MATLAB Code flow2d.m

```
1.    % The is pollutant flow across a lake.
2.    % The explicit finite difference method is used.
3.    clear;
4.    L = 1.0; % length in x direction
5.    W = 4.0; % length in y direction
6.    T = 10.; % final time
7.    maxk = 200; % number of time steps
8.    dt = T/maxk;
9.    nx = 10.; % number of steps in x direction
10.   dx = L/nx;
11.   ny = 20.; % number of steps in y direction
12.   dy = W/ny;
13.   velx = .1; % wind speed in x direction
14.   vely = .4; % wind speed in y direction
15.   decay = .1; %decay rate
16.   % Set initial conditions.
17.   for i = 1:nx+1
18.        x(i) =(i-1)*dx;
19.        for j = 1:ny+1
20.             y(j) =(j-1)*dy;
21.             u(i,j,1) = 0.;
22.        end
```

```
23.     end
24.     % Set upwind boundary conditions.
25.     for k=1:maxk+1
26.          time(k) = (k-1)*dt;
27.          for j=1:ny+1
28.               u(1,j,k) = .0;
29.          end
30.          for i=1:nx+1
31.               u(i,1,k) = (i<=(nx/2+1))*(k<26)
                          *5.0*sin(pi*x(i)*2)
                          +(i>(nx/2+1))*.1;
32.          end
33.     end
34.     %
35.     % Execute the explicit finite difference method.
36.     %
37.     for k=1:maxk
38.          for i=2:nx+1;
39.               for j=2:ny+1;
40.                    u(i,j,k+1) =(1 - velx*dt/dx
                               - vely*dt/dy - decay*dt)*u(i,j,k)
                               + velx*dt/dx*u(i-1,j,k)
                               + vely*dt/dy*u(i,j-1,k);
41.               end
42.          end
43.     end
44.     mesh(x,y,u(:,:,maxk)')
45.     % contour(x,y,u(:,:,maxk)')
```

Figure 1.5.4 is the concentration at the final time step as computed in flow2d.m. Figure 1.5.5 is sequence of mesh plots for the concentrations at various time steps. Note the vertical axis for the concentration is scaled so that the concentration plume decreases and moves in the direction of wind velocity (.1,.4). The MATLAB code mov2dflow.m generates a sequence of mesh plots.

MATLAB Code mov2dflow.m

```
1.     % This generates a sequence of 3D plots of concentration.
2.     flow2d;
3.     lim =[0 1 0 4 0 3];
4.     for k=1:5:200
5.          %contour(x,y,u(:,:,k)')
6.          mesh(x,y,u(:,:,k)')
7.          title ('concentration versus space at different times' )
8.          axis(lim);
9.          k = waitforbuttonpress;
10.    end
```

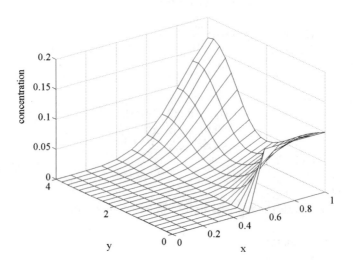

Figure 1.5.4: Concentration at the Final Time

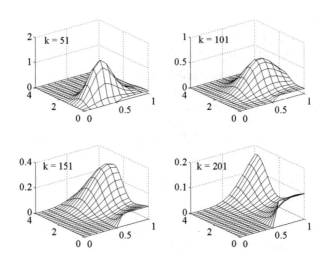

Figure 1.5.5: Concentrations at Different Times

1.5.6   Assessment

Diffusion of heat or the transfer of a pollutant may occur in nonrectangular domains. Certainly a rectangular lake is not realistic. Other discretization methods such as the finite element scheme are very useful in modeling more complicated geometric objects. Also, the assumption of the unknown depending on just two space variables may not be acceptable. Some discussion of three dimensional models is given in Sections 4.4-4.6, and in Sections 6.2-6.3 where there are three dimensional analogous codes heat3d.m and flow3d.m

1.5.7   Exercises

1.   Duplicate the calculations in heat2d.m. Use mesh and contour to view the temperatures at different times.
2.   In heat2d.m experiment with different time mesh sizes, $maxk = 150, 300,$ 450. Be sure to consider the stability constraint.
3.   In heat2d.m experiment with different space mesh sizes, $n = 10, 20$ and 40. Be sure to consider the stability constraint.
4.   In heat2d.m experiment with different thermal conductivities $K = cond =$ .01, .02 and .04. Be sure to make any adjustments to the time step so that the stability condition holds.
5.   Suppose heat is being generated at a rate of 3 units of heat per unit volume per unit time.
    (a).   How is the finite difference model for the 2D problem in equation (1.5.4) modified to account for this?
    (b).   Modify heat2d.m to implement this source of heat.
    (c).   Experiment with different values for the heat source $f = 0, 1, 2, 3.$
6.   In the 2D finite difference model in equation (1.5.4) and in the MATLAB code heat2d.m the space steps in the x and y directions were assumed to be equal to h.
    (a).   Modify these so that $\Delta x = dx$ and $\Delta y = dy$ are different.
    (b).   Experiment with different shaped fins that are not squares, that is, in lines 4-5 W and L may be different.
    (c).   Or, experiment in line 9 where n is replaced by nx and ny for different numbers of steps in the x and y directions so that the length of the space loops must change.
7.   Duplicate the calculations in flow2d.m. Use mesh and contour to view the temperatures at different times.
8.   In flow2d.m experiment with different time mesh sizes, $maxk = 100,$ 200, 400. Be sure to consider the stability constraint.
9.   In flow2d.m experiment with different space mesh sizes, $nx = 5, 10$ and 20. Be sure to consider the stability constraint.
10.   In flow2d.m experiment with different decay rates $dec = .01, .02$ and .04. Be sure to make any adjustments to the time step so that the stability condition holds.
11.   Experiment with the wind velocity in the MATLAB code flow2d.m.

(a).   Adjust the magnitudes of the velocity components and observe stability as a function of wind velocity.

(b).   If the wind velocity is not from the south and west, then the finite difference model in (1.5.10) will change. Let the wind velocity be from the north and west, say wind velocity = (.2, -.4). Modify the finite difference model.

(c).   Modify the MATLAB code flow2d.m to account for this change in wind direction.

12.   Suppose pollutant is being generated at a rate of 3 units of heat per unit volume per unit time.

(a).   How is the model for the 2D problem in equation (1.5.10) modified to account for this?

(b).   Modify flow2d.m to implement this source of pollution.

(c).   Experiment with different values for the heat source $f = 0, 1, 2, 3$.

1.6   Convergence Analysis

1.6.1   Introduction

Initial value problems have the form

$$u_t = f(t, u) \text{ and } u(0) = \text{given}. \tag{1.6.1}$$

The simplest cases can be solved by separation of variables, but in general they do not have closed form solutions. Therefore, one is forced to consider various approximation methods. In this section we study the simplest numerical method, the Euler finite difference method. We shall see that under appropriate assumptions the error made by this type of approximation is bounded by a constant times the step size.

1.6.2   Applied Area

Again consider a well stirred liquid such as a cup of coffee. Assume that the temperature is uniform with respect to space, but the temperature may be changing with respect to time. We wish to predict the temperature as a function of time given some initial observations about the temperature.

1.6.3   Model

A continuous model is Newton's *law of cooling* states that the rate of change of the temperature is proportional to the difference in the surrounding temperature and the temperature of the liquid

$$u_t = c(u_{sur} - u). \tag{1.6.2}$$

If $c = 0$, then there is perfect insulation, and the liquid's temperature must remain at its initial value. For large c the liquid's temperature will rapidly

approach the surrounding temperature. The closed form solution of this differential equation can be found by the separation of variables method and is, for
u_{sur} equal a constant,

$$u(t) = u_{sur} + (u(0) - u_{sur})e^{-ct}. \qquad (1.6.3)$$

If c is not given, then it can be found from a second observation such as $u(t_1) = u_1$. If u_{sur} is a function of t, one can still find a closed form solution provided the integrations steps are not too complicated.

1.6.4   Method

Euler's method involves the approximation of u_t by the finite difference

$$(u^{k+1} - u^k)/h$$

where $h = T/K$ and K is now the number of time steps, u^k is an approximation of $u(kh)$ and f is evaluated at (kh, u^k). If T is not finite, then h will be fixed and k may range over all of the positive integers. The differential equation (1.6.1) can be replaced by either

$$
\begin{aligned}
(u^{k+1} - u^k)/h &= f((k+1)h, u^{k+1}) \\
\text{or, } (u^{k+1} - u^k)/h &= f(kh, u^k).
\end{aligned}
\qquad (1.6.4)
$$

The choice in (1.6.4) is the simplest because it does not require the solution of a possibly nonlinear problem at each time step. The scheme given by (1.6.4) is called *Euler's method,* and it is a discrete model of the differential equation in (1.6.2). For the continuous Newton's law of cooling differential equation where $f(t, u) = c(u_{sur} - u)$ Euler's method is the same as the first order finite difference method for the discrete Newton's law of cooling.

The *improved Euler method* is given by the following two equations

$$
\begin{aligned}
(utemp - u^k)/h &= f(kh, u^k) & (1.6.5) \\
(u^{k+1} - u^k)/h &= 1/2(f(kh, u^k) + f((k+1)h, utemp)). & (1.6.6)
\end{aligned}
$$

Equation (1.6.5) gives a first estimate of the temperature at time kh, and then it is used in equation (1.6.6) where an average of the time derivative is computed. This is called improved because the errors for Euler's method are often bounded by a constant times the time step, while the errors for the improved Euler method are often bounded by a constant times the time step *squared.*

1.6.5   Implementation

One can easily use MATLAB to illustrate that as the time step decreases, the solution from the discrete models approaches the solution to the continuous model. This is depicted in both graphical and table form. In the MATLAB code eulerr.m we experiment with the number of time steps and fixed final time.

Newton's law of cooling for a constant surrounding temperature is considered so that the exact solution is known. The exact solution is compared with both the Euler and improved Euler approximation solutions.

In the MATLAB code eulerr.m lines 3-13 contain the input data. The arrays for the exact solution, Euler approximate solution and the improved Euler approximate solution are, respectively, uexact, ueul and uieul, and they are computed in time loop in lines 14-25. The output is given in lines 26-29 where the errors are given at the final time.

MATLAB Code eulerr.m

```
1.      % This code compares the discretization errors.
2.      % The Euler and improved Euler methods are used.
3.      clear;
4.      maxk = 5;       % number of time steps
5.      T = 10.0;       % final time
6.      dt = T/maxk;
7.      time(1) = 0;
8.      u0 = 200.;      % initial temperature
9.      c = 2./13.;     % insulation factor
10.     usur = 70.;       % surrounding temperature
11.     uexact(1) = u0;
12.     ueul(1) = u0;
13.     uieul(1) = u0;
14.     for k = 1:maxk       %time loop
15.         time(k+1) = k*dt;
16.         % exact solution
17.         uexact(k+1) = usur + (u0 - usur)*exp(-c*k*dt);
18.         % Euler numerical approximation
19.         ueul(k+1) = ueul(k) +dt*c*(usur - ueul(k));
20.         % improved Euler numerical approximation
21.         utemp = uieul(k) +dt*c*(usur - uieul(k));
22.         uieul(k+1)= uieul(k)
                        + dt/2*(c*(usur - uieul(k))+c*(usur - utemp));
23.         err_eul(k+1) = abs(ueul(k+1) - uexact(k+1));
24.         err_im_eul(k+1) = abs(uieul(k+1) - uexact(k+1));
25.     end
26.     plot(time, ueul)
27.     maxk
28.     err_eul_at_T = err_eul(maxk+1)
29.     err_im_eul_at_T = err_im_eul(maxk+1)
```

Figure 1.6.1 contains the plots of for the Euler method given in the arrays ueul for $maxk = 5, 10, 20$ and 40 times steps. The curve for $maxk = 5$ is not realistic because of the oscillations, but it does approach the surrounding temperature. The other three plots for all points in time increase towards the exact solution.

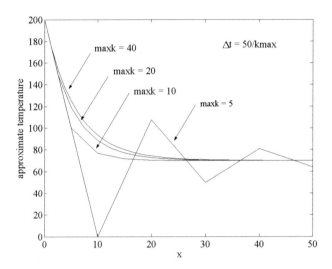

Figure 1.6.1: Euler Approximations

Table 1.6.1: Euler Errors at t = 10

Time Steps	Euler Error	Improved Euler Error
5	7.2378	0.8655
10	3.4536	0.1908
20	1.6883	0.0449
40	0.8349	0.0109

Another way of examining the error is to fix a time and consider the difference in the exact temperature and the approximate temperatures given by the Euler methods. Table 1.6.1 does this for time equal to 10. The Euler errors are cut in half whenever the number of time steps are doubled, that is, the Euler errors are bounded by a constant times the time step size. The improved Euler errors are cut in one quarter when the number of time steps are doubled, that is, the improved Euler errors are bounded by a constant times the time step size *squared*.

1.6.6 Assessment

In order to give an explanation of the discretization error, we must review the Mean Value theorem and an extension. The Mean Value theorem, like the intermediate value theorem, appears to be clearly true once one draws the picture associated with it. Drawing the picture does make some assumptions. For example, consider the function given by $f(x) = 1 - |x|$. Here there is a "corner" in the graph at $x = 0$, that is, $f(x)$ does not have a derivative at $x = 0$.

Theorem 1.6.1 *(Mean Value Theorem) Let $f : [a,b] \to \mathbb{R}$ be continuous on $[a,b]$. If f has a derivative at each point of (a,b), then there is a c between a and x such that $f'(c) = (f(b) - f(a))/(b - a)$.*

If b is replaced by x and we solve for $f(x)$ in $f'(c) = (f(x) - f(a))/(x - a)$, then provided $f(x)$ has a derivative

$$f(x) = f(a) + f'(c)(x - a)$$

for some c between a and x. Generally, one does not know the exact value of c, but if the derivative is bounded by M, then the following inequality holds

$$|f(x) - f(a)| \le M\,|x - a|.$$

An extension of this linear approximation to a quadratic approximation of $f(x)$ is stated in the next theorem.

Theorem 1.6.2 *(Extended Mean Value Theorem) If $f : [a,b] \to \mathbb{R}$ has two continuous derivatives on $[a,b]$, then there is a c between a and x such that*

$$f(x) = f(a) + f'(a)(x - a) + f''(c)(x - a)^2/2. \qquad (1.6.7)$$

Clearly Euler's method is a very inexpensive algorithm to execute. However, there are sizable errors. There are two types of errors:

$$Discretization\ error \equiv E_d^k = u^k - u(kh)$$

where u^k is from Euler's algorithm (1.6.4) with no roundoff error and $u(kh)$ is from the exact continuum solution (1.6.1).

$$Accumulation\ error \equiv E_r^k = U^k - u^k$$

where U^k is from Euler's algorithm, but with round errors.
The *overall error* contains both errors and is $E_r^k + E_d^k = U^k - u(kh)$. In Table 1.6.1 the discretization error for Euler's method is bounded by a constant times h, and the discretization error for the improved Euler method is bounded by a constant times h *squared*.

Now we will give the discretization error analysis for the Euler method applied to equation (1.6.2). The relevant terms for the error analysis are

$$u_t(kh) = c(u_{sur} - u(kh)) \qquad (1.6.8)$$
$$u^{k+1} = u^k + hc(u_{sur} - u^k) \qquad (1.6.9)$$

Use the extended Mean Value theorem on $u(kh + h)$ where a is replaced by kh and x is replaced by $kh + h$

$$u((k+1)h) = u(kh) + u_t(kh)h + u_{tt}(c_{k+1})h^2/2 \qquad (1.6.10)$$

Use the right side of (1.6.8) for $u_t(kh)$ in (1.6.10), and combine this with (1.6.9) to get

$$
\begin{aligned}
E_d^{k+1} &= u^{k+1} - u((k+1)h) \\
&= [u^k + hc(u_{sur} - u^k)] - \\
&\quad [u(kh) + c(u_{sur} - u(kh))h + u_{tt}(c_{k+1})h^2/2] \\
&= aE_d^k + b_{k+1}h^2/2 \\
\text{where } a &= 1 - ch \text{ and } b_{k+1} = -u_{tt}(c_{k+1}).
\end{aligned}
\tag{1.6.11}
$$

Suppose $a = 1 - ch > 0$ and $|b_{k+1}| \leq M$. Use the triangle inequality, a "telescoping" argument and the partial sums of the geometric series $1 + a + a^2 + \cdots + a^k = (a^{k+1} - 1)/(a - 1)$ to get

$$
\begin{aligned}
|E_d^{k+1}| &\leq a|E_d^k| + Mh^2/2 \\
&\leq a(a|E_d^{k-1}| + Mh^2/2) + Mh^2/2 \\
&\vdots \\
&\leq a^{k+1}|E_d^0| + (a^{k+1} - 1)/(a - 1)\ Mh^2/2.
\end{aligned}
\tag{1.6.12}
$$

Assume $E_d^0 = 0$ and use the fact that $a = 1 - ch$ with $h = T/K$ to obtain

$$
\begin{aligned}
|E_d^{k+1}| &\leq [(1 - cT/K)^K - 1]/(-ch)\ Mh^2/2 \\
&\leq 1/c\ Mh/2.
\end{aligned}
\tag{1.6.13}
$$

We have proved the following theorem, which is a special case of a more general theorem about Euler's method applied to ordinary differential equations of the form (1.6.1), see [4, chapter 5.2]

Theorem 1.6.3 *(Euler Error Theorem) Consider the continuous (1.6.2) and discrete (1.6.4) Newton cooling models. Let T be finite, $h = T/K$ and let solution of (1.6.2) have two derivatives on the time interval $[0, T]$. If the second derivative of the solution is bounded by M, the initial condition has no roundoff error and $1 - ch > 0$, then the discretization error is bounded by $(M/2c)h$.*

In the previous sections we consider discrete models for heat and pollutant transfer

$$
\begin{aligned}
\text{Pollutant Transfer} \quad &: \quad u_t = f - au_x - cu, \tag{1.6.14} \\
&\quad u(0, t) \text{ and } u(x, 0) \text{ given.} \\
\text{Heat Diffusion} \quad &: \quad u_t = f + (\kappa u_x)_x - cu, \tag{1.6.15} \\
&\quad u(0, t), u(L, t) \text{ and } u(x, 0) \text{ given.}
\end{aligned}
$$

The *discretization errors* for (1.6.14) and (1.6.15), where the solutions depend both on space and time, have the form

$$
\begin{aligned}
E_i^{k+1} &\equiv u_i^{k+1} - u(i\Delta x, (k+1)\Delta t) \\
\|E^{k+1}\| &\equiv \max_i |E_i^{k+1}|.
\end{aligned}
$$

Table 1.6.2: Errors for Flow

Δt	Δx	Flow Errors in (1.6.14)
1/10	1/20	0.2148
1/20	1/40	0.1225
1/40	1/60	0.0658
1/80	1/80	0.0342

Table 1.6.3: Errors for Heat

Δt	Δx	Heat Errors in (1.6.15)
1/50	1/5	$9.2079 \ 10^{-4}$
1/200	1/10	$2.6082 \ 10^{-4}$
1/800	1/20	$0.6630 \ 10^{-4}$
1/3200	1/40	$0.1664 \ 10^{-4}$

$u(i\Delta x, (k+1)\Delta t)$ is the exact solution, and u_i^{k+1} is the numerical or approximate solution. In the following examples the discrete models were from the explicit finite difference methods used in Sections 1.3 and 1.4.

Example for (1.6.14). Consider the MATLAB code flow1d.m (see flow1derr.m and equations (1.4.2-1.4.4)) that generates the numerical solution of (1.6.14) with $c = dec = .1$, $a = vel = .1$, $f = 0$, $u(0,t) = sin(2\pi(0 - vel\ t))$ and $u(x,0) = sin(2\pi x)$. It is compared over the time interval $t = 0$ to $t = T = 20$ and at $x = L = 1$ with the exact solution $u(x,t) = e^{-dec\ t}\ sin(2\pi(x - vel\ t))$. Note the error in Table 1.6.2 is proportional to $\Delta t + \Delta x$.

Example for (1.6.15). Consider the MATLAB code heat.m (see heaterr.m and equations (1.2.1)-(1.2.3)) that computes the numerical solution of (1.6.15) with $k = 1/\pi^2$, $c = 0$, $f = 0$, $u(0,t) = 0$, $u(1,t) = 0$ and $u(x,0) = sin(\pi x)$. It is compared at $(x,t) = (1/2, 1)$ with the exact solution $u(x,t) = e^{-t}sin(\pi x)$. Here the error in Table 1.6.3 is proportional to $\Delta t + \Delta x^2$.

In order to give an explanation of the discretization errors, one must use higher order *Taylor polynomial approximation*. The proof of this is similar to the extended mean value theorem. It asserts if $f : [a, b] \to \mathbb{R}$ has $n + 1$ continuous derivatives on $[a, b]$, then there is a c between a and x such that

$$\begin{aligned} f(x) \ = \ & f(a) + f^{(1)}(a)(x - a) + \cdots + f^{(n)}(a)/n!\ (x - a)^n \\ & + f^{(n+1)}(c)/(n + 1)!\ (x - a)^{n+1}. \end{aligned}$$

Theorem 1.6.4 *(Discretization Error for (1.6.14)) Consider the continuous model (1.6.14) and its explicit finite difference model. If a, c and $(1 - a\Delta t/\Delta x - \Delta t\, c)$ are nonnegative, and u_{tt} and u_{xx} are bounded on $[0, L] \times [0, T]$, then there are constants C_1 and C_2 such that*

$$\left\| E^{k+1} \right\| \leq (C_1 \Delta x + C_2 \Delta t) T .$$

Theorem 1.6.5 *(Discretization Error for (1.6.15)) Consider the continuous model (1.6.15) and its explicit finite difference model. If $c > 0, \kappa > 0, \alpha = (\Delta t/\Delta x^2)\kappa$ and $(1 - 2\alpha - \Delta t\, c) > 0$, and u_{tt} and u_{xxxx} are bounded on $[0, L] \times [0, T]$, then there are constants C_1 and C_2 such that*

$$\left\| E^{k+1} \right\| \leq (C_1 \Delta x^2 + C_2 \Delta t) T.$$

1.6.7 Exercises

1. Duplicate the calculations in Figure 1.6.1, and find the graphical solution when $maxk = 80$.

2. Verify the calculations in Table 1.6.1, and find the error when $maxk = 80$.

3. Assume the surrounding temperature initially is 70 and increases at a constant rate of one degree every ten minutes.

  (a). Modify the continuous model in (1.6.2) and find its solution via the MATLAB command desolve.

  (b). Modify the discrete model in (1.6.4).

4. Consider the time dependent surrounding temperature in problem 3.

  (a). Modify the MATLAB code eulerr.m to account for the changing surrounding temperature.

  (b). Experiment with different number of time steps with $maxk = 5$, 10, 20, 40 and 80.

5. In the proof of the Theorem 1.6.3 justify the (1.6.11) and $|b_{k+1}| \leq M$.

6. In the proof of the Theorem 1.6.3 justify the (1.6.12) and (1.6.13).

7. Modify Theorem 1.6.3 to account for the case where the surrounding temperature can depend on time, $u_{sur} = u_{sur}(t)$. What assumptions should be placed on $u_{sur}(t)$ so that the discretization error will be bounded by a constant times the step size?

8. Verify the computations in Table 1.6.14. Modify flow1d.m by inserting an additional line inside the time-space loops for the error (see flow1derr.m).

9. Verify the computations in Table 1.6.15. Modify heat.m by inserting an additional line inside the time-space loops for the error (see heaterr.m).

10. Consider a combined model for (1.6.14)-(1.6.15): $u_t = f + (\kappa u_x)_x - au_x - cu$. Formulate suitable boundary conditions, an explicit finite difference method, a MATLAB code and prove an error estimate.

Chapter 2

Steady State Discrete Models

This chapter considers the steady state solution to the heat diffusion model. Here boundary conditions that have derivative terms in them are applied to the cooling fin model, which will be extended to two and three space variables in the next two chapters. Variations of the Gauss elimination method are studied in Sections 2.3 and 2.4 where the block structure of the coefficient matrix is utilized. This will be very important for parallel solution of large algebraic systems. The last two sections are concerned with the analysis of two types of convergence: one with respect to discrete time and one with respect to the mesh size. Additional introductory references include Burden and Faires [4] and Meyer [16].

2.1 Steady State and Triangular Solves

2.1.1 Introduction

The next four sections will be concerned with solving the linear algebraic system

$$Ax = d \qquad (2.1.1)$$

where A is a given $n \times n$ matrix, d is a given column vector and x is a column vector to be found. In this section we will focus on the special case where A is a triangular matrix. Algebraic systems have many applications such as inventory management, electrical circuits, the steady state polluted stream and heat diffusion in a wire.

  Both the polluted stream and heat diffusion problems initially were formulated as time and space dependent problems, but for larger times the concentrations or temperatures depend less on time than on space. A time independent solution is called *steady state* or *equilibrium* solution, which can be modeled by

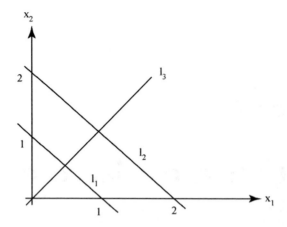

Figure 2.1.1: Infinite or None or One Solution(s)

systems of algebraic equations (2.1.1) with x being the steady state solution. Systems of the form $Ax = d$ can be derived from $u = Au+b$ via $(I-A)u = b$ and replacing u by x, b by d and $(I - A)$ by A.

There are several cases of (2.1.1), which are illustrated by the following examples.

Example 1. The algebraic system may not have a solution. Consider

$$\begin{bmatrix} 1 & 1 \\ 2 & 2 \end{bmatrix} = \begin{bmatrix} d_1 \\ d_2 \end{bmatrix}.$$

If $d = [1 \quad 2]^T$, then there are an infinite number of solutions given by points on the line l_1 in Figure 2.1.1. If $d = [1 \quad 4]^T$, then there are no solutions because the lines l_1 and l_2 are parallel. If the problem is modified to

$$\begin{bmatrix} 1 & 1 \\ -2 & 2 \end{bmatrix} = \begin{bmatrix} 1 \\ 0 \end{bmatrix},$$

then there will be exactly one solution given by the intersection of lines l_1 and l_3.

Example 2. This example illustrates a system with three equations with either no solution or a set of solutions that is a straight line in 3D space.

$$\begin{bmatrix} 1 & 1 & 1 \\ 0 & 0 & 3 \\ 0 & 0 & 3 \end{bmatrix} \begin{bmatrix} x_1 \\ x_2 \\ x_3 \end{bmatrix} = \begin{bmatrix} 1 \\ d_2 \\ 3 \end{bmatrix}$$

If $d_2 \neq 3$, then the second row or equation implies $3x_3 \neq 3$ and $x_1 \neq 1$. This contradicts the third row or equation, and hence, there is no solution to the system of equations. If $d_2 = 3$, then $x_3 = 1$ and x_2 is a free parameter. The first row or equation is $x_1 + x_2 + 1 = 1$ or $x_1 = -x_2$. The vector form of the solution is

$$
\begin{bmatrix} x_1 \\ x_2 \\ x_3 \end{bmatrix} = \begin{bmatrix} 0 \\ 0 \\ 1 \end{bmatrix} + x_2 \begin{bmatrix} -1 \\ 1 \\ 0 \end{bmatrix}.
$$

This is a straight line in 3D space containing the point $\begin{bmatrix} 0 & 0 & 1 \end{bmatrix}^T$ and going in the direction $\begin{bmatrix} -1 & 1 & 0 \end{bmatrix}^T$.

The easiest algebraic systems to solve have either diagonal or a triangular matrices.

Example 3. Consider the case where A is a diagonal matrix.

$$
\begin{bmatrix} 1 & 0 & 0 \\ 0 & 2 & 0 \\ 0 & 0 & 3 \end{bmatrix} \begin{bmatrix} x_1 \\ x_2 \\ x_3 \end{bmatrix} = \begin{bmatrix} 1 \\ 4 \\ 7 \end{bmatrix} \text{ whose solution is } \begin{bmatrix} x_1 \\ x_2 \\ x_3 \end{bmatrix} = \begin{bmatrix} 1/1 \\ 4/2 \\ 7/3 \end{bmatrix}.
$$

Example 4. Consider the case where A is a lower triangular matrix.

$$
\begin{bmatrix} 1 & 0 & 0 \\ 1 & 2 & 0 \\ 1 & 4 & 3 \end{bmatrix} \begin{bmatrix} x_1 \\ x_2 \\ x_3 \end{bmatrix} = \begin{bmatrix} 1 \\ 4 \\ 7 \end{bmatrix}.
$$

The first row or equation gives $x_1 = 1$. Use this in the second row or equation to get $1 + 2x_2 = 4$ and $x_2 = 3/2$. Put these two into the third row or equation to get $1(1) + 4(3/2) + 3x_3 = 7$ and $x_3 = 0$. This is known as a *forward* sweep.

Example 5. Consider the case where A is an upper triangular matrix

$$
\begin{bmatrix} 1 & -1 & 1 \\ 0 & 2 & 2 \\ 0 & 0 & 3 \end{bmatrix} \begin{bmatrix} x_1 \\ x_2 \\ x_3 \end{bmatrix} = \begin{bmatrix} 1 \\ 4 \\ 9 \end{bmatrix}.
$$

First, the last row or equation gives $x_3 = 3$. Second, use this in the second row or equation to get $2x_2 + 2(3) = 4$ and $x_2 = -1$. Third, put these two into the first row or equation to get $1(x_1) - 1(-1) + 3(3) = 1$ and $x_1 = -9$. This illustrates a *backward* sweep where the components of the matrix are retrieved by rows.

2.1.2   Applied Area

Consider a stream which initially has an *industrial spill* upstream. Suppose that at the farthest point upstream the river is being polluted so that the concentration is independent of time. Assume the flow rate of the stream is known and the chemical decay rate of the pollutant is known. We would like to determine the short and long term effect of this initial spill and upstream pollution.

The discrete model was developed in Section 1.4 for the concentration u_i^{k+1} approximation of $u(i\Delta x, (k+1)\Delta t))$.

$$
\begin{aligned}
u_i^{k+1} &= vel\ (\Delta t/\Delta x)u_{i-1}^k + (1 - vel\ (\Delta t/\Delta x) - \Delta t\ dec)u_i^k \\
i &= 1, ..., n-1 \text{ and } k = 0, ..., maxk - 1, \\
u_i^0 &= \text{given for } i = 1, ..., n-1 \text{ and} \\
u_0^k &= \text{given for } k = 1, ..., maxk.
\end{aligned}
$$

This discrete model should approximate the solution to the continuous space and time model

$$
\begin{aligned}
u_t &= -vel\ u_x - dec\ u, \\
u(x,0) &= \text{given and} \\
u(0,t) &= \text{given.}
\end{aligned}
$$

The steady state solution will be independent of time. For the discrete model this is

$$
0 = vel\ (\Delta t/\Delta x)u_{i-1} + (0 - vel\ (\Delta t/\Delta x) - \Delta t\ dec)u_i \quad (2.1.2)
$$
$$
u_0 = \text{given.} \quad (2.1.3)
$$

The *discrete steady state* model may be reformulated as in (2.1.1) where A is a lower triangular matrix. For example, if there are 3 unknown concentrations, then (2.1.2) must hold for $i = 1, 2$, and 3

$$
\begin{aligned}
0 &= vel\ (\Delta t/\Delta x)u_0 + (0 - vel\ (\Delta t/\Delta x) - \Delta t\ dec)u_1 \\
0 &= vel\ (\Delta t/\Delta x)u_1 + (0 - vel\ (\Delta t/\Delta x) - \Delta t\ dec)u_2 \\
0 &= vel\ (\Delta t/\Delta x)u_2 + (0 - vel\ (\Delta t/\Delta x) - \Delta t\ dec)u_3.
\end{aligned}
$$

Or, when $d = vel/\Delta x$ and $c = 0 - d - dec$, the vector form of this is

$$
\begin{bmatrix} c & 0 & 0 \\ d & c & 0 \\ 0 & d & c \end{bmatrix} \begin{bmatrix} u_1 \\ u_2 \\ u_3 \end{bmatrix} = \begin{bmatrix} du_0 \\ 0 \\ 0 \end{bmatrix}. \quad (2.1.4)
$$

If the velocity of the stream is negative so that the stream is moving from right to left, then $u(L,t)$ will be given and the resulting steady state discrete model will be upper triangular.

The *continuous steady state* model is

$$
0 = -vel\ u_x - dec\ u, \quad (2.1.5)
$$
$$
u(0) = \text{given.} \quad (2.1.6)
$$

The solution is $u(x) = u(0)e^{-(dec/vel)x}$. If the velocity of the steam is negative (moving from the right to the left), then the given concentration will be u_n where n is the size of matrix and the resulting matrix will be upper triangular.

2.1.3 Model

The general model will be an algebraic system (2.1.1) of n equations and n unknowns. We will assume the matrix has *upper triangular* form

$$A = [a_{ij}] \text{ where } a_{ij} = 0 \text{ for } i > j \text{ and } 1 \leq i, j \leq n.$$

The row numbers of the matrix are associated with i, and the column numbers are given by j. The component form of $Ax = d$ when A is upper triangular is for all i

$$a_{ii}x_i + \sum_{j>i} a_{ij}x_j = d_i. \tag{2.1.7}$$

One can take advantage of this by setting $i = n$, where the summation is now vacuous, and solve for x_n.

2.1.4 Method

The last equation in the component form is $a_{nn}x_n = d_n$, and hence, $x_n = d_n/a_{nn}$. The $(n-1)$ equation is $a_{n-1,n-1}x_{n-1} + a_{n-1,n}x_n = d_{n-1}$, and hence, we can solve for $x_{n-1} = (d_{n-1} - a_{n-1,n}x_n)/a_{n-1,n-1}$. This can be repeated, provided each a_{ii} is nonzero, until all x_j have been computed. In order to execute this on a computer, there must be two loops: one for the equation (2.1.7) (the i-loop) and one for the summation (the j-loop). There are two versions: the ij version with the i-loop on the outside, and the ji version with the j-loop on the outside. The ij version is a reflection of the backward sweep as in Example 5. Note the inner loop retrieves data from the array by jumping from one column to the next. In Fortran this is in stride n and can result in slower computation times. Example 6 illustrates the ji version where we subtract multiples of the columns of A, the order of the loops is interchanged, and the components of A are retrieved by moving down the columns of A.

Example 6. Consider the following 3×3 algebraic system

$$\begin{bmatrix} 4 & 6 & 1 \\ 0 & 1 & 1 \\ 0 & 0 & 4 \end{bmatrix} \begin{bmatrix} x_1 \\ x_2 \\ x_3 \end{bmatrix} = \begin{bmatrix} 100 \\ 10 \\ 20 \end{bmatrix}.$$

This product can also be viewed as linear combinations of the columns of the matrix

$$\begin{bmatrix} 4 \\ 0 \\ 0 \end{bmatrix} x_1 + \begin{bmatrix} 6 \\ 1 \\ 0 \end{bmatrix} x_2 + \begin{bmatrix} 1 \\ 1 \\ 4 \end{bmatrix} x_3 = \begin{bmatrix} 100 \\ 10 \\ 20 \end{bmatrix}.$$

First, solve for $x_3 = 20/4 = 5$. Second, subtract the last column times x_3 from both sides to reduce the dimension of the problem

$$\begin{bmatrix} 4 \\ 0 \\ 0 \end{bmatrix} x_1 + \begin{bmatrix} 6 \\ 1 \\ 0 \end{bmatrix} x_2 = \begin{bmatrix} 100 \\ 10 \\ 20 \end{bmatrix} - \begin{bmatrix} 1 \\ 1 \\ 4 \end{bmatrix} 5 = \begin{bmatrix} 95 \\ 5 \\ 0 \end{bmatrix}.$$

Third, solve for $x_2 = 5/1$. Fourth, subtract the second column times x_2 from both sides

$$\begin{bmatrix} 4 \\ 0 \\ 0 \end{bmatrix} x_1 = \begin{bmatrix} 95 \\ 5 \\ 0 \end{bmatrix} - \begin{bmatrix} 6 \\ 1 \\ 0 \end{bmatrix} 5 = \begin{bmatrix} 65 \\ 0 \\ 0 \end{bmatrix}.$$

Fifth, solve for $x_1 = 65/4$.

Since the following MATLAB codes for the ij and ji methods of an upper triangular matrix solve are very clear, we will not give a formal statement of these two methods.

2.1.5   Implementation

We illustrate two MATLAB codes for doing upper triangular solve with the ij (row) and the ji (column) methods. Then the MATLAB solver $x = A\backslash d$ and $inv(A) * d$ will be used to solve the steady state polluted stream problem.

In the code jisol.m lines 1-4 are the data for Example 6, and line 5 is the first step of the column version. The j-loop in line 6 moves the rightmost column of the matrix to the right side of the vector equation, and then in line 10 the next value of the solution is computed.

MATLAB Code jisol.m

```
1.    clear;
2.    A = [4 6 1;0 1 1;0 0 4]
3.    d = [100 10 20]'
4.    n = 3
5.    x(n) = d(n)/A(n,n);
6.    for j = n:-1:2
7.        for i = 1:j-1
8.            d(i) = d(i) - A(i,j)*x(j);
9.        end
10.   x(j-1) = d(j-1)/A(j-1,j-1);
11.   end
12.   x
```

In the code ijsol.m the i-loop in line 6 computes the partial row sum with respect to the j index, and this is done for each row i by the j-loop in line 8.

MATLAB Code ijsol.m

```
1.    clear;
2.    A = [4 6 1;0 1 1;0 0 4]
```

```
3.      d = [100 10 20]'
4.      n = 3
5.      x(n) = d(n)/A(n,n);
6.      for i = n:-1:1
7.          sum = d(i);
8.          for j = i+1:n
9.              sum = sum - A(i,j)*x(j);
10.         end
11.         x(i) = sum/A(i,i);
12.     end
13.  x
```

MATLAB can easily solve problems with n equations and n unknowns, and the coefficient matrix, A, does not have to be either upper or lower triangular. The following are two commands to do this, and these will be more completely described in the next section.

MATLAB Linear Solve A\d and inv(A)*d.

```
>A
A =
4 6 1
0 1 1
0 0 4

>d
d =
100
10
20

>x = A\d
x =
16.2500
5.0000
5.0000

>x = inv(A)*d
x =
16.2500
5.0000
5.0000
```

Finally, we return to the steady state polluted stream in (2.1.4). Assume $L = 1$, $\Delta x = L/3 = 1/3$, $vel = 1/3$, $dec = 1/10$ and $u(0) = 2/10$. The continuous steady state solution is $u(x) = (2/10)e^{-(3/10)x}$. We approximate this solution by either the discrete solution for large k, or the solution to the algebraic system. For just three unknowns the algebraic system in (2.1.4) with

$d = (1/3)/(1/3) = 1$ and $c = 0 - 1 - (1/10) = -1.1$ is easily solved for the approximate concentration at three positions in the stream.

>A = [1.1 0 0;-1 1.1 0;0 -1 1.1]
A =
1.1000 0 0
-1.0000 1.1000 0
0 -1.0000 1.1000

>d = [.2 0 0]'
d =
 0.2000
 0
 0

>A\d
ans =
0.1818
0.1653
0.1503

The above numerical solution is an approximation of continuous solution $u(x) = .2e^{-x}$ where $x_1 = 1\Delta x = 1/3$, $x_2 = 2\Delta x = 2/3$ and $x_3 = 3\Delta x = 1$ so that $.2e^{-.1} = .18096$, $.2e^{-.2} = .16375$ and $.2e^{-.3} = .14816$, respectively.

2.1.6   Assessment

One problem with the upper triangular solve algorithm may occur if the diagonal components of A, a_{ii}, are very small. In this case the floating point approximation may induce significant errors. Another instance is two equations which are nearly the same. For example, for two equations and two variables suppose the lines associated with the two equations are almost parallel. Then small changes in the slopes, given by either floating point or empirical data approximations, will induce big changes in the location of the intersection, that is, the solution. The following elementary theorem gives conditions on the matrix that will yield unique solutions.

Theorem 2.1.1 *(Upper Triangular Existence)   Consider $Ax = d$ where A is upper triangular ($a_{ij} = 0$ for $i > j$) and an $n \times n$ matrix. If all a_{ii} are not zero, then $Ax = d$ has a solution. Moreover, this solution is unique.*

Proof.   The derivation of the ij method for solving upper triangular algebraic systems established the existence part. In order to prove the solution is unique, let x and y be two solutions $Ax = d$ and $Ay = d$. Subtract these two and use the distributive property of matrix products $Ax - Ay = d - d$ so that $A(x - y) = 0$. Now apply the upper triangular solve algorithm with d replaced by 0 and x replaced by $x - y$. This implies $x - y = 0$ and so $x = y$. ∎

2.1.7   Exercises

1.   State an ij version of an algorithm for solving lower triangular problems.
2.   Prove an analogous existence and uniqueness theorem for lower triangular problems.
3.   Use the ij version to solve the following

$$
\begin{bmatrix}
1 & 0 & 0 & 0 \\
2 & 5 & 0 & 0 \\
-1 & 4 & 5 & 0 \\
0 & 2 & 3 & -2
\end{bmatrix}
\begin{bmatrix}
x_1 \\ x_2 \\ x_3 \\ x_4
\end{bmatrix}
=
\begin{bmatrix}
1 \\ 3 \\ 7 \\ 11
\end{bmatrix}.
$$

4.   Consider example 5 and use example 6 as a guide to formulate a ji (column) version of the solution for example 5.
5.   Use the ji version to solve the problem in 3.
6.   Write a MATLAB version of the ji method for a lower triangular solve. Use it to solve the problem in 3.
7.   Use the ij version and MATLAB to solve the problem in 3.
8.   Verify the calculations for the polluted stream problem. Experiment with different flow and decay rates. Observe stability and steady state solutions.
9.   Consider the steady state polluted stream problem with fixed $L = 1.0$, $vel = 1/3$ and $dec = 1/10$. Experiment with 4, 8 and 16 unknowns so that $\Delta x = 1/4, 1/8$ and$1/16$, respectively. Formulate the analogue of the vector equation (2.1.14) and solve it. Compare the solutions with the solution of the continuous model.
10.   Formulate a discrete model for the polluted stream problem when the velocity of the stream is negative.

2.2   Heat Diffusion and Gauss Elimination

2.2.1   Introduction

In most applications the coefficient matrix is not upper or lower triangular. By adding and subtracting multiples of the equations, often one can convert the algebraic system into an equivalent triangular system. We want to make this systematic so that these calculations can be done on a computer.

A first step is to reduce the notation burden. Note that the positions of all the x_i were always the same. Henceforth, we will simply delete them. The entries in the $n \times n$ matrix A and the entries in the $n \times 1$ column vector d may be combined into the $n \times (n+1)$ *augmented matrix*

$$[A \; d].$$

For example, the augmented matrix for the algebraic system

$$2x_1 + 6x_2 + 0x_3 = 12$$
$$0x_1 + 6x_2 + 1x_3 = 0$$
$$1x_1 - 1x_2 + 1x_3 = 0$$

is

$$[A\ d] = \begin{bmatrix} 2 & 6 & 0 & 12 \\ 0 & 6 & 1 & 0 \\ 1 & -1 & 1 & 0 \end{bmatrix}.$$

Each row of the augmented matrix represents the coefficients and the right side of an equation in the algebraic system.

The next step is to add or subtract multiples of rows to get all zeros in the lower triangular part of the matrix. There are three basic row operations:

(i).   interchange the order of two rows or equations,
(ii).   multiply a row or equation by a nonzero constant and
(iii).   add or subtract rows or equations.

In the following example we use a combination of (ii) and (iii), and note each row operation is equivalent to a multiplication by an *elementary matrix*, a matrix with ones on the diagonal and one nonzero off-diagonal component.

Example. Consider the above problem. First, subtract 1/2 of row 1 from row 3 to get a zero in the (3,1) position:

$$E_1[A\ d] = \begin{bmatrix} 2 & 6 & 0 & 12 \\ 0 & 6 & 1 & 0 \\ 0 & -4 & 1 & -6 \end{bmatrix} \text{ where } E_1 = \begin{bmatrix} 1 & 0 & 0 \\ 0 & 1 & 0 \\ -1/2 & 0 & 1 \end{bmatrix}.$$

Second, add 2/3 of row 2 to row 3 to get a zero in the (3,2) position:

$$E_2E_1[A\ d] = \begin{bmatrix} 2 & 6 & 0 & 12 \\ 0 & 6 & 1 & 0 \\ 0 & 0 & 5/3 & -6 \end{bmatrix} \text{ where } E_2 = \begin{bmatrix} 1 & 0 & 0 \\ 0 & 1 & 0 \\ 0 & 2/3 & 1 \end{bmatrix}.$$

Let $E = E_2E_1$, $U = EA$ and $\hat{d} = Ed$ so that $E[A\ d] = [U\ \hat{d}]$. Note U is upper triangular. Each elementary row operation can be reversed, and this has the form of a matrix inverse of each elementary matrix:

$$E_1^{-1} = \begin{bmatrix} 1 & 0 & 0 \\ 0 & 1 & 0 \\ 1/2 & 0 & 1 \end{bmatrix} \text{ and } E_1^{-1}E_1 = I = \begin{bmatrix} 1 & 0 & 0 \\ 0 & 1 & 0 \\ 0 & 0 & 1 \end{bmatrix},$$

$$E_2^{-1} = \begin{bmatrix} 1 & 0 & 0 \\ 0 & 1 & 0 \\ 0 & -2/3 & 1 \end{bmatrix} \text{ and } E_2^{-1}E_2 = I.$$

Note that $A = LU$ where $L = E_1^{-1}E_2^{-1}$ because by repeated use of the associative property

$$\begin{align}
(E_1^{-1}E_2^{-1})(EA) &= (E_1^{-1}E_2^{-1})((E_2E_1)A) \\
&= ((E_1^{-1}E_2^{-1})(E_2E_1))A \\
&= (E_1^{-1}(E_2^{-1}(E_2E_1)))A \\
&= (E_1^{-1}((E_2^{-1}E_2)E_1))A \\
&= (E_1^{-1}E_1)A \\
&= A.
\end{align}$$

The product $L = E_1E_2$ is a lower triangular matrix and $A = LU$ is called an *LU factorization of A.*

Definition. *An $n \times n$ matrix, A, has an inverse $n \times n$ matrix, A^{-1}, if and only if $A^{-1}A = AA^{-1} = I$, the $n \times n$ identity matrix.*

Theorem 2.2.1 *(Basic Properties) Let A be an $n \times n$ matrix that has an inverse:*

1. *A^{-1} is unique,*

2. *$x = A^{-1}d$ is a solution to $Ax = d$,*

3. *$(AB)^{-1} = B^{-1}A^{-1}$ provided B also has an inverse and*

4. *$A^{-1} = \begin{bmatrix} c_1 & c_2 & \cdots & c_n \end{bmatrix}$ has column vectors that are solutions to $Ac_j = e_j$ where e_j are unit column vectors with all zero components except the jth, which is equal to one.*

We will later discuss these properties in more detail. Note, given an inverse matrix one can solve the associated linear system. Conversely, if one can solve the linear problems in property 4 via Gaussian elimination, then one can find the inverse matrix. Elementary matrices can be used to find the LU factorizations and the inverses of L and U. Once L and U are known apply property 3 to find $A^{-1} = U^{-1}L^{-1}$. A word of caution is appropriate and also see Section 8.1 for more details. Not all matrices have inverses such as

$$A = \begin{bmatrix} 1 & 0 \\ 2 & 0 \end{bmatrix}.$$

Also, one may need to use permutations of the rows of A so that $PA = LU$ such as

$$\begin{align}
A &= \begin{bmatrix} 0 & 1 \\ 2 & 3 \end{bmatrix} \\
PA &= \begin{bmatrix} 0 & 1 \\ 1 & 0 \end{bmatrix}\begin{bmatrix} 0 & 1 \\ 2 & 3 \end{bmatrix} \\
&= \begin{bmatrix} 2 & 3 \\ 0 & 1 \end{bmatrix}.
\end{align}$$

2.2.2   Applied Area

We return to the heat conduction problem in a thin wire, which is thermally insulated on its lateral surface and has length L. Earlier we used the explicit method for this problem where the temperature depended on both time and space. In our calculations we observed, provided the stability condition held, the time dependent solution converges to time independent solution, which we called a steady state solution.

Steady state solutions correspond to models, which are also derived from Fourier's heat law. The difference now is that the change, with respect to time, in the heat content is zero. Also, the temperature is a function of just space so that $u_i \approx u(ih)$ where $h = L/n$.

$$
\begin{aligned}
\text{change in heat content} \quad = \quad & 0 \approx \text{(heat from the source)} \\
& +\text{(heat diffusion from the left side)} \\
& +\text{(heat diffusion from the right side)}.
\end{aligned}
$$

Let A be the cross section area of the thin wire and K be the thermal conductivity so that the approximation of the change in the heat content for the small volume Ah is

$$
0 = Ah\ \Delta t f + A\Delta t\ K(u_{i+1} - u_i)/h - A\Delta t\ K(u_i - u_{i-1})/h. \qquad (2.2.1)
$$

Now, divide by $Ah\ \Delta t$, let $\beta = K/h^2$, and we have the following $n-1$ equations for the $n-1$ unknown approximate temperatures u_i.

Finite Difference Equations for Steady State Heat Diffusion.

$$
\begin{aligned}
0 \quad &= \quad f + \beta(u_{i+1} + u_{i-1}) - 2\beta u_i \text{ where} && (2.2.2) \\
i \quad &= \quad 1, ..., n-1 \text{ and } \beta = K/h^2 \text{ and} \\
u_0 \quad &= \quad u_n = 0. && (2.2.3)
\end{aligned}
$$

Equation (2.2.3) is the temperature at the left and right ends set equal to zero. The discrete model (2.2.2)-(2.2.3) is an approximation of the continuous model (2.2.4)-(2.2.5). The partial differential equation (2.2.4) can be derived from (2.2.1) by replacing u_i by $u(ih)$, dividing by $Ah\ \Delta t$ and letting h and Δt go to zero.

Continuous Model for Steady State Heat Diffusion.

$$
\begin{aligned}
0 \quad &= \quad f + (Ku_x)_x \text{ and} && (2.2.4) \\
u(0) \quad &= \quad 0 = u(L). && (2.2.5)
\end{aligned}
$$

2.2.3   Model

The finite difference model may be written in matrix form where the matrix is a *tridiagonal* matrix. For example, if $n = 4$, then we are dividing the wire into four equal parts and there will be 3 unknowns with the end temperatures set equal to zero.

Tridiagonal Algebraic System with n = 4.

$$\begin{bmatrix} 2\beta & -\beta & 0 \\ -\beta & 2\beta & -\beta \\ 0 & -\beta & 2\beta \end{bmatrix} \begin{bmatrix} u_1 \\ u_2 \\ u_3 \end{bmatrix} = \begin{bmatrix} f_1 \\ f_2 \\ f_3 \end{bmatrix}.$$

Suppose the length of the wire is 1 so that $h = 1/4$, and the thermal conductivity is .001. Then $\beta = .016$ and if $f_i = 1$, then upon dividing all rows by β and using the augmented matrix notation we have

$$[A\ d] = \begin{bmatrix} 2 & -1 & 0 & 62.5 \\ -1 & 2 & -1 & 62.5 \\ 0 & -1 & 2 & 62.5 \end{bmatrix}.$$

Forward Sweep (put into upper triangular form):
    Add 1/2(row 1) to (row 2),

$$E_1[A\ d] = \begin{bmatrix} 2 & -1 & 0 & 62.5 \\ 0 & 3/2 & -1 & (3/2)62.5 \\ 0 & -1 & 2 & 62.5 \end{bmatrix} \text{ where } E_1 = \begin{bmatrix} 1 & 0 & 0 \\ 1/2 & 1 & 0 \\ 0 & 0 & 1 \end{bmatrix}.$$

Add 2/3(row 2) to (row 3),

$$E_2 E_1[A\ d] = \begin{bmatrix} 2 & -1 & 0 & 62.5 \\ 0 & 3/2 & -1 & (3/2)62.5 \\ 0 & 0 & 4/3 & (2)62.5 \end{bmatrix} \text{ where } E_2 = \begin{bmatrix} 1 & 0 & 0 \\ 0 & 1 & 0 \\ 0 & 2/3 & 1 \end{bmatrix}.$$

Backward Sweep (solve the triangular system):

$$\begin{aligned} u_3 &= (2)62.5(3/4) = 93.75, \\ u_2 &= ((3/2)62.5 + 93.75)(2/3) = 125 \text{ and} \\ u_1 &= (62.5 + 125)/2 = 93.75. \end{aligned}$$

The above solutions of the discrete model should be an approximation of the continuous model $u(x)$ where $x = 1\Delta x, 2\Delta x$ and $3\Delta x$. Note the LU factorization of the 3×3 coefficient A has the form

$$\begin{aligned} A &= (E_2 E_1)^{-1} U \\ &= E_1^{-1} E_2^{-1} U \\ &= \begin{bmatrix} 1 & 0 & 0 \\ -1/2 & 1 & 0 \\ 0 & 0 & 1 \end{bmatrix} \begin{bmatrix} 1 & 0 & 0 \\ 0 & 1 & 0 \\ 0 & -2/3 & 1 \end{bmatrix} \begin{bmatrix} 2 & -1 & 0 \\ 0 & 3/2 & -1 \\ 0 & 0 & 4/3 \end{bmatrix} \\ &= \begin{bmatrix} 1 & 0 & 0 \\ -1/2 & 1 & 0 \\ 0 & -2/3 & 1 \end{bmatrix} \begin{bmatrix} 2 & -1 & 0 \\ 0 & 3/2 & -1 \\ 0 & 0 & 4/3 \end{bmatrix} \\ &= LU. \end{aligned}$$

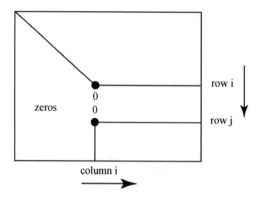

Figure 2.2.1: Gaussian Elimination

2.2.4   Method

The general Gaussian elimination method requires forming the augmented matrix, a forward sweep to convert the problem to upper triangular form, and a backward sweep to solve this upper triangular system. The row operations needed to form the upper triangular system must be done in a systematic way:

(i).   Start with column 1 and row 1 of the augmented matrix. Use an appropriate multiple of row 1 and subtract it from row i to get a zero in the $(i,1)$ position in column 1 with $i > 1$.

(ii).   Move to column 2 and row 2 of the new version of the augmented matrix. In the same way use row operations to get zero in each $(i, 2)$ position of column 2 with $i > 2$.

(iii).   Repeat this until all the components in the lower left part of the subsequent augmented matrices are zero.

This is depicted in the Figure 2.2.1 where the (i, j) component is about to be set to zero.

Gaussian Elimination Algorithm.

```
        define the augmented matrix [A d]
        for j = 1,n-1                 (forward sweep)
            for i = j+1,n
                add multiple of (row j) to (row i) to get
                                a zero in the (i,j) position
            endloop
        endloop
        for i = n,1                   (backward sweep)
            solve for xᵢ using row i
        endloop.
```

The above description is not very complete. In the forward sweep more details and special considerations with regard to roundoff errors are essential. The row operations in the inner loop may not be possible without some permutation of the rows, for example,

$$A = \begin{bmatrix} 0 & 1 \\ 2 & 3 \end{bmatrix}.$$

More details about this can be found in Section 8.1. The backward sweep is just the upper triangular solve step, and two versions of this were studied in the previous section. The number of floating point operations needed to execute the forward sweep is about equal to $n^3/3$ where n is the number of unknowns. So, if the number of unknowns doubles, then the number of operations will increase by a factor of eight!

2.2.5   Implementation

MATLAB has a number of intrinsic procedures which are useful for illustration of Gaussian elimination. These include lu, inv, A\d and others. The LU factorization of A can be used to solve $Ax = d$ because $Ax = (LU)x = L(Ux) = d$. Therefore, first solve $Ly = d$ and second solve $Ux = y$. If both L and U are known, then the solve steps are easy lower and upper triangular solves.

MATLAB and lu, inv and A\d

```
>A = [2 -1 0;-1 2 -1;0 -1 2]
>d = [62.5 62.5 62.5]'
>sol = A\d
sol =
93.7500
125.0000
93.750

>[L U] = lu(A)
L =
 1.0000 0 0
-0.5000 1.0000 0
 0 -0.6667 1.0000
U =
2.0000 -1.0000 0
0 1.5000 -1.0000
0 0 1.3333

>L*U
ans =
 2 -1 0
-1 2 -1
 0 -1 2

>y = L\d
y =
```

```
       62.5000
       93.7500
       125.0000

       >x =U\y
       x =
        93.7500
       125.0000
        93.7500

       >inv(A)
       ans =
        0.7500 0.5000 0.2500
        0.5000 1.0000 0.5000
        0.2500 0.5000 0.7500

       >inv(U)*inv(L)
       ans =
        0.7500 0.5000 0.2500
        0.5000 1.0000 0.5000
        0.2500 0.5000 0.7500
```

Computer codes for these calculations have been worked on for many decades. Many of these codes are stored, updated and optimized for particular computers in netlib (see http://www.netlib.org). For example LU factorizations and the upper triangular solves can be done by the LAPACK subroutines sgetrf() and sgetrs() and also sgesv(), see the user guide [1].

The next MATLAB code, heatgelm.m, solves the 1D steady state heat diffusion problem for a number of different values of n. Note that numerical solutions converge to $u(ih)$ where $u(x)$ is the continuous model and h is the step size. Lines 1-5 input the basic data of the model, and lines 6-16 define the right side, d, and the coefficient matrix, A. Line 17 converts the d to a column vector and prints it, and line 18 prints the matrix. The solution is computed in line 19 and printed.

MATLAB Code heatgelm.m

```
1.     clear
2.     n = 3
3.     h = 1./(n+1);
4.     K = .001;
5.     beta = K/(h*h);
6.     A= zeros(n,n);
7.     for i=1:n
8.           d(i) = sin(pi*i*h)/beta;
9.           A(i,i) = 2;
10.          if i<n
11.               A(i,i+1) = -1;
```

```
12.          end;
13.          if i>1
14.              A(i,i-1) = -1;
15.          end;
16.      end
17.      d = d'
18.      A
19.      temp = A\d
```

Output for n = 3:
    temp =
    75.4442
    106.6942
    75.4442

Output for n = 7:
    temp =
    39.2761
    72.5728
    94.8209
    102.6334
    94.8209
    72.5728
    39.2761

2.2.6 Assessment

The above model for heat conduction depends upon the mesh size, h, but as the mesh size h goes to zero there will be little difference in the computed solutions. For example, in the MATLAB output, the component i of temp is the approximate temperature at ih where $h = 1/(n+1)$. The approximate temperatures at the center of the wire are 106.6942 for $n = 3$, 102.6334 for $n = 7$ and 101.6473 for $n = 15$. The continuous model is $-(.001u_x)_x = sin(\pi x)$ with $u(0) = 0 = u(1)$, and the solution is $u(x) = (1000/\pi^2)sin(\pi x)$. So, $u(1/2) = 1000/\pi^2 = 101.3212$, which is approached by the numerical solutions as n increases. An analysis of this will be given in Section 2.6.

The four basic properties of inverse matrices need some justification.

Proof that the inverse is unique:
    Let B and C be inverses of A so that $AB = BA = I$ and $AC = CA = I$. Subtract these matrix equations and use the distributive property

$$AB - AC = I - I = 0$$
$$A(B - C) = 0.$$

Since B is an inverse of A and use the associative property,

$$
\begin{aligned}
B(A(B - C)) &= B0 = 0 \\
(BA)(B - C) &= 0 \\
I(B - C) &= 0.
\end{aligned}
$$

Proof that $A^{-1}d$ is a solution of $Ax = d$:
Let $x = A^{-1}d$ and again use the associative property

$$
A(A^{-1}d) = (AA^{-1})d = Id = d.
$$

Proofs of properties 3 and 4 are also a consequence of the associative property.

2.2.7   Exercises

1.   Consider the following algebraic system

$$
\begin{aligned}
1x_1 + 2x_2 + 3x_3 &= 1 \\
-1x_1 + 1x_2 - 1x_3 &= 2 \\
2x_1 + 4x_2 + 3x_3 &= 3.
\end{aligned}
$$

(a).   Find the augmented matrix.
(b).   By hand calculations with row operations and elementary matrices find E so that $EA = U$ is upper triangular.
(c).   Use this to find the solution, and verify your calculations using MATLAB.
2.   Use the MATLAB code heatgelm.m and experiment with the mesh sizes, by using $n = 11, 21$ and 41, in the heat conduction problem and verify that the computed solution converges as the mesh goes to zero, that is, $u_i - u(ih)$ goes to zero as h goes to zero
3.   Prove property 3 of Theorem 2.2.1.
4.   Prove property 4 of Theorem 2.2.1.
5.   Prove that the solution of $Ax = d$ is unique if A^{-1} exists.

2.3   Cooling Fin and Tridiagonal Matrices

2.3.1   Introduction

In the thin wire problem we derived a tridiagonal matrix, which was generated from the finite difference approximation of the differential equation. It is very common to obtain either similar tridiagonal matrices or more complicated matrices that have blocks of tridiagonal matrices. We will illustrate this by a sequence of models for a cooling fin. This section is concerned with a very efficient version of the Gaussian elimination algorithm for the solution of

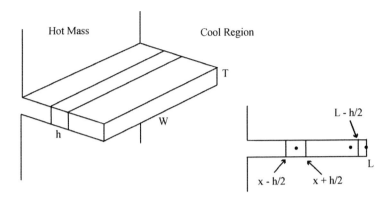

Figure 2.3.1: Thin Cooling Fin

tridiagonal algebraic systems. The full version of a Gaussian elimination algorithm for n unknowns requires order $n^3/3$ operations and order n^2 storage locations. By taking advantage of the number of zeros and their location, the Gaussian elimination algorithm for tridiagonal systems can be reduced to order $5n$ operations and order $8n$ storage locations!

2.3.2 Applied Area

Consider a hot mass, which must be cooled by transferring heat from the mass to a cooler surrounding region. Examples include computer chips, electrical amplifiers, a transformer on a power line, or a gasoline engine. One way to do this is to attach *cooling fins* to this mass so that the surface area that transmits the heat will be larger. We wish to be able to model heat flow so that one can determine whether or not a particular configuration will sufficiently cool the mass.

In order to start the modeling process, we will make some assumptions that will simplify the model. Later we will return to this model and reconsider some of these assumptions. First, assume no time dependence and the temperature is approximated by a function of only the distance from the mass to be cooled. Thus, there is diffusion in only one direction. This is depicted in Figure 2.3.1 where x is the direction perpendicular to the hot mass.

Second, assume the heat lost through the surface of the fin is similar to Newton's law of cooling so that for a slice of the lateral surface

$$
\begin{aligned}
\text{heat loss through a slice} \ &= \ (\text{area})(\text{time interval})c(u_{sur} - u) \\
&= \ h(2W + 2T)\,\Delta t\, c(u_{sur} - u).
\end{aligned}
$$

Here u_{sur} is the surrounding temperature, and the c reflects the ability of the fin's surface to transmit heat to the surrounding region. If c is near zero, then little heat is lost. If c is large, then a larger amount of heat is lost through the lateral surface.

Third, assume heat diffuses in the x direction according to *Fourier's heat law* where K is the thermal conductivity. For interior volume elements with $x < L = 1$,

$$
\begin{aligned}
0 \approx\ & \text{(heat through lateral surface)} \\
& +\text{(heat diffusing through front)} \\
& -\text{(heat diffusing through back)} \\
=\ & h\ (2W + 2T)\ \Delta t\ c(u_{sur} - u(x)) \\
& +TW\ \Delta t\ Ku_x(x + h/2) \\
& -TW\ \Delta t\ Ku_x(x - h/2).
\end{aligned} \tag{2.3.1}
$$

For the tip of the fin with $x = L$, we use $Ku_x(L) = c(u_{sur} - u(L))$ and

$$
\begin{aligned}
0 \approx\ & \text{(heat through lateral surface of tip)} \\
& +\text{(heat diffusing through front)} \\
& -\text{(heat diffusing through back)} \\
=\ & (h/2)(2W + 2T)\ \Delta t\ c(u_{sur} - u(L)) \\
& +TW\ \Delta t\ c(u_{sur} - u(L)) \\
& -TW\ \Delta t\ Ku_x(L - h/2).
\end{aligned} \tag{2.3.2}
$$

Note, the volume element near the tip of the fin is one half of the volume of the interior elements.

These are only approximations because the temperature changes continuously with space. In order to make these approximations in (2.3.1) and (2.3.2) more accurate, we divide by $h\ \Delta t\ TW$ and let h go to zero

$$
0 = (2W + 2T)/(TW)\ c(u_{sur} - u) + (Ku_x)_x. \tag{2.3.3}
$$

Let $C \equiv ((2W + 2T)/(TW))\ c$ and $f \equiv Cu_{sur}$. The *continuous* model is given by the following differential equation and two boundary conditions.

$$
\begin{aligned}
-(Ku_x)_x + Cu &= f, & (2.3.4) \\
u(0) &= \text{given and} & (2.3.5) \\
Ku_x(L) &= c(u_{sur} - u(L)). & (2.3.6)
\end{aligned}
$$

The boundary condition in (2.3.6) is often called a *derivative or flux or Robin* boundary condition.. If $c = 0$, then no heat is allowed to pass through the right boundary, and this type of boundary condition is often called a *Neumann* boundary condition.. If c approaches infinity and the derivative remains bounded, then (2.3.6) implies $u_{sur} = u(L)$. When the value of the function is given at the boundary, this is often called the *Dirichlet* boundary condition.

2.3.3 Model

The above derivation is useful because (2.3.1) and (2.3.2) suggest a way to discretize the continuous model. Let u_i be an approximation of $u(ih)$ where $h = L/n$. Approximate the derivative $u_x(ih + h/2)$ by $(u_{i+1} - u_i)/h$. Then equations (2.3.2) and (2.3.3) yield the *finite difference* approximation, a discrete model, of the continuum model (2.3.4)-(2.3.6).

Let u_0 be given and let $1 \leq i < n$:

$$-[K(u_{i+1} - u_i)/h - K(u_i - u_{i-1})/h] + hCu_i = hf(ih). \tag{2.3.7}$$

Let $i = n$:

$$-[c(u_{sur} - u_n) - K(u_n - u_{n-1})/h] + (h/2)Cu_n = (h/2)f(nh). \tag{2.3.8}$$

The discrete system (2.3.7) and (2.3.8) may be written in matrix form. For ease of notation we let $n = 4$, multiply (2.3.7) by h and (2.3.8) by $2h$, $B \equiv 2K + h^2 C$ so that there are 4 equations and 4 unknowns:

$$
\begin{aligned}
Bu_1 - Ku_2 &= h^2 f_1 + Ku_0, \\
-Ku_1 + Bu_2 - Ku_3 &= h^2 f_2, \\
-Ku_2 + Bu_3 - Ku_4 &= h^2 f_3 \text{ and} \\
-2Ku_3 + (B + 2hc)u_4 &= h^2 f_4 + 2chu_{sur}.
\end{aligned}
$$

The matrix form of this is $AU = F$ where A is, in general, $n \times n$ matrix and U and F are $n \times 1$ column vectors. For $n = 4$ we have

$$
A = \begin{bmatrix}
B & -K & 0 & 0 \\
-K & B & -K & 0 \\
0 & -K & B & -K \\
0 & 0 & -2K & B + 2ch
\end{bmatrix}
$$

$$
\text{where } U = \begin{bmatrix} u_1 \\ u_2 \\ u_3 \\ u_4 \end{bmatrix} \text{ and } F = \begin{bmatrix} h^2 f_1 + Ku_0 \\ h^2 f_2 \\ h^2 f_3 \\ h^2 f_4 + 2chu_{sur} \end{bmatrix}.
$$

2.3.4 Method

The solution can be obtained by either using the *tridiagonal (Thomas)* algorithm, or using a solver that is provided with your computer software. Let us consider the tridiagonal system $Ax = d$ where A is an $n \times n$ matrix and x and d are $n \times 1$ column vectors. We assume the matrix A has components as indicated in

$$
A = \begin{bmatrix}
a_1 & c_1 & 0 & 0 \\
b_2 & a_2 & c_2 & 0 \\
0 & b_3 & a_3 & c_3 \\
0 & 0 & b_4 & a_4
\end{bmatrix}.
$$

In previous sections we used the Gaussian elimination algorithm, and we noted the matrix could be factored into two matrices $A = LU$. Assume A is tridiagonal so that L has nonzero components only in its diagonal and subdiagonal, and U has nonzero components only in its diagonal and superdiagonal. For the above 4×4 matrix this is

$$
\begin{bmatrix}
a_1 & c_1 & 0 & 0 \\
b_2 & a_2 & c_2 & 0 \\
0 & b_3 & a_3 & c_3 \\
0 & 0 & b_4 & a_4
\end{bmatrix}
=
\begin{bmatrix}
\alpha_1 & 0 & 0 & 0 \\
b_2 & \alpha_2 & 0 & 0 \\
0 & b_3 & \alpha_3 & 0 \\
0 & 0 & b_4 & \alpha_4
\end{bmatrix}
\begin{bmatrix}
1 & \gamma_1 & 0 & 0 \\
0 & 1 & \gamma_2 & 0 \\
0 & 0 & 1 & \gamma_3 \\
0 & 0 & 0 & 1
\end{bmatrix}.
$$

The plan of action is (i) solve for α_i and γ_i in terms of a_i, b_i and c_i by matching components in the above matrix equation, (ii) solve $Ly = d$ and (iii) solve $Ux = y$.

Step (i): For $i = 1$, $a_1 = \alpha_1$ and $c_1 = \alpha_1 \gamma_1$. So, $\alpha_1 = a_1$ and $\gamma_1 = c_1/a_1$. For $2 \le i \le n-1$, $a_i = b_i \gamma_{i-1} + \alpha_i$ and $c_i = \alpha_i \gamma_i$. So, $\alpha_i = a_i - b_i \gamma_{i-1}$ and $\gamma_i = c_i/\alpha_i$. For $i = n$, $a_n = b_n \gamma_{n-1} + \alpha_n$. So, $\alpha_n = a_n - b_n \gamma_{n-1}$. These steps can be executed provided the α_i are not zero or too close to zero!

Step (ii): Solve $Ly = d$.
$$y_1 = d_1/\alpha_1 \text{ and for } i = 2, ..., n \quad y_i = (d_i - b_i y_{i-1})/\alpha_i.$$
Step (iii): Solve $Ux = y$.
$$x_n = y_n \text{ and for } i = n - 1, ..., 1 \quad x_i = y_i - \gamma_i x_{i+1}.$$

The loops for steps (i) and (ii) can be combined to form the following very important algorithm.

Tridiagonal Algorithm.

```
α(1) = a(1), γ(1) = c(1)/a(1) and y(1) = d(1)/a(1)
for i = 2, n
    α(i) = a(i)- b(i)*γ(i-1)
    γ(i) = c(i)/α(i)
    y(i) = (d(i) - b(i)*y(i-1))/α(i)
endloop
x(n) = y(n)
for i = n - 1,1
    x(i) = y(i) -γ(i)*x(i+1)
endloop.
```

2.3.5   Implementation

In this section we use a MATLAB user defined function trid.m and the tridiagonal algorithm to solve the finite difference equations in (2.3.7) and (2.3.8). The function $trid(n, a, b, c, d)$ has input n and the column vectors a, b, c. The output is the solution of the tridiagonal algebraic system. In the MATLAB code fin1d.m lines 7-20 enter the basic data for the cooling fin. Lines 24-34 define the column vectors in the variable list for trid.m. Line 38 is the call to trid.m.

The output can be given as a table, see line 44, or as a graph, see line 55. Also, the heat balance is computed in lines 46-54. Essentially, this checks to see if the heat entering from the hot mass is equal to the heat lost off the lateral and tip areas of the fin. More detail about this will be given later. In the trid.m function code lines 8-12 do the forward sweep where the LU factors are computed and the $Ly = d$ solve is done. Lines 13-16 do the backward sweep to solve $Ux = y$.

MATLAB Codes fin1d.m and trid.m

```
1.      % This is a model for the steady state cooling fin.
2.      % Assume heat diffuses in only one direction.
3.      % The resulting algebraic system is solved by trid.m.
4.      %
5.      % Fin Data.
6.      %
7.      clear
8.      n = 40
9.      cond = .001;
10.     csur = .001;
11.     usur = 70.;
12.     uleft = 160.;
13.     T = .1;
14.     W = 10.;
15.     L = 1.;
16.     h = L/n;
17.     CC = csur*2.*(W+T)/(T*W);
18.     for i = 1:n
19.             x(i) = h*i;
20.     end
21.     %
22.     % Define Tridiagonal Matrix
23.     %
24.     for i = 1:n-1
25.             a(i) = 2*cond+h*h*CC;
26.             b(i) = -cond;
27.             c(i) = -cond;
28.             d(i) = h*h*CC*usur;
29.     end
30.     d(1) = d(1) + cond*uleft;
31.     a(n) = 2.*cond + h*h*CC + 2.*h*csur;
32.     b(n) = -2.*cond;
33.     d(n) = h*h*CC*usur + 2.*csur*usur*h;
34.     c(n) = 0.0;
35.     %
36.     % Execute Tridiagonal Algorithm
```

```
37.     %
38.     u = trid(n,a,b,c,d)
39.     %
40.     % Output as a Table or Graph
41.     %
42.     u = [uleft u];
43.     x = [0 x];
44.     % [x u];
45.     %      Heat entering left side of fin from hot mass
46.     heatenter = T*W*cond*(u(2)-u(1))/h
47.     heatouttip = T*W*csur*(usur-u(n+1));
48.     heatoutlat =h*(2*T+2*W)*csur*(usur-u(1))/2;
49.     for i=2:n
50.         heatoutlat=heatoutlat+h*(2*T+2*W)*csur*(usur-u(i));
51.     end
52.     heatoutlat=heatoutlat+h*(2*T+2*W)*csur*(usur-u(n+1))/2;
53.     heatout = heatouttip + heatoutlat
54.     errorinheat = heatenter-heatout
55.     plot(x,u)
```

```
1.      function x = trid(n,a,b,c,d)
2.      alpha = zeros(n,1);
3.      gamma = zeros(n,1);
4.      y = zeros(n,1);
5.      alpha(1) = a(1);
6.      gamma(1) = c(1)/alpha(1);
7.      y(1) = d(1)/alpha(1);
8.      for i = 2:n
9.          alpha(i) = a(i) - b(i)*gamma(i-1);
10.         gamma(i) = c(i)/alpha(i);
11.         y(i) = (d(i) - b(i)*y(i-1))/alpha(i);
12.     end
13.     x(n) = y(n);
14.     for i = n-1:-1:1
15.         x(i) = y(i) - gamma(i)*x(i+1);
16.     end
```

In Figure 2.3.2 the graphs of temperature versus space are given for variable $c = csur$ in (2.3.4) and (2.3.6). For larger c the solution or temperature should be closer to the surrounding temperature, 70. Also, for larger c the derivative at the left boundary is very large, and this indicates, via the Fourier heat law, that a large amount of heat is flowing from the hot mass into the right side of the fin. The heat entering the fin from the left should equal the heat leaving the fin through the lateral sides and the right tip; this is called *heat balance*.

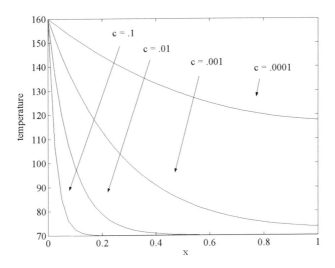

Figure 2.3.2: Temperature for c = .1, .01, .001, .0001

2.3.6 Assessment

In the derivation of the model for the fin we made several assumptions. If the thickness T of the fin is too large, there will be a varying temperature with the vertical coordinate. By assuming the W parameter is large, one can neglect any end effects on the temperature of the fin. Another problem arises if the temperature varies over a large range in which case the thermal conductivity K will be temperature dependent. We will return to these problems.

Once the continuum model is agreed upon and the finite difference approximation is formed, one must be concerned about an appropriate mesh size. Here an analysis much the same as in the previous chapter can be given. In more complicated problems several computations with decreasing mesh sizes are done until little variation in the numerical solutions is observed.

Another test for correctness of the mesh size and the model is to compute the *heat balance* based on the computations. The heat balance simply states the heat entering from the hot mass must equal the heat leaving through the fin. One can derive a formula for this based on the steady state continuum model (2.3.4)-(2.3.6). Integrate both sides of (2.3.4) to give

$$\int_0^L 0 dx = \int_0^L ((2W + 2T)/(TW)c(u_{sur} - u) + (Ku_x)_x) dx$$

$$0 = \int_0^L ((2W + 2T)/(TW)c(u_{sur} - u)) dx + Ku_x(L) - Ku_x(0).$$

Next use the boundary condition (2.3.6) and solve for $Ku_x(0)$

$$Ku_x(0) \quad = \quad \int_0^L ((2W + 2T)/(TW)c(u_{sur} - u))dx$$
$$+c(u_{sur} - u(L)) \tag{2.3.9}$$

In the MATLAB code fin1d.m lines 46-54 approximate both sides of (2.3.9) where the integration is done by the trapezoid rule and both sides are multiplied by the cross section area, TW. A large difference in these two calculations indicates significant numerical errors. For $n = 40$ and smaller $c = .0001$, the difference was small and equaled 0.0023. For $n = 40$ and large $c = .1$, the difference was about 50% of the approximate heat loss from the fin! However, larger n significantly reduces this difference, for example when $n = 320$ and large $c = .1$, then heat_enter = 3.7709, heat_out = 4.0550

The tridiagonal algorithm is not always applicable. Difficulties will arise if the α_i are zero or near zero. The following theorem gives conditions on the components of the tridiagonal matrix so that the tridiagonal algorithm works very well.

Theorem 2.3.1 *(Existence and Stability) Consider the tridiagonal algebraic system. If $|a_1| > |c_1| > 0$, $|a_i| > |b_i| + |c_i|$, $c_i \neq 0$, $b_i \neq 0$ and $1 < i < n$, $|a_n| > |c_n| > 0$, then*

1. $0 < |a_i| - |b_i| < |\alpha_i| < |a_i| + |b_i|$ *for* $1 \leq i \leq n$ *(avoids division by small numbers) and*

2. $|\gamma_i| < 1$ *for* $1 \leq i \leq n$ *(the stability in the backward solve loop).*

Proof.   The proof uses mathematical induction on n. Set $i = 1$: $b_1 = 0$ and $|\alpha_1| = |a_1| > 0$ and $|\gamma_1| = |c_1|/|a_1| < 1$.
Set $i > 1$ and assume it is true for $i - 1$: $\alpha_i = a_i - b_i\gamma_{i-1}$ and $\gamma_i = c_i/\alpha_i$. So, $a_i = b_i\gamma_{i-1} + \alpha_i$ and $|a_i| \leq |b_i||\gamma_{i-1}| + |\alpha_i| < |b_i|1 + |\alpha_i|$. Then $|\alpha_i| > |a_i| - |b_i| \geq |c_i| > 0$. Also, $|\alpha_i| = |a_i - b_i\gamma_{i-1}| \leq |a_i| + |b_i||\gamma_{i-1}| < |a_i| + |b_i|1$. $|\gamma_i| = |c_i|/|\alpha_i| < |c_i|/(|a_i| - |b_i|) \leq 1$. ∎

2.3.7   Exercises

1.   By hand do the tridiagonal algorithm for $3x_1 - x_2 = 1$, $-x_1 + 4x_2 - x_3 = 2$ and $-x_2 + 2x_3 = 3$.
2.   Show that the tridiagonal algorithm fails for the following problem $x_1 - x_2 = 1$, $-x_1 + 2x_2 - x_3 = 1$ and $-x_2 + x_3 = 1$.
3.   In the derivation of the tridiagonal algorithm we combined some of the loops. Justify this.
4.   Use the code fin1d.m and verify the calculations in Figure 2.3.2. Experiment with different values of $T = .05, .10, .15$ and $.20$. Explain your results and evaluate the accuracy of the model.

5.   Find the exact solution of the fin problem and experiment with different mesh sizes by using $n = 10, 20, 40$ and 80. Observe convergence of the discrete solution to the continuum solution. Examine the heat balance calculations.

6.   Modify the above model and code for a tapered fin where $T = .2(1-x) + .1x$.

7.   Consider the steady state axially symmetric heat conduction problem $0 = rf + (Kru_r)_r$, $u(r_0) = given$ and $u(R_0) = given$. Assume $0 < r_0 < R_0$. Find a discrete model and the solution to the resulting algebraic problems.

2.4   Schur Complement

2.4.1   Introduction

In this section we will continue to discuss Gaussian elimination for the solution of $Ax = d$. Here we will examine a block version of Gaussian elimination. This is particularly useful for two reasons. First, this allows for efficient use of the computer's memory hierarchy. Second, when the algebraic equation evolves from models of physical objects, then the decomposition of the object may match with the blocks in the matrix A. We will illustrate this for steady state heat diffusion models with one and two space variables, and later for models with three space variables.

2.4.2   Applied Area

In the previous section we discussed the steady state model of diffusion of heat in a *cooling fin*. The continuous model has the form of an ordinary differential equation with given temperature at the boundary that joins the hot mass. If there is heat diffusion in two directions, then the model will be more complicated, which will be more carefully described in the next chapter. The objective is to solve the resulting algebraic system of equations for the approximate temperature as a function of more than one space variable.

2.4.3   Model

The continuous models for steady state heat diffusion are a consequence of the Fourier heat law applied to the directions of heat flow. For simplicity assume the temperature is given on all parts of the boundary. More details are presented in Chapter 4.2 where the steady state cooling fin model for diffusion in two directions is derived.

Continuous Models:
   Diffusion in 1D. Let $u = u(x) = $ temperature on an interval.

$$0 \;=\; f + (Ku_x)_x \text{ and} \tag{2.4.1}$$
$$u(0), u(L) \;=\; given. \tag{2.4.2}$$

Diffusion in 2D. Let $u = u(x, y) =$ temperature on a square.

$$0 = f + (Ku_x)_x + (Ku_y)_y \text{ and} \qquad (2.4.3)$$
$$u = \text{given on the boundary.} \qquad (2.4.4)$$

The discrete models can be either viewed as discrete versions of the Fourier heat law, or as finite difference approximations of the continuous models.

Discrete Models:

Diffusion in 1D. Let u_i approximate $u(ih)$ with $h = L/n$.

$$0 = f + \beta(u_{i+1} + u_{i-1}) - \beta 2u_i \qquad (2.4.5)$$
$$\text{where } i = 1, ..., n - 1 \text{ and } \beta = K/h^2 \text{ and}$$
$$u_0, u_n = \text{given.} \qquad (2.4.6)$$

Diffusion in 2D. Let u_{ij} approximate $u(ih, jh)$ with $h = L/n = \Delta x = \Delta y$.

$$0 = f + \beta(u_{i+1,j} + u_{i-1,j}) - \beta 2u_{i,j} +$$
$$\beta(u_{i,j+1} + u_{i,j-1}) - \beta 2u_{i,j} \qquad (2.4.7)$$
$$\text{where } i, j = 1, ..., n - 1 \text{ and } \beta = K/h^2 \text{ and}$$
$$u_{0,j}, u_{n,j}, u_{i,0}, u_{i,n} = \text{given.} \qquad (2.4.8)$$

The matrix version of the discrete 1D model with $n = 6$ is as follows. This 1D model will have 5 unknowns, which we list in classical order from left to right. The matrix A will be 5×5 and is derived from (2.4.5) by dividing both sides by $\beta = K/h^2$.

$$\begin{bmatrix} 2 & -1 & & & \\ -1 & 2 & -1 & & \\ & -1 & 2 & -1 & \\ & & -1 & 2 & -1 \\ & & & -1 & 2 \end{bmatrix} \begin{bmatrix} u_1 \\ u_2 \\ u_3 \\ u_4 \\ u_5 \end{bmatrix} = (1/\beta) \begin{bmatrix} f_1 \\ f_2 \\ f_3 \\ f_4 \\ f_5 \end{bmatrix}$$

The matrix version of the discrete 2D model with $n = 6$ will have $5^2 = 25$ unknowns. Consequently, the matrix A will be 25×25. The location of its components will evolve from line (2.4.7) and will depend on the ordering of the unknowns u_{ij}. The classical method of ordering is to start with the bottom grid row $(j = 1)$ and move from left $(i = 1)$ to right $(i = n - 1)$ so that

$$u = \begin{bmatrix} U_1^T & U_2^T & U_3^T & U_4^T & U_5^T \end{bmatrix}^T \text{ with } U_j = \begin{bmatrix} u_{1j} & u_{2j} & u_{3j} & u_{4j} & u_{5j} \end{bmatrix}^T$$

is a grid row j of unknowns. The final grid row corresponds to $j = n - 1$. So, it is reasonable to think of A as a 5×5 *block matrix* where each block is 5×5 and corresponds to a grid row. With careful writing of the equations in (2.4.7) one can derive A as

$$
\begin{bmatrix}
B & -I & & & \\
-I & B & -I & & \\
& -I & B & -I & \\
& & -I & B & -I \\
& & & -I & B
\end{bmatrix}
\begin{bmatrix}
U_1 \\
U_2 \\
U_3 \\
U_4 \\
U_5
\end{bmatrix}
= (1/\beta)
\begin{bmatrix}
F_1 \\
F_2 \\
F_3 \\
F_4 \\
F_5
\end{bmatrix}
\text{ where}
$$

$$
B =
\begin{bmatrix}
4 & -1 & & & \\
-1 & 4 & -1 & & \\
& -1 & 4 & -1 & \\
& & -1 & 4 & -1 \\
& & & -1 & 4
\end{bmatrix}
\text{ and } I =
\begin{bmatrix}
1 & & & & \\
& 1 & & & \\
& & 1 & & \\
& & & 1 & \\
& & & & 1
\end{bmatrix}.
$$

2.4.4   Method

In the above 5×5 block matrix it is tempting to try a block version of Gaussian elimination. The first block row could be used to eliminate the $-I$ in the block (2,1) position (block row 2 and block column 1). Just multiply block row 1 by B^{-1} and add the new block row 1 to block row 2 to get

$$
\begin{bmatrix} 0 & (B - B^{-1}) & -I & 0 & 0 \end{bmatrix}
$$

where the 0 represents a 5×5 *zero matrix*. If all the inverse matrices of any subsequent block matrices on the diagonal exist, then one can continue this until all blocks in the lower block part of A have been modified to 5×5 zero matrices.

In order to make this more precise, we will consider just a 2×2 block matrix where the diagonal blocks are square but may not have the same dimension

$$
A = \begin{bmatrix} B & E \\ F & C \end{bmatrix}.
\tag{2.4.9}
$$

In general A will be $n \times n$ with $n = k + m$, B is $k \times k$, C is $m \times m$, E is $k \times m$ and F is $m \times k$. For example, in the above 5×5 block matrix we may let $n = 25$, $k = 5$ and $m = 20$ and

$$
C =
\begin{bmatrix}
B & -I & & \\
-I & B & -I & \\
& -I & B & -I \\
& & -I & B
\end{bmatrix}
\text{ and } E = F^T = \begin{bmatrix} -I & 0 & 0 & 0 \end{bmatrix}.
$$

If B has an inverse, then we can multiply block row 1 by FB^{-1} and subtract it from block row 2. This is equivalent to multiplication of A by a *block elementary matrix* of the form

$$
\begin{bmatrix}
I_k & 0 \\
-FB^{-1} & I_m
\end{bmatrix}.
$$

If $Ax = d$ is viewed in block form, then

$$
\begin{bmatrix} B & E \\ F & C \end{bmatrix} \begin{bmatrix} X_1 \\ X_2 \end{bmatrix} = \begin{bmatrix} D_1 \\ D_2 \end{bmatrix}. \tag{2.4.10}
$$

The above block elementary matrix multiplication gives

$$
\begin{bmatrix} B & E \\ 0 & C - FB^{-1}E \end{bmatrix} \begin{bmatrix} X_1 \\ X_2 \end{bmatrix} = \begin{bmatrix} D_1 \\ D_2 - FB^{-1}D_1 \end{bmatrix}. \tag{2.4.11}
$$

So, if the block upper triangular matrix is nonsingular, then this last block equation can be solved.

The following basic properties of square matrices play an important role in the solution of (2.4.10). These properties follow directly from the definition of an inverse matrix.

Theorem 2.4.1 (*Basic Matrix Properties*) *Let B and C be square matrices that have inverses. Then the following equalities hold:*

1. $\begin{bmatrix} B & 0 \\ 0 & C \end{bmatrix}^{-1} = \begin{bmatrix} B^{-1} & 0 \\ 0 & C^{-1} \end{bmatrix}$,

2. $\begin{bmatrix} I_k & 0 \\ F & I_m \end{bmatrix}^{-1} = \begin{bmatrix} I_k & 0 \\ -F & I_m \end{bmatrix}$,

3. $\begin{bmatrix} B & 0 \\ F & C \end{bmatrix} = \begin{bmatrix} B & 0 \\ 0 & C \end{bmatrix} \begin{bmatrix} I_k & 0 \\ C^{-1}F & I_m \end{bmatrix}$ and

4. $\begin{bmatrix} B & 0 \\ F & C \end{bmatrix}^{-1} = \begin{bmatrix} I_k & 0 \\ C^{-1}F & I_m \end{bmatrix}^{-1} \begin{bmatrix} B & 0 \\ 0 & C \end{bmatrix}^{-1} = \begin{bmatrix} B^{-1} & 0 \\ -C^{-1}FB^{-1} & C^{-1} \end{bmatrix}$.

Definition. *Let A have the form in (2.4.9) and B be nonsingular. The Schur complement of B in A is $C - FB^{-1}E$.*

Theorem 2.4.2 (*Schur Complement Existence*) *Consider A as in (2.4.10). If both B and the Schur complement of B in A are nonsingular, then A is nonsingular. Moreover, the solution of $Ax = d$ is given by using a block upper triangular solve of (2.4.11).*

The choice of the blocks B and C can play a very important role. Often the choice of the physical object, which is being modeled, suggests the choice of B and C. For example, if the heat diffusion in a thin wire is being modeled, the unknowns associated with B might be the unknowns on the left side of the thin wire and the unknowns associated with C would then be the right side. Another alternative is to partition the wire into three parts: a small center and a left and right side; this might be useful if the wire was made of two types of materials. A somewhat more elaborate example is the model of airflow over an aircraft. Here we might partition the aircraft into wing, rudder, fuselage and "connecting" components. Such partitions of the physical object or the matrix are called *domain decompositions*.

2.4.5   Implementation

MATLAB will be used to illustrate the Schur complement, domain decomposition and different ordering of the unknowns. The classical ordering of the unknowns can be changed so that the "solve" or "inverting" of B or its Schur complement is a minimal amount of work.

1D Heat Diffusion with n = 6 (5 unknowns).

Classical order of unknowns u_1, u_2, u_3, u_4, u_5 gives the coefficient matrix

$$A = \begin{bmatrix} 2 & -1 & & & \\ -1 & 2 & -1 & & \\ & -1 & 2 & -1 & \\ & & -1 & 2 & -1 \\ & & & -1 & 2 \end{bmatrix}.$$

Domain decomposition order of unknowns is $u_3; u_1, u_2; u_4, u_5$. In order to form the new coefficient matrix A', list the equations in the new order. For example, the equation for the third unknown is $-u_2 + 2u_3 - u_4 = (1/\beta)f_3$, and so, the first row of the new coefficient matrix should be $\begin{bmatrix} 2 & 0 & -1 & -1 & 0 \end{bmatrix}$. The other rows in the new coefficient matrix are found in a similar fashion so that

$$A' = \begin{bmatrix} 2 & & -1 & -1 & \\ & 2 & -1 & & \\ -1 & -1 & 2 & & \\ -1 & & & 2 & -1 \\ & & & -1 & 2 \end{bmatrix}.$$

Here $B = [2]$ and C is block diagonal. In the following MATLAB calculations note that B is easy to invert and that the Schur complement is more complicated than the C matrix.

```
>b = [2];
>e = [0 -1 -1 0];
>f = e';
>c = [2 -1 0 0;-1 2 0 0;0 0 2 -1;0 0 -1 2];
>a = [b e;f c]
a =
2 0 -1 -1 0
0 2 -1 0 0
-1 -1 2 0 0
-1 0 0 2 -1
0 0 0 -1 2

>schurcomp = c - f*inv(b)*e      % 4x4 tridiagonal matrix
schurcomp =
2.0 -1.0 0 0
-1.0 1.5 -0.5 0
```

0 -0.5 1.5 -1.0
0 0 -1.0 2.

```
>d1 = [1];
>d2 = [1 1 1 1]';
>dd2 = d2 - f*inv(b)*d1
dd2 =
1.0000
1.5000
1.5000
1.0000

>x2 = schurcomp\dd2      % block upper triangular solve
x2 =
2.5000
4.0000
4.0000
2.5000

>x1 = inv(b)*(d1 - e*x2)
x1 =
4.5000

>x = a\[d1 d2']'
x =
4.5000
2.5000
4.0000
4.0000
2.5000
```

Domain decomposition order of unknowns is $u_1, u_2; u_4, u_5; u_3$ so that the new coefficient matrix is

$$A'' = \begin{bmatrix} 2 & -1 & & & \\ -1 & 2 & & & -1 \\ & & 2 & -1 & -1 \\ & & -1 & 2 & \\ & -1 & -1 & & 2 \end{bmatrix}.$$

Here $C = [2]$ and B is block diagonal. The Schur complement of B will be 1×1 and is easy to invert. Also, B is easy to invert because it is block diagonal. The following MATLAB calculations illustrate this.

```
>f = [ 0 -1 -1 0];
>e = f';
>b = [2 -1 0 0;-1 2 0 0;0 0 2 -1;0 0 -1 2];
>c = [2];
>a = [ b e;f c]
```

```
a =
2 -1 0 0 0
-1 2 0 0 -1
0 0 2 -1 -1
0 0 -1 2 0
0 -1 -1 0 2

>schurcomp = c -f*inv(b)*e      % 1x1 matrix
schurcomp =
0.6667

>d1 = [1 1 1 1]';
>d2 = [1];
>dd2 = d2 -f*inv(b)*d1
dd2 =
3

>x2 = schurcomp\dd2      % block upper triangular solve
x2 =
4.5000

>x1 = inv(b)*(d1 - e*x2)
x1 =
2.5000
4.0000
4.0000
2.5000

>x = inv(a)*[d1' d2]'
x =
2.5000
4.0000
4.0000
2.5000
4.5000
```

2D Heat Diffusion with n = 6 (25 unknowns).

Here we will use domain decomposition where the third grid row is listed last, and the first, second, fourth and fifth grid rows are listed first in this order. Each block is 5×5 for the 5 unknowns in each grid row, and i is a 5×5 identity matrix

$$A'' = \begin{bmatrix} b & -i & & & \\ -i & b & & & -i \\ & & b & -i & -i \\ & & -i & b & \\ & -i & -i & & b \end{bmatrix} \quad \text{where}$$

$$b = \begin{bmatrix} 4 & -1 & & & \\ -1 & 4 & -1 & & \\ & -1 & 4 & -1 & \\ & & -1 & 4 & -1 \\ & & & -1 & 4 \end{bmatrix}.$$

The B will be the block 4×4 matrix and $C = b$. The B matrix is block diagonal and is relatively easy to invert. The C matrix and the Schur complement of B are 5×5 matrices and will be easy to invert or "solve". With this type of domain decomposition the Schur complement matrix will be small, but it will have mostly nonzero components. This is illustrated by the following MATLAB calculations.

```
>clear
>b = [4 -1 0 0 0;-1 4 -1 0 0; 0 -1 4 -1 0; 0 0 -1 4 -1;0 0 0 -1 4];
>ii = -eye(5);
>z = zeros(5);
>B = [b ii z z;ii b z z; z z b ii; z z ii b];
>f = [z ii ii z];
>e = f';
>C = b;
>schurcomp = C - f*inv(B)*e      % full 5x5 matrix
schurcomp =
 3.4093 -1.1894 -0.0646 -0.0227 -0.0073
-1.1894  3.3447 -1.2121 -0.0720 -0.0227
-0.0646 -1.2121  3.3374 -1.2121 -0.0646
-0.0227 -0.0720 -1.2121  3.3447 -1.1894
-0.0073 -0.0227 -0.0646 -1.1894  3.4093

>whos
Name Size Bytes Class
 B       20x20 3200 double array
 C       5x5 200 double array
 b       5x5 200 double array
 e       20x5 800 double array
 f       5x20 800 double array
 ii      5x5 200 double array
 schurcomp 5x5 200 double array
 z       5x5 200 double array
```

2.4.6   Assessment

Heat and mass transfer models usually involve transfer in more than one direction. The resulting discrete models will have structure similar to the 2D heat
diffusion model. There are a number of zero components that are arranged in
very nice patterns, which are often block tridiagonal. Here domain decomposition and the Schur complement will continue to help reduce the computational
burden.

The proof of the Schur complement theorem is a direct consequence of using
a block elementary row operation to get a zero matrix in the block row 2 and
column 1 position

$$\begin{bmatrix} I_k & 0 \\ -FB^{-1} & I_m \end{bmatrix} \begin{bmatrix} B & E \\ F & C \end{bmatrix} = \begin{bmatrix} B & E \\ 0 & C - FB^{-1}E \end{bmatrix}.$$

Thus

$$\begin{bmatrix} B & E \\ F & C \end{bmatrix} = \begin{bmatrix} I_k & 0 \\ FB^{-1} & I_m \end{bmatrix} \begin{bmatrix} B & E \\ 0 & C - FB^{-1}E \end{bmatrix}.$$

Since both matrices on the right side have inverses, the left side, A, has an
inverse.

2.4.7   Exercises

1.   Use the various orderings of the unknowns and the Schur complement to
solve $Ax = d$ where

$$A = \begin{bmatrix} 2 & -1 & & & \\ -1 & 2 & -1 & & \\ & -1 & 2 & -1 & \\ & & -1 & 2 & -1 \\ & & & -1 & 2 \end{bmatrix} \text{ and } d = \begin{bmatrix} 1 \\ 2 \\ 3 \\ 4 \\ 5 \end{bmatrix}.$$

2.   Consider the above 2D heat diffusion model for 25 unknowns. Suppose
d is a 25×1 column vector whose components are all equal to 10. Use the Schur
complement with the third grid row of unknowns listed last to solve $Ax = d$.

3.   Repeat problem 2 but now list the third grid row of unknowns first.

4.   Give the proofs of the four basic properties in Theorem 2.4.1.

5.   Find the inverse of the block upper triangular matrix

$$\begin{bmatrix} B & E \\ 0 & C \end{bmatrix}.$$

6.   Use the result in problem 5 to find the inverse of

$$\begin{bmatrix} B & E \\ 0 & C - FB^{-1}E \end{bmatrix}.$$

7.   Use the result in problem 6 and the proof of the Schur complement theorem to find the inverse of

$$\begin{bmatrix} B & E \\ F & C \end{bmatrix}.$$

2.5   Convergence to Steady State

2.5.1   Introduction

In the applications to heat and mass transfer the discrete time-space dependent models have the form

$$u^{k+1} = Au^k + b.$$

Here u^{k+1} is a sequence of column vectors, which could represent approximate temperature or concentration at time step $k + 1$. Under stability conditions on the time step the time dependent solution may "converge" to the solution of the discrete steady state problem

$$u = Au + b.$$

In Chapter 1.2 one condition that ensured this was when the matrix products A^k "converged" to the zero matrix, then u^{k+1} "converges" to u. We would like to be more precise about the term "converge" and to show how the stability conditions are related to this "convergence."

2.5.2   Vector and Matrix Norms

There are many different norms, which are a "measure" of the length of a vector. A common norm is the *Euclidean norm*

$$\|x\|_2 \equiv (x^T x)^{\frac{1}{2}}.$$

Here we will only use the infinity norm. Any real valued function of $x \in \mathbb{R}^n$ that satisfies the properties 1-3 of subsequent Theorem 2.5.1 is called a *norm.*

Definition. *The infinity norm of the $n \times 1$ column vector $x = [x_i]$ is a real number*

$$\|x\| \equiv \max_i |x_i|.$$

The infinity norm of an $n \times n$ matrix $Ax = [a_{ij}]$ is

$$\|A\| \equiv \max_i \sum_j |a_{ij}|.$$

Example.

$$\text{Let } x \;=\; \begin{bmatrix} -1 \\ 6 \\ -9 \end{bmatrix} \text{ and } A = \begin{bmatrix} 1 & 3 & -4 \\ 1 & 3 & 1 \\ 3 & 0 & 5 \end{bmatrix}.$$

$$\|x\| \;=\; \max\{1, 6, 9\} = 9 \text{ and } \|A\| = \max\{8, 5, 8\} = 8.$$

Theorem 2.5.1 *(Basic Properties of the Infinity Norm) Let A and B be $n \times n$ matrices and $x, y \in \mathbb{R}^{n}$. Then*

1. $\|x\| \geq 0$, *and* $\|x\| = 0$ *if and only if* $x = 0$,

2. $\|x + y\| \leq \|x\| + \|y\|$,

3. $\|\alpha x\| \leq |\alpha| \|x\|$ *where* α *is a real number,*

4. $\|Ax\| \leq \|A\| \|x\|$ *and*

5. $\|AB\| \leq \|A\| \|B\|$.

Proof. The proofs of 1-3 are left as exercises. The proof of 4 uses the definitions of the infinity norm and the matrix-vector product.

$$\|Ax\| = \max_i \left| \sum_j a_{ij} x_j \right|$$
$$\leq \max_i \sum_j |a_{ij}| \cdot |x_j|$$
$$\leq (\max_i \sum_j |a_{ij}|) \cdot (\max_j |x_j|) = \|A\| \|x\|.$$

The proof of 5 uses the definition of a matrix-matrix product.

$$\|AB\| \equiv \max_i \sum_j \left| \sum_k a_{ik} b_{kj} \right|$$
$$\leq \max_i \sum_j \sum_k |a_{ik}| |b_{kj}|$$
$$= \max_i \sum_k |a_{ik}| \sum_j |b_{kj}|$$
$$\leq (\max_i \sum_k |a_{ik}|)(\max_k \sum_j |b_{kj}|)$$
$$= \|A\| \|B\|$$

Property 5 can be generalized to any number of matrix products. ∎

Definition. *Let x^k and x be vectors. x^k converges to x if and only if each component of x_i^k converges to x_i. This is equivalent to $\|x^k - x\| = \max_i |x_i^k - x_i|$ converges to zero.*

Like the geometric series of single numbers the iterative scheme $x^{k+1} = Ax^k + b$ can be expressed as a summation via recursion

$$
\begin{aligned}
x^{k+1} &= Ax^k + b \\
&= A(Ax^{k-1} + b) + b \\
&= A^2 x^{k-1} + Ab + b \\
&= A^2(Ax^{k-2} + b) + Ab + b \\
&= A^3 x^{k-2} + (A^2 + A^1 + I)b \\
&\quad \vdots \\
&= A^{k+1} x^0 + (A^k + \cdots + I)b. \quad (2.5.1)
\end{aligned}
$$

Definition. *The summation $I + \cdots + A^k$ and the series $I + \cdots + A^k + \cdots$ are generalizations of the geometric partial sums and series, and the latter is often referred to as the von Neumann series.*

In Section 1.2 we showed if A^k converges to the zero matrix, then $x^{k+1} = Ax^k + b$ must converge to the solution of $x = Ax + b$, which is also a solution of $(I - A)x = b$. If $I - A$ has an inverse, equation (2.5.1) suggests that the von Neumann series must converge to the inverse of $I - A$. If the norm of A is less than one, then these are true.

Theorem 2.5.2 *(Geometric Series for Matrices) Consider the scheme $x^{k+1} = Ax^k + b$. If the norm of A is less than one, then*

1. $x^{k+1} = Ax^k + b$ *converges to $x = Ax + b$,*

2. $I - A$ *has an inverse and*

3. $I + \cdots + A^k$ *converges to the inverse of $I - A$.*

Proof. For the proof of 1 subtract $x^{k+1} = Ax^k + b$ and $x = Ax + b$ to get by recursion or "telescoping"

$$x^{k+1} - x = A(x^k - x)$$
$$\vdots$$
$$= A^{k+1}(x^0 - x). \qquad (2.5.2)$$

Apply properties 4 and 5 of the vector and matrix norms with $B = A^k$ so that after recursion

$$\left\|x^{k+1} - x\right\| \leq \left\|A^{k+1}\right\| \left\|x^0 - x\right\|$$
$$\leq \left\|A\right\| \left\|A^k\right\| \left\|x^0 - x\right\|$$
$$\vdots$$
$$\leq \left\|A\right\|^{k+1} \left\|x^0 - x\right\|. \qquad (2.5.3)$$

Because the norm of A is less than one, the right side must go to zero. This forces the norm of the error to go to zero.

For the proof of 2 use the following result from matrix algebra: $I - A$ has an inverse if and only if $(I - A)x = 0$ implies $x = 0$. Suppose x is not zero and $(I - A)x = 0$. Then $Ax = x$. Apply the norm to both sides of $Ax = x$ and use property 4 to get

$$\|x\| = \|Ax\| \leq \|A\| \, \|x\| \qquad (2.5.4)$$

Because x is not zero, its norm must not be zero. So, divide both sides by the norm of x to get $1 \leq \|A\|$, which is a contradiction to the assumption of the theorem.

For the proof of 3 use the associative and distributive properties of matrices so that

$$(I - A)(I + A + \cdots + A^k)$$
$$= I(I + A + \cdots + A^k) - A(I + A + \cdots + A^k)$$
$$= I - A^{k+1}.$$

Multiply both sides by the inverse of $I - A$ to get

$$(I + A + \cdots + A^k) = (I - A)^{-1}(I - A^{k+1})$$
$$= (I - A)^{-1}I - (I - A)^{-1}A^{k+1}$$
$$(I + A + \cdots + A^k) - (I - A)^{-1} = -(I - A)^{-1}A^{k+1}.$$

Apply property 5 of the norm

$$\left\| (I + A + \cdots + A^k) - (I - A)^{-1} \right\| = \left\| -(I - A)^{-1}A^{k+1} \right\|$$
$$\leq \left\| -(I - A)^{-1} \right\| \left\| A^{k+1} \right\|$$
$$\leq \left\| -(I - A)^{-1} \right\| \left\| A \right\|^{k+1}.$$

Since the norm is less than one the right side must go to zero. Thus, the partial sums must converge to the inverse of $I - A$. ■

2.5.3   Application to the Cooling Wire

Consider a *cooling wire* as discussed in Section 1.3 with some heat loss through the lateral surface. Assume this heat loss is directly proportional to the product of change in time, the lateral surface area and to the difference in the surrounding temperature and the temperature in the wire. Let c_{sur} be the proportionality constant, which measures insulation. Let r be the radius of the wire so that the lateral surface area of a small wire segment is $2\pi rh$. If u_{sur} is the surrounding temperature of the wire, then the heat loss through the small lateral area is $c_{sur} \, \Delta t \, 2\pi rh \, (u_{sur} - u_i)$ where u_i is the approximate temperature. Additional heat loss or gain from a source such as electrical current and from left and right diffusion gives a discrete model where $\alpha \equiv (\Delta t/h^2)(K/\rho c)$

$$\begin{aligned}
u_i^{k+1} &= (\Delta t/\rho c)(f + c_{sur}(2/r)u_{sur}) + \alpha(u_{i+1}^k + u_{i-1}^k) \\
&\quad + (1 - 2\alpha - (\Delta t/\rho c)c_{sur}(2/r))u_i^k \qquad\qquad (2.5.5)
\end{aligned}$$

$$\text{for } i = 1, ..., n - 1 \text{ and } k = 0, ..., maxk - 1.$$

For $n = 4$ there are three unknowns and the equations in (2.5.5) for $i = 1, 2$ and 3 may be written in matrix form. These three scalar equations can be written as one 3D vector equation $u^{k+1} = Au^k + b$ where

$$u^k = \begin{bmatrix} u_1^k \\ u_2^k \\ u_3^k \end{bmatrix}, \quad b = (\Delta t/\rho c)F \begin{bmatrix} 1 \\ 1 \\ 1 \end{bmatrix} \text{ and}$$

$$A = \begin{bmatrix} 1 - 2\alpha - d & \alpha & 0 \\ \alpha & 1 - 2\alpha - d & \alpha \\ 0 & \alpha & 1 - 2\alpha - d \end{bmatrix} \text{ with}$$

$$F \equiv f + c_{sur}(2/r)u_{sur} \text{ and } d \equiv (\Delta t/\rho c)c_{sur}(2/r).$$

Stability Condition for (2.5.5).

$$1 - 2\alpha - d > 0 \text{ and } \alpha > 0.$$

When the stability condition holds, then the norm of the above 3×3 matrix is

$$\max\{|1 - 2\alpha - d| + |\alpha| + |0|, |\alpha| + |1 - 2\alpha - d| + |\alpha|,$$
$$|0| + |1 - 2\alpha - d| + |\alpha|\}$$
$$= \max\{1 - 2\alpha - d + \alpha, \alpha + 1 - 2\alpha - d + \alpha,$$
$$1 - 2\alpha - d + \alpha\}$$
$$= \max\{1 - \alpha - d, 1 - d, 1 - \alpha - d\}$$
$$= 1 - d < 1.$$

2.5.4 Application to Pollutant in a Stream

Let the concentration u at $(i\Delta x, k\Delta t)$ be approximated by u_i^k where $\Delta t = T/maxk, \Delta x = L/n$ and L is the length of the stream. The model will have the general form

$$\text{change in amount} \approx \text{(amount entering from upstream)}$$
$$- \text{(amount leaving to downstream)}$$
$$- \text{(amount decaying in a time interval)}.$$

As in Section 1.4 this eventually leads to the discrete model

$$u_i^{k+1} = vel(\Delta t/\Delta x)u_{i-1}^k + (1 - vel(\Delta t/\Delta x) - \Delta t \, dec)u_i^k \quad (2.5.6)$$
$$i = 1, ..., n - 1 \text{ and } k = 0, ..., maxk - 1.$$

For $n = 3$ there are three unknowns and equations, and (2.5.6) with $i = 1, 2,$ and 3 can be written as one 3D vector equation $u^{k+1} = Au^k + b$ where

$$\begin{bmatrix} u_1^{k+1} \\ u_2^{k+1} \\ u_3^{k+1} \end{bmatrix} = \begin{bmatrix} c & 0 & 0 \\ d & c & 0 \\ 0 & d & c \end{bmatrix} \begin{bmatrix} u_1^k \\ u_2^k \\ u_3^k \end{bmatrix} + \begin{bmatrix} du_0^k \\ 0 \\ 0 \end{bmatrix}$$

where $d \equiv vel\,(\Delta t/\Delta x)$ and $c \equiv 1 - d - dec\,\Delta t$.

Stability Condition for (2.5.6).

$$1 - d - dec\,\Delta t \text{ and } vel, dec > 0.$$

When the stability condition holds, then the norm of the 3×3 matrix is given by

$$\max\{|c| + |0| + |0|, |d| + |c| + |0|, |0| + |d| + |c|\}$$
$$= \max\{1 - d - dec\,\Delta t, d + 1 - d - dec\,\Delta t$$
$$, d + 1 - d - dec\,\Delta t\}$$
$$= 1 - dec\,\Delta t < 1.$$

2.5.5 Exercises

1. Find the norms of the following

$$x = \begin{bmatrix} 1 \\ -7 \\ 0 \\ 3 \end{bmatrix} \text{ and } A = \begin{bmatrix} 4 & -5 & 3 \\ 0 & 10 & -1 \\ 11 & 2 & 4 \end{bmatrix}.$$

2. Prove properties 1-3 of the infinity norm.
3. Consider the array

$$A = \begin{bmatrix} 0 & .3 & -.4 \\ .4 & 0 & .2 \\ .3 & .1 & 0 \end{bmatrix}.$$

(a). Find the infinity norm of A.
(b). Find the inverse of $I - A$.
(c). Use MATLAB to compute A^k for $k = 2, 3, \cdots, 10$.
(d). Use MATLAB to compute the partial sums $I + A + \cdots + A^k$.
(e). Compare the partial sums in (d) with the inverse of $I - A$ in (b).

4. Consider the application to a cooling wire. Let $n = 5$. Find the matrix and determine when its infinity norm will be less than one.

5. Consider the application to pollution of a stream. Let $n = 4$. Find the matrix and determine when its infinity norm will be less than one.

2.6 Convergence to Continuous Model

2.6.1 Introduction

In the past sections we considered differential equations whose solutions were dependent on space but not time. The main physical illustration of this was heat transfer. The simplest continuous model is a boundary value problem

$$-(Ku_x)_x + Cu = f \text{ and} \tag{2.6.1}$$
$$u(0), u(1) = \text{given.} \tag{2.6.2}$$

Here $u = u(x)$ could represent temperature and K is the thermal conductivity, which for small changes in temperature K can be approximated by a constant. The function f can have many forms: (i). $f = f(x)$ could be a heat source such as electrical resistance in a wire, (ii). $f = c(u_{sur} - u)$ from Newton's law of cooling, (iii). $f = c(u_{sur}^4 - u^4)$ from Stefan's radiative cooling or (iv). $f \approx f(a) + f'(a)(u - a)$ is a linear Taylor polynomial approximation. Also, there are other types of boundary conditions, which reflect how fast heat passes through the boundary.

In this section we will illustrate and give an analysis for the convergence of the discrete steady state model to the continuous steady state model. This differs from the previous section where the convergence of the discrete time-space model to the discrete steady state model was considered.

2.6.2   Applied Area

The derivation of (2.6.1) for steady state one space dimension heat diffusion is based on the empirical Fourier heat law. In Section 1.3 we considered a time dependent model for heat diffusion in a wire. The steady state continuous model is

$$-(Ku_x)_x + (2c/r)u = f + (2c/r)u_{sur}. \tag{2.6.3}$$

A similar model for a cooling fin was developed in Chapter 2.3

$$-(Ku_x)_x + ((2W + 2T)/(TW))cu = f. \tag{2.6.4}$$

2.6.3   Model

If K, C and f are constants, then the closed form solution of (2.6.1) is relatively easy to find. However, if they are more complicated or if we have diffusion in two and three dimensional space, then closed form solutions are harder to find. An alternative is the finite difference method, which is a way of converting continuum problems such as (2.6.1) into a finite set of algebraic equations. It uses numerical derivative approximation for the second derivative. First, we break the space into n equal parts with $x_i = ih$ and $h = 1/n$. Second, we let $u_i \approx u(ih)$ where $u(x)$ is from the continuum solution, and u_i will come from the finite difference (or discrete) solution. Third, we approximate the second derivative by

$$u_{xx}(ih) \approx [(u_{i+1} - u_i)/h - (u_i - u_{i-1})/h]/h. \tag{2.6.5}$$

The finite difference method or discrete model approximation to (2.6.1) is for $0 < i < n$

$$-K[(u_{i+1} - u_i)/h - (u_i - u_{i-1})/h]/h + Cu_i = f_i = f(ih). \tag{2.6.6}$$

This gives $n - 1$ equations for $n - 1$ unknowns. The end points $u_0 = u(0)$ and $u_n = u(1)$ are given as in $f(x)$.

   The discrete system (2.6.6) may be written in matrix form. For ease of notation we multiply (2.6.6) by h^2, let $B \equiv 2K + h^2C$, and $n = 5$ so that there are 4 equations and 4 unknowns

$$
\begin{aligned}
Bu_1 - Ku_2 &= h^2 f_1 + Ku_0, \\
-Ku_1 + Bu_2 - Ku_3 &= h^2 f_2, \\
- Ku_2 + Bu_3 - Ku_4 &= h^2 f_3 \text{ and} \\
-Ku_3 + Bu_4 &= h^2 f_4 + Ku_5.
\end{aligned}
$$

The matrix form of this is

$$AU = F \text{ where} \tag{2.6.7}$$

A is, in general, an $(n-1) \times (n-1)$ matrix, and U and F are $(n-1) \times 1$ column vectors. For example, for $n = 5$ we have a tridiagonal algebraic system

$$A = \begin{bmatrix} B & -K & 0 & 0 \\ -K & B & -K & 0 \\ 0 & -K & B & -K \\ 0 & 0 & -K & B \end{bmatrix},$$

$$U = \begin{bmatrix} u_1 \\ u_2 \\ u_3 \\ u_4 \end{bmatrix} \text{ and } F = \begin{bmatrix} h^2 f_1 + K u_0 \\ h^2 f_2 \\ h^2 f_3 \\ h^2 f_4 + K u_5 \end{bmatrix}.$$

2.6.4 Method

The solution can be obtained by either using the tridiagonal algorithm, or using a solver that is provided with your computer software. When one considers two and three space models, the coefficient matrix will become larger and more complicated. In these cases one may want to use a block tridiagonal solver, or an iterative method such as the classical *successive over relaxation (SOR)* approximation, see Sections 3.1 and 3.2.

2.6.5 Implementation

The user defined MATLAB function $bvp(n, cond, r, c, usur, f)$ defines the tridiagonal matrix, the right side and calls the MATLAB function trid(), which executes the tridiagonal algorithm. We have experimented with different radii, r, of the wire and different mesh sizes, $\Delta x = 1/n$. The user defined MATLAB function trid() is the same as in Section 2.3.

The parameter list of six numbers in the function file bvp.m and lines 2-10, define the diagonals in the tridiagonal matrix. The right side, which is stored in the vector d in line 9, use f replaced by a function of x, xx(i) in line 5. The function file trid.m is called in line 11, and it outputs an $n-1$ vector called *sol*. Then in lines 12-13 the xx and *sol* vectors are augmented to include the left and right boundaries. One could think of bvp as a mapping from $\mathbb{R}^6$ to $\mathbb{R}^{2(n+1)}$.

MATLAB Code bvp.m

```
1.     function [xx, sol] = bvp(n,cond,r,c,usur,f)
2.     h = 1/n;
3.     C = (2/r)*c;
4.     for i = 1:n-1
5.         xx(i) = i*h;
6.         a(i) = cond*2 + C*h*h;
7.         b(i) = -cond;
8.         c(i) = -cond;
9.         d(i) = h*h*(f + C*usur);
```

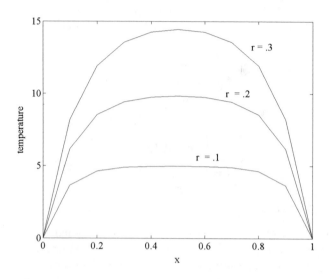

Figure 2.6.1: Variable r = .1, .2 and .3

10.   end
11.   sol = trid(n-1,a,b,c,d);
12.   xx = [0 xx 1.];
13.   sol = [0 sol 0.];

The following calculations vary the radii $r = .1, .2$ and $.3$ while fixing $n = 10, cond = .001, c = .01, usur = 0$ and $f = 1$. In Figure 2.6.1 the lowest curve corresponds to the approximate temperature for the smallest radius wire:

[xx1 uu1]=bvp(10,.001,.1,.01,0,1)
[xx2 uu2]=bvp(10,.001,.2,.01,0,1)
[xx3,uu3]=bvp(10,.001,.3,.01,0,1)
plot(xx1,uu1,xx2,uu2,xx3,uu3).

The following calculations vary the $n = 4, 8$ and 16 while fixing $cond = .001$, $r = .3, c = .01, usur = 0$ and $f = 1$. In Figure 2.6.2 the approximations as a function of n appear to be converging:

[xx4 uu4]=bvp(4,.001,.3,.01,0,1)
[xx8 uu8]=bvp(8,.001,.3,.01,0,1)
[xx16,uu16]=bvp(16,.001,.3,.01,0,1)
plot(xx4,uu4,xx8,uu8,xx16,uu16).

2.6.6   Assessment

In the above models of heat diffusion, the thermal conductivity was held constant relative to the space and temperature. If the temperature varies over a

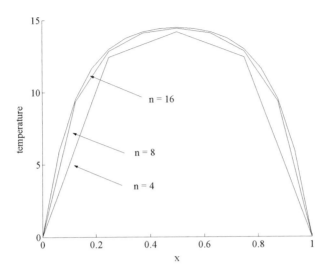

Figure 2.6.2: Variable n = 4, 8 and 16

large range, the thermal conductivity will show significant changes. Also, the electrical resistance will vary with temperature, and hence, the heat source, f, may be a function of temperature.

Another important consideration for the heat in a wire model is the possibility of the temperature being a function of the radial space variable, that is, as the radius increases, the temperature is likely to vary. Hence, a significant amount of heat will also diffuse in the radial direction.

A third consideration is the choice of mesh size, h. Once the algebraic system has been solved, one wonders how close the numerical solution of the finite difference method (2.6.6), the discrete model, is to the solution of the differential equation (2.6.1), the continuum model. We want to analyze the

$$discretization \ error \ \equiv \ E_i = u_i - u(ih). \tag{2.6.8}$$

Neglect any roundoff errors. As in Section 1.6 use the Taylor polynomial approximation with $n = 3$, and the fact that $u(x)$ satisfies (2.6.1) at $a = ih$ and $x = a \pm h$ to get

$$
\begin{aligned}
&K(u((i-1)h) - 2u(ih) + u((i+1)h))/h^2 \\
&= \ Cu(ih) - f(ih) + Ku^{(4)}(c_i)/12 \ h^2.
\end{aligned} \tag{2.6.9}
$$

The finite difference method (2.6.6) gives

$$-K(u_{i-1} - 2u_i + u_{i+1})/h^2 + Cu_i = f(ih). \tag{2.6.10}$$

Table 2.6.1: Second Order Convergence

n(h = 1/n)	$\|E\| * 10^{-4}$	$\|E\| / h^2$
10	17.0542	0.1705
20	04.2676	0.1707
40	01.0671	0.1707
80	00.2668	0.1707
160	00.0667	0.1707

Add equations (2.6.9) and (2.6.10) and use the definition of E_i to obtain

$$K(-E_{i-1} + 2E_i - E_{i+1})/h^2 + CE_i = Ku^{(4)}(c_i)/12\ h^2.$$

Or,

$$(2K/h^2 + C)E_i = KE_{i+1}/h^2 + KE_{i-1}/h^2 + Ku^{(4)}(c_i)/12\ h^2. \qquad (2.6.11)$$

Let $\|E\| = max_i|E_i|$ and $M_4 = max|u^{(4)}(x)|$ where x is in $[0, 1]$.
Then for all i equation (2.6.11) implies

$$(2K/h^2 + C)|E_i| \leq 2K/h^2\ \|E\| + KM_4/12\ h^2.$$

This must be true for the i that gives the maximum $\|E\|$, and therefore,

$$
\begin{aligned}
(2K/h^2 + C)\,\|E\| &\leq 2K/h^2\,\|E\| + KM_4/12\ h^2. \\
C\,\|E\| &\leq KM_4/12\ h^2. \qquad (2.6.12)
\end{aligned}
$$

We have just proved the next theorem.

Theorem 2.6.1 *(Finite Difference Error)   Consider the solution of (2.6.1) and the associated finite difference system (2.6.6). If the solution of (2.6.1) has four continuous derivatives on [0,1], then for $M_4 = max|u^{(4)}(x)|$ where x is in [0,1]*

$$\|E\| = max_i|u_i - u(ih)| \leq (KM_4/(12C))\ h^2.$$

Example. This example illustrates the second order convergence of the finite difference method, which was established in the above theorem. Consider (2.6.1) with $K = C = 1$ and $f(x) = 10x(1 - x)$. The exact solution is $u(x) = c_1 e^x + c_2 e^{-x} + 10(x(1.-x)-2.)$ where the constants are chosen so that $u(0) = 0$ and $u(1) = 0$. See the MATLAB code bvperr.m and the second column in Table 2.6.1 for the error, which is proportional to the *square* of the space step, $\Delta x = h$. For small h the error will decrease by one-quarter when h is decreased by one-half, and this is often called *second order convergence* of the finite difference solution to the continuous solution.

2.6.7   Exercises

1.   Experiment with the thin wire model. Examine the effects of varying $cond = .005, .001$ and $.0005$.

2.   Experiment with the thin wire model. Examine the effects of varying $c = .1, .01$ and $.001$.

3.   Find the exact solution for the calculations in Table 2.6.1, and verify the quadratic convergence of the finite difference method.

4.   Justify equation (2.6.9).

5.   Consider (2.6.1) but with a new boundary condition at $x = 1$ in the form $Ku_x(1) = (1 - u(1))$. Find the new algebraic system associated with the finite difference method.

6.   In exercise 5 find the exact solution and generate a table of errors, which is similar to Table 2.6.1.

7.   In exercises 5 and 6 prove a theorem, which is similar to the finite difference error theorem, Theorem 2.6.1.

Chapter 3

Poisson Equation Models

This chapter is the extension from one to two dimensional steady state space models. The solution of the discrete versions of these can be approximated by various iterative methods, and here the successive over-relaxation and conjugate gradient methods will be implemented. Three application areas are diffusion in two directions, ideal and porous fluid flows in two directions, and the deformation of the steady state membrane problem. The model for the membrane problem requires the shape of the membrane to minimize the potential energy, and this serves to motivate the formulation of the conjugate gradient method. The classical iterative methods are described in G. D. Smith [23] and Burden and Faires [4]

3.1 Steady State and Iterative Methods

3.1.1 Introduction

Models of heat flow in more than one direction will generate large and non-tridiagonal matrices. Alternatives to the full version of Gaussian elimination, which requires large storage and number of operations, are the iterative methods. These usually require less storage, but the number of iterations needed to approximate the solution can vary with the tolerance parameter of the particular method. In this section we present the most elementary iterative methods: Jacobi, Gauss-Seidel and successive over-relaxation (SOR). These methods are useful for sparse (many zero components) matrices where the nonzero patterns are very systematic. Other iterative methods such as the preconditioned conjugate gradient (PCG) or generalized minimum residual (GMRES) are particularly useful, and we will discuss these later in this chapter and in Chapter 9.

3.1.2   Applied Area

Consider the cooling fin problem from the previous chapter, but here we will use the iterative methods to solve the algebraic system. Also we will study the effects of varying the parameters of the fin such as thickness, T, and width, W. In place of solving the algebraic problem by the tridiagonal algorithm as in Section 2.3, the solution will be found iteratively. Since we are considering a model with diffusion in one direction, the coefficient matrix will be tridiagonal. So, the preferred method is the tridiagonal algorithm. Here the purpose of using iterative methods is to simply introduce them so that their application to models with more than one direction can be solved.

3.1.3   Model

Let $u(x)$ be the temperature in a cooling fin with only significant diffusion in one direction. Use the notation in Section 2.3 with $C = ((2W + 2T)/(TW))c$ and $f = Cu_{sur}$. The continuous model is given by the following differential equation and two boundary conditions.

$$-(Ku_x)_x + Cu = f, \tag{3.1.1}$$
$$u(0) = \text{given and} \tag{3.1.2}$$
$$Ku_x(L) = c(u_{sur} - u(L)). \tag{3.1.3}$$

The boundary condition in (3.1.3) is often called a *derivative or flux or Robin* boundary condition.

Let u_i be an approximation of $u(ih)$ where $h = L/n$. Approximate the derivative $u_x(ih + h/2)$ by $(u_{i+1} - u_i)/h$. Then equations (3.1.1) and (3.3.3) yield the finite difference approximation, a discrete model, of the continuous model.

Let u_0 be given and let $1 \le i < n$:

$$-[K(u_{i+1} - u_i)/h - K(u_i - u_{i-1})/h] + hCu_i = hf(ih). \tag{3.1.4}$$

Let $i = n$:

$$-[c(u_{sur} - u_n) - K(u_n - u_{n-1})/h] + (h/2)Cu_n = (h/2)f(nh). \tag{3.1.5}$$

For ease of notation we let $n = 4$, multiply (3.1.4) by h and (3.1.5) by $2h$, and let $B \equiv 2K + h2C$ so that there are 4 equations and 4 unknowns:

$$
\begin{aligned}
Bu_1 - Ku_2 &= h^2 f_1 + Ku_0, \\
-Ku_1 + Bu_2 - Ku_3 &= h^2 f_2, \\
-Ku_2 + Bu_3 - Ku_4 &= h^2 f_3 \text{ and} \\
-2Ku_3 + (B + 2hc)u_4 &= h^2 f_4 + 2chu_{sur}
\end{aligned}
$$

The matrix form of this is $AU = F$ where A is, in general, an $n \times n$ tridiagonal matrix and U and F are $n \times 1$ column vectors.

3.1.4 Method

In order to motivate the definition of these iterative algorithms, consider the following 3×3 example with $u_0 = 0$, $u_4 = 0$ and

$$-u_{i-1} + 3u_i - u_{i+1} = 1 \text{ for } i = 1, 2 \text{ and } 3.$$

Since the diagonal component is the largest, an approximation can be made by letting u_{i-1} and u_{i+1} be either previous guesses or calculations, and then computing the new u_i from the above equation.

Jacobi Method: Let $u^0 = [0, 0, 0]$ be the initial guess. The formula for the next iteration for node i is

$$u_i^{m+1} = (1 + u_{i-1}^m + u_{i+1}^m)/3.$$

$u^1 = [(1+0)/3, (1+0)/3, (1+0)/3] = [1/3, 1/3, 1/3]$
$u^2 = [(1+1/3)/3, (1+1/3+1/3)/3, (1+1/3)/3] = [4/9, 5/9, 4/9]$
$u^3 = [14/27, 17/27, 14/27].$

One repeats this until there is little change for all the nodes i.

Gauss-Seidel Method: Let $u^0 = [0, 0, 0]$ be the initial guess. The formula for the next iteration for node i is

$$u_i^{m+1} = (1 + u_{i-1}^{m+1} + u_{i+1}^m)/3.$$

$u^1 = [(1+0)/3, (1+1/3+0)/3, (1+4/9)/3] = [9/27, 12/27, 13/27]$
$u^2 = [(1+12/27)/3, (1+39/81+13/27)/3, (1+53/81)/3]$
$u^3 = [117/243, 159/243, 134/243].$

Note, the $m+1$ on the right side. This method varies from the Jacobi method because the most recently computed values are used. Repeat this until there is little change for all the nodes i.

The Gauss-Seidel algorithm usually converges much faster than the Jacobi method. Even though we can define both methods for any matrix, the methods may or may not converge. Even if it does converge, it may do so very slowly and have little practical use. Fortunately, for many problems similar to heat conduction, these methods and their variations do effectively converge. One variation of the Gauss-Seidel method is the successive over-relaxation (SOR) method, which has an acceleration parameter ω. Here the choice of the parameter ω should be between 1 and 2 so that convergence is as rapid as possible. For very special matrices there are formulae for such optimal ω, but generally the optimal ω are approximated by virtue of experience. Also the initial guess should be as close as possible to the solution, and in this case one may rely on the nature of the solution as dictated by the physical problem that is being modeled.

Jacobi Algorithm.

> for m = 0, maxit
> > for i = 1,n
> > $$x_i^{m+1} = (d_i - \sum_{j \neq i} a_{ij} x_j^m)/a_{ii}$$
> > endloop
> > test for convergence
> endloop.

SOR Algorithm (Gauss-Seidel for $\omega = 1.0$).

> for m = 0, maxit
> > for i = 1,n
> > $$x_i^{m+1/2} = (d_i - \sum_{j<i} a_{ij} x_j^{m+1} - \sum_{j>i} a_{ij} x_j^m)/a_{ii}$$
> > $$x_i^{m+1} = (1-\omega)x_i^m + \omega\, x_i^{m+1/2}$$
> > endloop
> > test for convergence
> endloop.

There are a number of *tests for convergence*. One common test is to determine if at each node the absolute value of two successive iterates is less than some given small number. This does not characterize convergence! Consider the following sequence of partial sums of the harmonic series

$$x^m = 1 + 1/2 + 1/3 + \cdots + 1/m.$$

Note $x^{m+1} - x^m = 1/(m+1)$ goes to zero and x^m goes to infinity. So, the above convergence test could be deceptive.

Four common tests for possible convergence are *absolute* error, *relative* error, *residual* error and *relative residual* error. Let $r(x^{m+1}) = d - Ax^{m+1}$ be the residual, and let x^m be approximations of the solution for $Ax = d$. Let $\|*\|$ be a suitable norm and let $\epsilon_i > 0$ be suitably small error tolerances. The *absolute*, *relative*, *residual* and *relative residual* errors are, respectively,

$$\|x^{m+1} - x^m\| < \epsilon_1,$$
$$\|x^{m+1} - x^m\|/\|x^m\| < \epsilon_2.$$
$$\|r(x^{m+1})\| < \epsilon_3 \text{ and}$$
$$\|r(x^{m+1})\|/\|d\| < \epsilon_4.$$

Often a combination of these is used to determine when to terminate an iterative method.

In most applications of these iterative methods the matrix is sparse. Consequently, one tries to use the zero pattern to reduce the computations in the summations. It is very important not to do the parts of the summation where the components of the matrix are zero. Also, it is not usually necessary to store all the computations. In Jacobi's algorithm one needs two $n \times 1$ column vectors, and in the SOR algorithm only one $n \times 1$ column vector is required.

3.1.5 Implementation

The cooling fin problem of the Section 2.3 is reconsidered with the tridiagonal algorithm replaced by the SOR iterative method. Although SOR converges much more rapidly than Jacobi, one should use the tridiagonal algorithm for tridiagonal problems. Some calculations are done to illustrate convergence of the SOR method as a function of the SOR parameter, ω. Also, numerical experiments with variable thickness, T, of the fin are done.

The MATLAB code fin1d.m, which was described in Section 2.3, will be used to call the following user defined MATLAB function sorfin.m. In fin1d.m on line 38 $u = trid(n, a, b, c, d)$ should be replaced by $[u, m, w] = sorfin(n, a, b, c, d)$, where the solution will be given by the vector u, m is the number of SOR steps required for convergence and w is the SOR parameter. In sorfin.m the accuracy of the SOR method will be controlled by the tolerance or error parameter, *eps* on line 7, and by the SOR parameter, w on line 8. The initial guess is given in lines 10-12. The SOR method is executed in the while loop in lines 13-39 where m is the loop index. The counter for the number of nodes that satisfy the error test is initialized in line 14 and updated in lines 20, 28 and 36. SOR is done for the left node in lines 15-21, for the interior nodes in lines 22-30 and for the right node in lines 31-37. In all three cases the $m + 1$ iterate of the unknowns over-writes the mth iterate of the unknowns. The error test requires the absolute value of the difference between successive iterates to be less than *eps*. When *numi* equals n, the while loop will be terminated. The while loop will also be terminated if the loop counter m is too large, in this case larger than $maxm = 500$.

MATLAB Code sorfin.m

```
1.    %
2.    % SOR Algorithm for Tridiagonal Matrix
3.    %
4.    function [u, m, w] =sorfin(n,a,b,c,d)
5.    maxm = 500;      % maximum iterations
6.    numi = 0;       % counter for nodes satisfying error
7.    eps = .1;       % error tolerance
8.    w = 1.8;       % SOR parameter
9.    m = 1;
10.   for i =1:n
11.        u(i) = 160.;      % initial guess
12.   end
13.   while ((numi<n)*(m<maxm))      % begin SOR loop
14.        numi = 0;
15.        utemp = (d(1) -c(1)*u(2))/a(1);      % do left node
16.        utemp = (1.-w)*u(1) + w*utemp;
17.        error = abs(utemp - u(1)) ;
18.        u(1) = utemp;
19.        if (error<eps)
```

Table 3.1.1: Variable SOR Parameter

SOR Parameter	Iterations for Conv.
1.80	178
1.85	133
1.90	077
1.95	125

```
20.                  numi = numi +1;
21.          end
22.          for i=2:n-1      % do interior nodes
23.                  utemp = (d(i) -b(i)*u(i-1) - c(i)*u(i+1))/a(i);
24.                  utemp = (1.-w)*u(i) + w*utemp;
25.                  error = abs(utemp - u(i));
26.                  u(i) = utemp;
27.                  if (error<eps)
28.                          numi = numi +1;
29.                  end
30.          end
31.          utemp = (d(n) -b(n)*u(n-1))/a(n);      % do right node
32.          utemp = (1.-w)*u(n) + w*utemp;
33.          error = abs(utemp - u(n)) ;
34.          u(n) = utemp ;
35.          if (error<eps)
36.                  numi = numi +1;      % exit if all nodes "converged"
37.          end
38.          m = m+1;
39.  end
```

The calculations in Table 3.1.1 are from an experiment with the SOR parameter where $n = 40$, $eps = 0.01$, $cond = 0.001$, $csur = 0.0001$, $usur = 70$, $W = 10$ and $L = 1$. The number of iterations that were required to reach the error test are recorded in column two where it is very sensitive to the choice of the SOR parameter.

Figure 3.1.1 is a graph of temperature versus space for variable thickness T of the fin. If T is larger, then the temperature of the cooling fin will be larger. Or, if T is smaller, then the temperature of the cooling fin will be closer to the cooler surrounding temperature, which in this case is $usur = 70$.

3.1.6 Assessment

Previously, we noted some shortcomings of the cooling fin model with diffusion in only one direction. The new models for such problems will have more complicated matrices. They will not be tridiagonal, but they will still have large diagonal components relative to the other components in each row of the

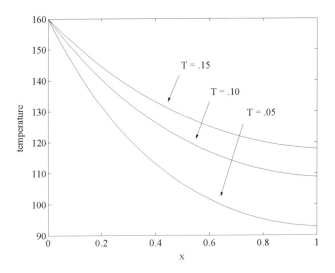

Figure 3.1.1: Cooling Fin with T = .05, .10 and .15

matrix. This property is very useful in the analysis of whether or not iterative methods converge to a solution of an algebraic system.

Definition. *Let* $A = [a_{ij}]$. *A is called strictly diagonally dominant if and only if for all i*

$$|a_{ii}| > \sum_{j \neq i} |a_{ij}|.$$

Examples.

1. The 3×3 example in the beginning of this section is strictly diagonally dominant

$$\begin{bmatrix} 3 & -1 & 0 \\ -1 & 3 & -1 \\ 0 & -1 & 3 \end{bmatrix}.$$

2. The matrix from the cooling fin is strictly diagonally dominant matrices because $B = 2K + h2C$

$$A = \begin{bmatrix} B & -K & 0 & 0 \\ -K & B & -K & 0 \\ 0 & -K & B & -K \\ 0 & 0 & -2K & B + 2ch \end{bmatrix}.$$

The next section will contain the proof of the following theorem and more examples that are not tridiagonal.

Theorem 3.1.1 *(Existence Theorem)   Consider $Ax = d$ and assume A is strictly diagonally dominant. If there is a solution, it is unique. Moreover, there is a solution.*

Theorem 3.1.2 *(Jacobi and Gauss-Seidel Convergence)   Consider $Ax = d$. If A is strictly diagonally dominant, then for all x^0 both the Jacobi and the Gauss-Seidel algorithms will converge to the solution.*

Proof.   Let x^{m+1} be from the Jacobi iteration and $Ax = d$. The component forms of these are

$$a_{ii}x_i^{m+1} = d_i - \sum_{j \neq i} a_{ij}x_j^m$$
$$a_{ii}x_i = d_i - \sum_{j \neq i} a_{ij}x_j.$$

Let the error at node i be

$$e_i^{m+1} = x_i^{m+1} - x_i$$

Subtract the above to get

$$a_{ii}e_i^{m+1} = 0 - \sum_{j \neq i} a_{ij}e_j^m$$
$$e_i^{m+1} = 0 - \sum_{j \neq i} \frac{a_{ij}}{a_{ii}}e_j^m$$

Use the triangle inequality

$$\left|e_i^{m+1}\right| = \left|\sum_{j \neq i} \frac{a_{ij}}{a_{ii}}e_i^m\right| \leq \sum_{j \neq i} \left|\frac{a_{ij}}{a_{ii}}\right| \left|e_j^m\right|.$$

$$\left\|e^{m+1}\right\| \equiv \max_i \left|e_i^{m+1}\right| \leq \max_i \sum_{j \neq i} \left|\frac{a_{ij}}{a_{ii}}\right| \left|e_j^m\right|$$
$$\leq (\max_i \sum_{j \neq i} \left|\frac{a_{ij}}{a_{ii}}\right|) \left\|e^m\right\|.$$

Because A is strictly diagonally dominant,

$$r = \max_i \sum_{j \neq i} \left|\frac{a_{ij}}{a_{ii}}\right| < 1.$$

$$\left\|e^{m+1}\right\| \leq r \left\|e^m\right\| \leq r(r \left\|e^{m-1}\right\|) \leq \cdots \leq r^{m+1} \left\|e^0\right\|$$

Since r < 1, the norm of the error must go to zero.   ∎

3.1.7 Exercises

1.   By hand do two iterations of the Jacobi and Gauss-Seidel methods for
the 3×3 example

$$
\begin{bmatrix} 3 & 1 & 0 \\ 1 & 3 & 1 \\ 0 & 1 & 3 \end{bmatrix}
\begin{bmatrix} x_1 \\ x_2 \\ x_3 \end{bmatrix}
=
\begin{bmatrix} 1 \\ 2 \\ 3 \end{bmatrix}.
$$

2.   Use the SOR method for the cooling fin and verify the calculations in
Table 3.1.1. Repeat the calculations but now use $n = 20$ and 80 as well as
$n = 40$.

3.   Use the SOR method for the cooling fin and experiment with the parameters $eps = .1, .01$ and $.001$. For a fixed $n = 40$ and eps find by numerical
experimentation the ω such that the number of iterations required for convergence is a minimum.

4.   Use the SOR method on the cooling fin problem and vary the width
$W = 5, 10$ and 20. What effect does this have on the temperature?

5.   Prove the Gauss-Seidel method converges for strictly diagonally dominant
matrices.

6.   The Jacobi algorithm can be described in matrix form by

$$
\begin{aligned}
x^{m+1} &= D^{-1}(L+U)x^m + D^{-1}d \text{ where} \\
A &= D - (L+U), \\
D &= diag(A).
\end{aligned}
$$

Assume A is strictly diagonally dominant.

(a).   Show $\left\| D^{-1}(L+U) \right\| < 1$.

(b).   Use the results in Section 2.5 to prove convergence of the Jacobi
algorithm.

3.2 Heat Transfer in 2D Fin and SOR

3.2.1 Introduction

This section contains a more detailed description of heat transfer models with
diffusion in two directions. The SOR algorithm is used to solve the resulting
algebraic problems. The models generate block tridiagonal algebraic systems,
and block versions of SOR and the tridiagonal algorithms will be described.

3.2.2 Applied Area

In the previous sections we considered a thin and long cooling fin so that one
could assume heat diffusion is in only one direction moving normal to the mass
to be cooled. If the fin is thick (large T) or if the fin is not long (small W), then
the temperature will significantly vary as one moves in the z or y directions.

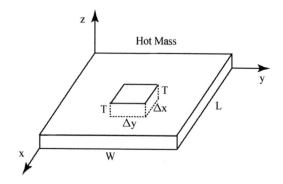

Figure 3.2.1: Diffusion in Two Directions

In order to model the *2D fin*, assume temperature is given along the 2D boundary and that the thickness T is small. Consequently, there will be diffusion in just the x and y directions. Consider a small mass within the plate, depicted in Figure 3.2.1, whose volume is $(\Delta x \Delta y T)$. This volume will have heat sources or sinks via the two $(\Delta x T)$ surfaces, two $(\Delta y T)$ surfaces, and two $(\Delta x \Delta y)$ surfaces as well as any internal heat equal to $f(heat/(vol.\ time))$. The top and bottom surfaces will be cooled by a Newton like law of cooling to the surrounding region whose temperature is u_{sur}. The steady state heat model with diffusion through the four vertical surfaces will be given by the *Fourier heat law* applied to each of the two directions

$$
\begin{aligned}
0 \approx\ & f(x,y)(\Delta x \Delta y T)\ \Delta t \\
& +(2\Delta x \Delta y)\ \Delta t c\ (u_{sur} - u) \\
& +(\Delta x T)\ \Delta t\ (Ku_y(x, y + \Delta y/2) - Ku_y(x, y - \Delta y/2)) \\
& +(\Delta y T)\Delta t\ (Ku_x(x + \Delta x/2, y) - Ku_x(x - \Delta x/2, y)).\quad (3.2.1)
\end{aligned}
$$

This approximation gets more accurate as Δx and Δy go to zero. So, divide by $(\Delta x \Delta y T)\Delta t$ and let Δx and Δy go to zero. This gives a partial differential equation (3.2.2).

Steady State 2D Diffusion.

$$
\begin{aligned}
0 =\ & f(x,y) + (2c/T)(u_{sur} - u) \\
& +(Ku_x(x,y))_x + (Ku_y(x,y))_y \\
& \text{for } (x,y) \text{ in } (0,L) \times (0,W), \\
& f(x,y) \text{ is the internal heat source,} \\
& K \text{ is the thermal conductivity and} \\
& c \text{ is the heat transfer coefficient.}
\end{aligned}
\qquad (3.2.2)
$$

3.2.3 Model

The partial differential equation (3.2.2) is usually associated with boundary conditions, which may or may not have derivatives of the solution. For the present, we will assume the *temperature is zero on the boundary* of $(0, L) \times (0, W)$, $L = W = 1, K = 1, f(x, y) = 0$ and $T = 2$. So, equation (3.2.2) simplifies to

$$-u_{xx} - u_{yy} + cu = cu_{sur}. \tag{3.2.3}$$

Let $u_{i,j}$ be the approximation of $u(ih, jh)$ where $h = dx = dy = 1.0/n$. Approximate the second order partial derivatives by the centered finite differences, or use (3.2.1) with similar approximations to $Ku_y(x, y + \Delta y/2) \approx K(u_{i,j+1} - u_{ij})/h$.

Finite Difference Model of (3.2.3).

$$-[(u_{i+1,j} - u_{i,j})/h - (u_{i,j} - u_{i-1,j})/h]/h$$
$$-[(u_{i,j+1} - u_{i,j})/h - (u_{i,j} - u_{i,j-1})/h]/h$$
$$+cu_{i,j} = cu_{sur} \text{ where } 1 \leq i, j \leq n - 1. \tag{3.2.4}$$

There are $(n - 1)^2$ equations for $(n - 1)^2$ unknowns $u_{i,j}$. One can write equation (3.2.4) as fixed point equations by multiplying both sides by h^2, letting $f = cu_{sur}$ and solving for $u_{i,j}$

$$u_{i,j} = (h^2 f_{ij} + u_{i,j-1} + u_{i-1,j} + u_{i+1,j} + u_{i,j+1})/(4 + ch^2). \tag{3.2.5}$$

3.2.4 Method

The point version of SOR for the problem in (3.2.4) or (3.2.5) can be very efficiently implemented because we know exactly where and what the nonzero components are in each row of the coefficient matrix. Since the unknowns are identified by a grid pair (i, j), the SOR computation for each unknown will be done within two nested loops. The classical order is given by having the i-loop inside and the j-loop on the outside. In this application of the SOR algorithm the lower sum is $u(i, j-1) + u(i-1, j)$ and the upper sum is $u(i+1, j) + u(i, j+1)$.

SOR Algorithm for (3.2.5) with f = cu$_{sur}$.

```
for m = 0, maxit
   for j = 2, n
      for i = 2,n
         utemp=(h*h*f(i,j) + u(i,j-1) + u(i-1,j)
                 + u(i+1,j ) + u(i,j+1))/(4+c*h*h)
         u(i,j)=(1-ω)*u(i,j)+ω*utemp
      endloop
   endloop
   test for convergence
endloop.
```

The finite difference model in (3.2.5) can be put into matrix form by multiplying the equations (3.2.5) by h^2 and listing the unknowns starting with the smallest y values (smallest j) and moving from the smallest to the largest x values (largest i). The first grid row of unknowns is denoted by $U_1 = [\ u_{11}\quad u_{21}\quad u_{31}\quad u_{41}\ ]^T$ for $n = 5$. Hence, the block form of the above system with boundary values set equal to zero is

$$
\begin{bmatrix}
B & -I & & \\
-I & B & -I & \\
& -I & B & -I \\
& & -I & B
\end{bmatrix}
\begin{bmatrix}
U_1 \\ U_2 \\ U_3 \\ U_4
\end{bmatrix}
= h^2
\begin{bmatrix}
F_1 \\ F_2 \\ F_3 \\ F_4
\end{bmatrix}
\quad \text{where}
$$

$$
B =
\begin{bmatrix}
4 + h^2 c & -1 & & \\
-1 & 4 + h^2 c & -1 & \\
& -1 & 4 + h^2 c & -1 \\
& & -1 & 4 + h^2 c
\end{bmatrix},
$$

$$
U_j =
\begin{bmatrix}
u_{1j} \\ u_{2j} \\ u_{3j} \\ u_{4j}
\end{bmatrix},\
F_j =
\begin{bmatrix}
f_{1j} \\ f_{2j} \\ f_{3j} \\ f_{4j}
\end{bmatrix}
\quad \text{with } f_{ij} = f(ih, jh).
$$

The above block tridiagonal system is a special case of the following block tridiagonal system where all block components are $N \times N$ matrices ($N = 4$). The block system has N^2 blocks, and so there are N^2 unknowns. If the full version of the Gaussian elimination algorithm was used, it would require approximately $(N^2)^3/3 = N^6/3$ operations

$$
\begin{bmatrix}
A_1 & C_1 & & \\
B_2 & A_2 & C_2 & \\
& B_3 & A_3 & C_3 \\
& & B_4 & A_4
\end{bmatrix}
\begin{bmatrix}
X_1 \\ X_2 \\ X_3 \\ X_4
\end{bmatrix}
=
\begin{bmatrix}
D_1 \\ D_2 \\ D_3 \\ D_4
\end{bmatrix}.
$$

Or, for $X_0 = 0 = X_5$ and $i = 1, 2, 3, 4$

$$
B_i X_{i-1} + A_i X_i + C_i X_{i+1} = D_i. \tag{3.2.6}
$$

In the block tridiagonal algorithm for (3.2.6) the entries are either $N \times N$ matrices or $N \times 1$ column vectors. The "divisions" for the "point" tridiagonal algorithm must be replaced by matrix solves, and one must be careful to preserve the proper order of matrix multiplication. The derivation of the following is similar to the derivation of the point form.

Block Tridiagonal Algorithm for (3.2.6).

> $\alpha(1)$ = A(1), solve $\alpha(1)$*g(1)= C(1) and solve $\alpha(1)$*Y(1) = D(1)
> for i = 2, N
>> $\alpha(i)$ = A(i)- B(i)*g(i-1)

```
        solve α(i)*g(i) = C(i)
        solve α(i)*Y(i) = D(i) - B(i)*Y(i-1)
    endloop
    X(N) = Y(N)
    for i = N - 1,1
        X(i) = Y(i) - g(i)*X(i+1)
    endloop.
```

The block or line version of the SOR algorithm also requires a matrix solve step in place of a "division." Note the matrix solve has a point tridiagonal matrix for the problem in (3.2.4).

Block SOR Algorithm for (3.2.6).

```
    for m = 1,maxit
        for i = 1,N
            solve A(i)*Xtemp = D(i) - B(i)*X(i-1) - C(i)*X(i+1)
            X(i) = (1-w)*X(i) + w*Xtemp
        endloop
        test for convergence
    endloop.
```

3.2.5 Implementation

The point SOR algorithm for a cooling plate, which has a fixed temperature at its boundary, is relatively easy to code. In the MATLAB function file sor2d.m, there are two input parameters for n and w, and there are outputs for w, *soriter* (the number of iterations needed to "converge") and the array u (approximate temperature array). The boundary temperatures are fixed at 200 and 70 as given by lines 7-10 where $ones(n + 1)$ is an $(n + 1) \times (n + 1)$ array whose components are all equal to 1, and the values of u in the interior nodes define the initial guess for the SOR method. The surrounding temperature is defined in line 6 to be 70, and so the steady state temperatures should be between 70 and 200. Lines 14-30 contain the SOR loops. The unknowns for the interior nodes are approximated in the nested loops beginning in lines 17 and 18. The counter *numi* indicates the number of nodes that satisfy the error tolerance, and *numi* is initialized in line 15 and updated in lines 22-24. If *numi* equals the number of unknowns, then the SOR loop is exited, and this is tested in lines 27-29. The outputs are given in lines 31-33 where $meshc(x, y, u\prime)$ generates a surface and contour plot of the approximated temperature. A similar code is sor2d.f90 written in Fortran 9x.

MATLAB Code sor2d.m

```
1.    function [w,soriter,u] = sor2d(n,w)
2.    h = 1./n;
3.    tol =.1*h*h;
```

```
4.      maxit = 500;
5.      c = 10.;
6.      usur = 70.;
7.      u = 200.*ones(n+1);      % initial guess and hot boundary
8.      u(n+1,:) = 70;      % cooler boundaries
9.      u(:,1) = 70;
10.     u(:,n+1) = 70
11.     f = h*h*c*usur*ones(n+1);
12.     x =h*(0:1:n);
13.     y = x;
14.     for iter =1:maxit      % begin SOR iterations
15.          numi = 0;
16.          for j = 2:n      % loop over all unknowns
17.               for i = 2:n
18.                    utemp = (f(i,j)  + u(i,j-1) + u(i-1,j) +
                                  u(i+1,j) + u(i,j+1))/(4.+h*h*c);
19.                    utemp = (1. - w)*u(i,j) + w*utemp;
20.                    error = abs(utemp - u(i,j));
21.                    u(i,j) = utemp;
22.                    if error<tol      % test each node for convergence
23.                         numi = numi + 1;
24.                    end
25.               end
26.          end
27.          if numi==(n-1)*(n-1)      % global convergence test
28.               break;
29.          end
30.     end
31.     w
32.     soriter = iter
33.     meshc(x,y,u')
```

The graphical output in Figure 3.2.2 is for $c = 10.0$, and one can see the plate has been cooled to a lower temperature. Also, we have graphed the temperature by indicating the equal temperature curves or contours. For 39^2 unknowns, error tolerance $tol = 0.01h^2$ and the SOR parameter $\omega = 1.85$, it took 121 iterations to converge. Table 3.2.1 records numerical experiments with other choices of ω, and this indicates that ω near 1.85 gives convergence in a minimum number of SOR iterations.

3.2.6   Assessment

In the first 2D heat diffusion model we kept the boundary conditions simple. However, in the 2D model of the cooling fin one should consider the heat that passes through the edge portion of the fin. This is similar to what was done for the cooling fin with diffusion in only the x direction. There the heat flux

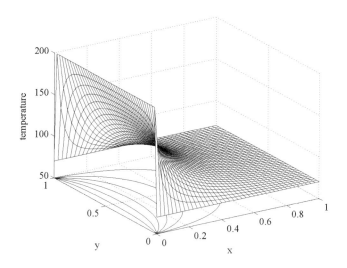

Figure 3.2.2: Temperature and Contours of Fin

Table 3.2.1: Convergence and SOR Parameter

SOR Parameter	Iter. for Conv.
1.60	367
1.70	259
1.80	149
1.85	121
1.90	167

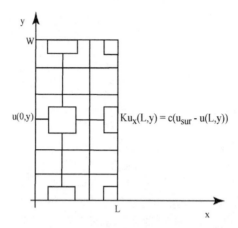

Figure 3.2.3: Cooling Fin Grid

at the tip was given by the boundary condition $Ku_x(L) = c(u_{sur} - u(L))$. For the 2D steady state cooling fin model we have similar boundary conditions on the edge of the fin that is away from the mass to be cooled.

The finite difference model must have additional equations for the cells near the edge of the fin. So, in Figure 3.2.3 there are 10 equations for the (Δx Δy) cells (interior), 5 equations for the ($\Delta x/2\ \Delta y$) cells (right), 4 equations for the ($\Delta x\ \Delta y/2$) cells (bottom and top) and 2 equations for the ($\Delta x/2\ \Delta y/2$) cells (corner). For example, the cells at the rightmost side with ($\Delta x/2\ \Delta y$) the finite difference equations are for $i = nx$ and $1 < j < ny$. The other portions of the boundary are similar, and for all the details the reader should examine the Fortran code fin2d.f90.

Finite Difference Model of (3.2.2) where i = nx and 1 < j < ny.

$$
\begin{aligned}
0 \ =\ & (2\Delta x/2\ \Delta y)c(u_{sur} - u_{nx,j}) \\
& + (\Delta x/2\ T)[K(u_{nx,j+1} - u_{nx,j})/\Delta y - K(u_{nx,j} - u_{nx,j-1})/\Delta y] \\
& + (\Delta y T)[c(u_{sur} - u_{nx,j}) - K(u_{nx,j} - u_{nx-1,j})/\Delta x]. \qquad (3.2.7)
\end{aligned}
$$

Another question is related to the existence of solutions. Note the diagonal components of the coefficient matrix are much larger than the off diagonal components. For each row the diagonal component is strictly larger than the sum of the absolute value of the off diagonal components. In the finite difference model for (3.2.4) the diagonal component is $4/h^2 + c$ and the four off diagonal components are $-1/h^2$. So, like the 1D cooling fin model the 2D cooling fin model has a strictly diagonally dominant coefficient matrix.

Theorem 3.2.1 (*Existence Theorem*)   *Consider* $Ax = d$ *and assume A is strictly diagonally dominant. If there is a solution, it is unique. Moreover, there is a solution.*

Proof.   Let x and y be two solutions. Then $A(x - y) = 0$. If $x - y$ is not a zero column vector, then we can choose i so that $|x_i - y_i| = max_j|x_j - y_j| > 0$.

$$a_{ii} (x_i - y_i) + \sum_{j \neq i} a_{ij} (x_j - y_j) = 0.$$

Divide by $x_i - y_i$ and use the triangle inequality to contradict the strict diagonal dominance. Since the matrix is square, the existence follows from the uniqueness. ∎

3.2.7   Exercises

1.   Consider the MATLAB code for the point SOR algorithm.

   (a). Verify the calculations in Table 3.2.1.

   (b). Experiment with the convergence parameter *tol* and observe the number of iterations required for convergence.

2.   Consider the MATLAB code for the point SOR algorithm. Experiment with the SOR parameter ω and observe the number of iterations for convergence. Do this for $n = 5, 10, 20$ and 40 and find the ω that gives the smallest number of iterations for convergence.

3.   Consider the MATLAB code for the point SOR algorithm. Let $c = 0$, $f(x, y) = 2\pi^2 sin(\pi x)sin(\pi y)$ and require u to be zero on the boundary of $(0, 1) \times (0, 1)$.

   (a). Show the solution is $u(x, y) = sin(\pi x)sin(\pi y)$.

   (b). Compare it with the numerical solution with $n = 5, 10, 20$ and 40. Observe the error is of order h^2.

   (c). Why have we used a convergence test with $tol = eps * h * h$?

4.   In the MATLAB code modify the boundary conditions so that at $u(0, y) = 200$, $u(x, 0) = u(x, 1) = u(1, y) = 70$. Experiment with n and ω.

5.   Implement the block tridiagonal algorithm for problem 3.

6.   Implement the block SOR algorithm for problem 3.

7.   Use Theorem 3.2.1 to show the block diagonal matrix from (3.2.4) for the block SOR is nonsingular.

8.   Use the Schur complement as in Section 2.4 and Theorem 3.2.1 to show the alpha matrices in the block tridiagonal algorithm applied to (3.2.4) are nonsingular.

3.3   Fluid Flow in a 2D Porous Medium

3.3.1   Introduction

In this and the next section we present two fluid flow models in 2D: flow in a porous media and ideal fluids. Both these models are similar to steady state 2D heat diffusion. The porous media flow uses an empirical law called Darcy's law, which is similar to Fourier's heat law. An application of this model to groundwater management will be studied.

3.3.2   Applied Area

In both applications assume the velocity of the fluid is $(u(x,y), v(x,y), 0)$, that is, it is a *2D steady state fluid flow*. In flows for both a porous medium and ideal fluid it is useful to be able to give a mathematical description of compressibility of the fluid. The compressibility of the fluid can be quantified by the divergence of the velocity. In 2D the *divergence of (u,v) is* $u_x + v_y$. This indicates how much mass enters a small volume in a given unit of time. In order to understand this, consider the small thin rectangular mass as depicted in Figure 3.3.1 with density ρ and approximate $u_x + v_y$ by finite differences. Let Δt be the change in time so that $u(x + \Delta x, y)\Delta t$ approximates the change in the x direction of the mass leaving (for $u(x + \Delta x, y) > 0$) the front face of the volume $(\Delta x \Delta y T)$.

$$
\begin{aligned}
\text{change in mass} \quad &= \quad \text{sum via four vertical faces of } (\Delta x \Delta y T) \\
&= \quad \rho T \, \Delta y \, (u(x + \Delta x, y) - u(x, y))\Delta t \\
&\quad + \rho T \, \Delta x \, (v(x, y + \Delta y) - v(x, y))\Delta t. \quad\quad (3.3.1)
\end{aligned}
$$

Divide by $(\Delta x \Delta y T)\Delta t$ and let Δx and Δy go to zero to get

$$
\text{rate of change of mass per unit volume} = \rho(u_x + v_y). \quad\quad (3.3.2)
$$

If the fluid is *incompressible*, then $u_x + v_y = 0$.

Consider a shallow saturated porous medium with at least one well. Assume the region is in the xy-plane and that the water moves towards the well in such a way that the velocity vector is in the xy-plane. At the top and bottom of the xy region assume there is no flow through these boundaries. However, assume there is ample supply from the left and right boundaries so that the pressure is fixed. The problem is to determine the flow rates of well(s), location of well(s) and number of wells so that there is still water to be pumped out. If a cell does not contain a well and is in the interior, then $u_x + v_y = 0$. If there is a well in a cell, then $u_x + v_y < 0$.

3.3.3   Model

Both porous and ideal flow models have a partial differential equation similar to that of the 2D heat diffusion model, but all three have different boundary

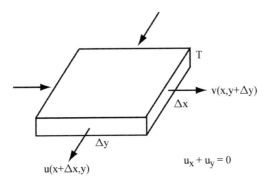

Figure 3.3.1: Incompressible 2D Fluid

conditions. For porous fluid flow problems, boundary conditions are either a given function along part of the boundary, or a zero derivative for the other parts of the boundary. The motion of the fluid is governed by an empirical law called Darcy's law.

Darcy Law.

$$(u, v) \quad = \quad -K(h_x, h_y) \text{ where} \tag{3.3.3}$$
$$h \text{ is the hydraulic head pressure and}$$
$$K \text{ is the hydraulic conductivity.}$$

The hydraulic conductivity depends on the pressure. However, if the porous medium is saturated, then it can be assumed to be constant. Next couple Darcy's law with the divergence of the velocity to get the following partial differential equation for the pressure.

$$u_x + v_y = -(Kh_x)_x - (Kh_y)_y = f. \tag{3.3.4}$$

Groundwater Fluid Flow Model.

$$-(Kh_x)_x - (Kh_y)_y \quad = \quad \begin{cases} 0 & , (x, y) \notin well \\ -R & , (x, y) \in well \end{cases} \quad (x, y) \in (0, L) \times (0, W),$$
$$Kh_y \quad = \quad 0 \text{ for } y = 0 \text{ and } y = W \text{ and}$$
$$h \quad = \quad h_\infty \text{ for } x = 0 \text{ and } x = L.$$

3.3.4   Method

We will use the finite difference method coupled with the SOR iterative scheme. For the $(\Delta x \Delta y)$ cells in the interior this is similar to the 2D heat diffusion problem. For the portions of the boundary where a derivative is set equal to

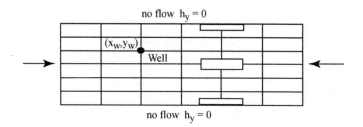

no flow $h_y = 0$

no flow $h_y = 0$

Figure 3.3.2: Groundwater 2D Porous Flow

zero on a half cell $(\Delta x/2 \ \Delta y)$ or $(\Delta x \ \Delta y/2)$ as in Figure 3.3.2, we insert some additional code inside the SOR loop. For example, consider the groundwater model where $h_y = 0$ at $y = W$ on the half cell $(\Delta x \ \Delta y/2)$. The finite difference equations for $u = h$, $dx = \Delta x$ and $dy = \Delta y$ in (3.3.5) and corresponding line of SOR code in (3.3.6) are

$$
\begin{aligned}
0 \ = \ & -[(0) - (u(i,j) - u(i,j-1))/dy]/(dy/2) \\
& -[(u(i+1,j) - u(i,j))/dx - \\
& \qquad (u(i,j) - u(i-1,j))/dx]/dx
\end{aligned} \tag{3.3.5}
$$

$$
\begin{aligned}
utemp \ = \ & ((u(i+1,j) + u(i-1,j))/(dx * dx) \\
& +2 * u(i,j-1)/(dy * dy))/(2/(dx * dx) + 2/(dy * dy)) \\
u(i,j) \ = \ & (1-w) * u(i,j) + w * utemp.
\end{aligned} \tag{3.3.6}
$$

3.3.5 Implementation

In the following MATLAB code por2d.m the SOR method is used to solve the discrete model for steady state saturated 2D groundwater porous flow. Lines 1-44 initialize the data. It is interesting to experiment with *eps* in line 6, the SOR parameter ww in line 7, nx and ny in lines 9,10, the location and flow rates of the wells given in lines 12-16, and the size of the flow field given in lines 28,29. In line 37 R_well is calibrated to be independent of the mesh. The SOR iteration is done in the while loop in lines 51-99. The bottom nodes in lines 73-85 and top nodes in lines 86-97, where there are no flow boundary conditions, must be treated differently from the interior nodes in lines 53-71. The locations of the two wells are given by the if statements in lines 58-63. The output is in lines 101-103 where the number of iterations for convergence and the SOR parameter are printed, and the surface and contour plots of the pressure are graphed by the MATLAB command meshc(x,y,u'). A similar code is por2d.f90 written in Fortran 9x.

MATLAB Code por2d.m

```
1.    % Steady state saturated 2D porous flow.
2.    % SOR is used to solve the algebraic system.
3.    % SOR parameters
4.    clear;
5.    maxm = 500;
6.    eps = .01;
7.    ww = 1.97;
8.    % Porous medium data
9.    nx = 50;
10.   ny = 20;
11.   cond = 10.;
12.   iw = 15;
13.   jw = 12;
14.   iwp = 32;
15.   jwp = 5;
16.   R_well = -250.;
17.   uleft = 100. ;
18.   uright = 100.;
19.   for j=1:ny+1
20.        u(1,j) = uleft;
21.        u(nx+1,j) = uright;
22.   end
23.   for j =1:ny+1
24.        for i = 2:nx
25.             u(i,j) = 100.;
26.        end
27.   end
28.   W = 1000.;
29.   L = 5000.;
30.   dx = L/nx;
31.   rdx = 1./dx;
32.   rdx2 = cond/(dx*dx);
33.   dy = W/ny;
34.   rdy = 1./dy;
35.   rdy2 = cond/(dy*dy);
36.   % Calibrate R_well to be independent of the mesh
37.   R_well = R_well/(dx*dy);
38.   xw = (iw)*dx;
39.   yw = (jw)*dy;
40.   for i = 1:nx+1
41.        x(i) = dx*(i-1);
42.   end
43.   for j = 1:ny+1
```

```
44.            y(j) = dy*(j-1);
45.        end
46.        % Execute SOR Algorithm
47.        nunkno = (nx-1)*(ny+1);
48.        m = 1;
49.        numi = 0;
50.        while ((numi<nunkno)*(m<maxm))
51.            numi = 0;
52.            % Interior nodes
53.            for j = 2:ny
54.                for i=2:nx
55.                    utemp = rdx2*(u(i+1,j)+u(i-1,j));
56.                    utempp = utemp + rdy2*(u(i,j+1)+u(i,j-1));
57.                    utemp = utempp/(2.*rdx2 + 2.*rdy2);
58.                    if ((i==iw)*(j==jw))
59.                        utemp=(utempp+R_well)/(2.*rdx2+2.*rdy2);
60.                    end
61.                    if ((i==iwp)*(j==jwp))
62.                        utemp =(utempp+R_well)/(2.*rdx2+2.*rdy2);
63.                    end
64.                    utemp = (1.-ww)*u(i,j) + ww*utemp;
65.                    error = abs(utemp - u(i,j)) ;
66.                    u(i,j) = utemp;
67.                    if (error<eps)
68.                        numi = numi +1;
69.                    end
70.                end
71.            end
72.            % Bottom nodes
73.            j = 1;
74.            for i=2:nx
75.                utemp = rdx2*(u(i+1,j)+u(i-1,j));
76.                utemp = utemp + 2.*rdy2*(u(i,j+1));
77.                utemp = utemp/(2.*rdx2 + 2.*rdy2 );
78.                utemp = (1.-ww)*u(i,j) + ww*utemp;
79.                error = abs(utemp - u(i,j)) ;
80.                u(i,j) = utemp;
81.                    if (error<eps)
82.                numi = numi +1;
83.                    end
84.            end
85.            % Top nodes
86.            j = ny+1;
87.            for i=2:nx
88.                utemp = rdx2*(u(i+1,j)+u(i-1,j));
```

```
89.                    utemp = utemp + 2.*rdy2*(u(i,j-1));
90.                    utemp = utemp/(2.*rdx2 + 2.*rdy2);
91.                    utemp = (1.-ww)*u(i,j) + ww*utemp;
92.                    error = abs(utemp - u(i,j));
93.                    u(i,j) = utemp;
94.                    if (error<eps)
95.                         numi = numi +1;
96.                    end
97.               end
98.               m = m+1;
99.          end
100.      % Output to Terminal
101.      m
102.      ww
103.      meshc(x,y,u')
```

The graphical output is given in Figure 3.3.3 where there are two wells and the pressure drops from 100 to around 45. This required 199 SOR iterations, and SOR parameter $w = 1.97$ was found by numerical experimentation. This numerical approximation may have significant errors due either to the SOR convergence criteria $eps = .01$ in line 6 being too large or to the mesh size in lines 9 and 10 being too large. If $eps = .001$, then 270 SOR iterations are required and the solution did not change by much. If $eps = .001$, ny is doubled from 20 to 40, and the jw and jwp are also doubled so that the wells are located in the same position in space, then 321 SOR iterations are computed and little difference in the graphs is noted. If the flow rate at both wells is increased from 250. to 500., then the pressure should drop. Convergence was attained in 346 SOR iterations for $eps = .001$, $nx = 50$ and $ny = 40$, and the graph shows the pressure at the second well to be negative, which indicates the well has gone dry!

3.3.6   Assessment

This porous flow model has enough assumptions to rule out many real applications. For groundwater problems the soils are usually not fully saturated, and the hydraulic conductivity can be highly nonlinear or vary with space according to the soil types. Often the soils are very heterogeneous, and the soil properties are unknown. Porous flows may require 3D calculations and irregularly shaped domains. The good news is that the more complicated models have many subproblems, which are similar to our present models from heat diffusion and fluid flow in saturated porous media.

3.3.7   Exercises

1.   Consider the groundwater problem. Experiment with the choice of w and eps. Observe the number of iterations required for convergence.

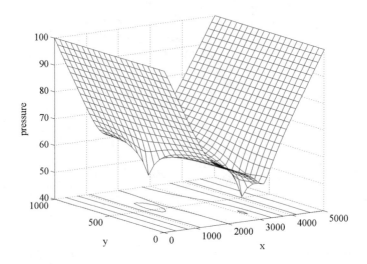

Figure 3.3.3: Pressure for Two Wells

2.   Experiment with the mesh sizes nx and ny, and convince yourself the discrete problem has converged to the continuous problem.
3.   Consider the groundwater problem. Experiment with the physical parameters $K = cond$, W, L and pump rate $R = R\_well$.
4.   Consider the groundwater problem. Experiment with the number and location of the wells.

3.4   Ideal Fluid Flow

3.4.1   Introduction

An *ideal fluid* is a steady state flow in 2D that is incompressible and has no circulation. In this case the velocity of the fluid can be represented by a stream function, which is a solution of a partial differential equation that is similar to the 2D heat diffusion and 2D porous flow models. The applied problem could be viewed as a first model for flow of a shallow river about an object, and the numerical solution will also be given by a variation of the previous SOR MATLAB codes.

3.4.2   Applied Area

Figure 3.4.1 depicts the flow about an obstacle. Because the fluid is not compressible, it must significantly increase its speed in order to pass near the obstacle. This can cause severe erosion of the nearby soil. The problem is to

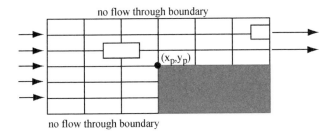

Figure 3.4.1: Ideal Flow About an Obstacle

determine these velocities of the fluid given the upstream velocity and the location and shape of the obstacle.

3.4.3 Model

We assume the velocity is a 2D steady state incompressible fluid flow. The *incompressibility* of the fluid can be characterized by the divergence of the velocity

$$u_x + v_y = 0. \tag{3.4.1}$$

The circulation or rotation of a fluid can be described by the curl of the velocity vector. In 2D the *curl of (u, v)* is $v_x - u_y$. Also the discrete form of this gives some insight to its meaning. Consider the loop about the rectangular region given in Figure 3.4.2. Let A be the cross-sectional area in this loop. The momentum of the vertical segment of the right side is $\rho A \Delta y\, v(x + \Delta x, y)$. The circulation or angular momentum of the loop about the tube with cross-section area A and density ρ is

$$\rho A \Delta y (v(x + \Delta x, y) - v(x, y)) - \rho A \Delta x (u(x, y + \Delta y) - u(x, y)).$$

Divide by $\rho(A \Delta y \Delta x)$ and let Δx and Δy go to zero to get $v_x - u_y$. If there is no circulation, then this must be zero. The fluid is called *irrotational* if

$$v_x - u_y = 0. \tag{3.4.2}$$

An ideal 2D steady state fluid flow is defined to be incompressible and irrotational so that both equations (3.4.1) and (3.4.2) hold. One can use the incompressibility condition and Green's theorem (more on this later) to show that there is a *stream function*, Ψ, such that

$$(\Psi_x, \Psi_y) = (-v, u). \tag{3.4.3}$$

The irrotational condition and (3.4.3) give

$$v_x - u_y = (-\Psi_x)_x - (\Psi_y)_y = 0.$$

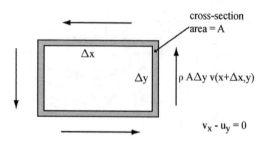

Figure 3.4.2: Irrotational 2D Flow $v_x - u_y = 0$

If the velocity upstream to the left in Figure 3.4.1 is $(u_0, 0)$, then let $\Psi = u_0\, y$. If the velocity at the right in Figure is $(u, 0)$, then let $\Psi_x = 0$. Equation (3.4.4) and these boundary conditions give the ideal fluid flow model.

We call Ψ a stream function because the curves $(x(\tau), y(\tau))$ defined by setting Ψ to a constant are parallel to the velocity vectors (u, v). In order to see this, let $\Psi(x(\tau), y(\tau)) = c$, compute the derivative with respect to τ and use the chain rule

$$
\begin{aligned}
0 &= \frac{d}{d\tau}\Psi(x(\tau), y(\tau)) \\
&= \Psi_x \frac{dx}{d\tau} + \Psi_y \frac{dy}{d\tau} \\
&= (-v, u) \cdot (\frac{dx}{d\tau}, \frac{dy}{d\tau}).
\end{aligned}
$$

Since $(u, v) \cdot (-v, u) = 0$, (u, v) and the tangent vector to the curve given by $\Psi(x(\tau), y(\tau)) = c$ must be parallel.

Ideal Flow Around an Obstacle Model.

$$
\begin{aligned}
-\Psi_{xx} - \Psi_{yy} &= 0 \text{ for } (x, y) \in (0, L) \times (0, W), \\
\Psi &= u_0 y \text{ for } x = 0\ (u = u_0), \\
\Psi &= u_0 W \text{ for } y = W\ (v = 0) \\
\Psi &= 0 \text{ for } y = 0 \text{ or } (x, y) \text{ on the obstacle and} \\
\Psi_x &= 0 \text{ for } x = L\ (v = 0).
\end{aligned}
$$

3.4.4   Method

Use the finite difference method coupled with the SOR iterative scheme. For the $(\Delta x \Delta y)$ cells in the interior this is similar to the 2D heat diffusion problem. For the portions of the boundary where a derivative is set equal to zero on a half cell $(\Delta x/2\ \Delta y)$ as in Figure 3.4.1, insert some additional code inside the SOR loop. In the obstacle model where $\Psi_x = 0$ at $x = L$ we have half cells $(\Delta x/2$

Δy). The *finite difference* equation in equation (3.4.5) and corresponding line in (3.4.6) of SOR code with $u = \Psi$, $dx = \Delta x$ and $dy = \Delta y$ are

$$
\begin{aligned}
0 = \; & -[(0)/dx - (u(i,j) - u(i-1,j))/dx]/dx/2 \\
& -[(u(i,j+1) - u(i,j))/dy \\
& \quad - (u(i,j) - u(i,j-1))/dy]/dy \quad (3.4.4)
\end{aligned}
$$

$$
\begin{aligned}
utemp = \; & (2 * u(i-1,j)/(dx*dx) + \\
& (u(i,j+1) + u(i,j-1))/(dy*dy)) \\
& /(2/(dx*dx) + 2/(dy*dy)) \\
u(i,j) = \; & (1-w)*u(i,j) + w*utemp. \quad (3.4.5)
\end{aligned}
$$

3.4.5 Implementation

The MATLAB code ideal2d.m has a similar structure as por2d.m, and also it uses the SOR scheme to approximate the solution to the algebraic system associated with the ideal flow about an obstacle. The obstacle is given by a darkened rectangle in Figure 3.4.1, and can be identified by indicating the indices of the point (ip, jp) as is done in lines 11,12. Other input data is given in lines 4-39. The SOR scheme is executed using the while loop in lines 46-90. The SOR calculations for the various nodes are done in three groups: the interior bottom nodes in lines 48-62, the interior top nodes in lines 62-75 and the right boundary nodes in lines 76-88. Once the SOR iterations have been completed, the output in lines 92-94 prints the number of SOR iterations, the SOR parameter and the contour graph of the stream line function via the MATLAB command contour(x,y,u').

MATLAB Code ideal2d.m

```
1.    % This code models flow around an obstacle.
2.    % SOR iterations are used to solve the system.
3.    % SOR parameters
4.    clear;
5.    maxm = 1000;
6.    eps = .01;
7.    ww = 1.6;
8.    % Flow data
9.    nx = 50;
10.   ny = 20;
11.   ip = 40;
12.   jp = 14;
13.   W = 100.;
14.   L = 500.;
15.   dx = L/nx;
16.   rdx = 1./dx;
```

```
17.     rdx2 = 1./(dx*dx);
18.     dy = W/ny;
19.     rdy = 1./dy;
20.     rdy2 = 1./(dy*dy);
21.     % Define Boundary Conditions
22.     uo = 1.;
23.          for j=1:ny+1
24.     u(1,j) = uo*(j-1)*dy;
25.     end
26.     for i = 2:nx+1
27.          u(i,ny+1) = uo*W;
28.     end
29.     for j =1:ny
30.          for i = 2:nx+1
31.               u(i,j) = 0.;
32.          end
33.     end
34.     for i = 1:nx+1
35.          x(i) = dx*(i-1);
36.     end
37.     for j = 1:ny+1
38.          y(j) = dy*(j-1);
39.     end
40.     %
41.     % Execute SOR Algorithm
42.     %
43.     unkno = (nx)*(ny-1) - (jp-1)*(nx+2-ip);
44.     m = 1;
45.     numi = 0;
46.     while ((numi<unkno)*(m<maxm))
47.          numi = 0;
48.          % Interior Bottom Nodes
49.          for j = 2:jp
50.               for i=2:ip-1
51.                    utemp = rdx2*(u(i+1,j)+u(i-1,j));
52.                    utemp = utemp + rdy2*(u(i,j+1)+u(i,j-1));
53.                    utemp = utemp/(2.*rdx2 + 2.*rdy2);
54.                    utemp = (1.-ww)*u(i,j) + ww*utemp;
55.                    error = abs(utemp - u(i,j));
56.                    u(i,j) = utemp;
57.                    if (error<eps)
58.                         numi = numi +1;
59.                    end
60.               end
61.          end
```

```
62.              % Interior Top Nodes
63.              for j = jp+1:ny
64.                  for i=2:nx
65.                      utemp = rdx2*(u(i+1,j)+u(i-1,j));
66.                      utemp = utemp + rdy2*(u(i,j+1)+u(i,j-1));
67.                      utemp = utemp/(2.*rdx2 + 2.*rdy2);
68.                      utemp = (1.-ww)*u(i,j) + ww*utemp;
69.                      error = abs(utemp - u(i,j)) ;
70.                      u(i,j) = utemp;
71.                      if (error<eps)
72.                          numi = numi +1;
73.                      end
74.                  end
75.              end
76.              % Right Boundary Nodes
77.              i = nx+1;
78.              for j = jp+1:ny
79.                  utemp = 2*rdx2*u(i-1,j);
80.                  utemp = utemp + rdy2*(u(i,j+1)+u(i,j-1));
81.                  utemp = utemp/(2.*rdx2 + 2.*rdy2);
82.                  utemp = (1.-ww)*u(i,j) + ww*utemp;
83.                  error = abs(utemp - u(i,j));
84.                  u(i,j) = utemp;
85.                  if (error<eps)
86.                      numi = numi +1;
87.                  end
88.              end
89.              m = m +1;
90.          end
91.          % Output to Terminal
92.          m
93.          ww
94.          contour(x,y,u')
```

The obstacle model uses the parameters $L = 500, W = 100$ and $u_0 = 1$. Since $u_0 = 1$, the stream function must equal $1y$ in the upstream position, the left side of Figure 3.4.1. The x component of the velocity is $u_0 = 1$, and the y component of the velocity will be zero. The graphical output gives the contour lines of the stream function. Since these curves are much closer near the exit, the right side of the figure, the x component of the velocity must be larger above the obstacle. If the obstacle is made smaller, then the exiting velocity will not be as large.

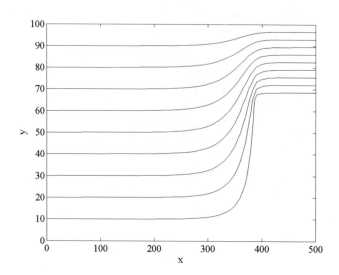

Figure 3.4.3: Flow Around an Obstacle

3.4.6   Assessment

This ideal fluid flow model also has enough assumptions to rule out many real applications. Often there is circulation in flow of water, and therefore, the irrotational assumption is only true for slow moving fluids in which circulation does not develop. Air is a compressible fluid. Fluid flows may require 3D calculations and irregularly shaped domains. Fortunately, the more complicated models have many subproblems which are similar to our present models from heat diffusion, fluid flow in saturated porous medium and ideal fluid flow.

The existence of stream functions such that $(\Psi_x, \Psi_y) = (-v, u)$ needs to be established. Recall the conclusion of Green's Theorem where Ω is a simply connected region in 2D with boundary C given by functions with piecewise continuous first derivatives

$$\oint_C P dx + Q dy = \iint_\Omega Q_x - P_y dx dy. \qquad (3.4.6)$$

Suppose $u_x + v_y = 0$ and let $Q = u$ and $P = -v$. Since $Q_x - P_y = (u)_x - (-v)_y = 0$, then the line integral about a closed curve will always be zero. This means that the line integral will be independent of the path taken between two points.

Define the stream function to be the line integral of $(P, Q) = (-v, u)$ starting at some (x_0, y_0) and ending at (x, y). This is single valued because the line integral is independent of the path taken. In order to show $(\Psi_x, \Psi_y) = (-v, u)$,

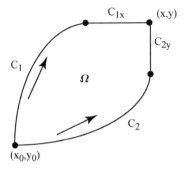

Figure 3.4.4: Two Paths to (x,y)

consider the special path $C_1 + C_{1x}$ in Figure 3.4.4 and show $\Psi_x = -v = P$

$$
\begin{aligned}
\Psi_x &= \frac{d}{dx} \int_{C_1} Pdx + Qdy + \frac{d}{dx} \int_{C_{1x}} Pdx + Qdy \\
&= 0 + \frac{d}{dx} \int_{C_{1x}} Pdx + Q0 \\
&= P.
\end{aligned}
$$

The proof that $\Psi_y = u = Q$ is similar and uses the path $C_2 + C_{2y}$ in Figure 3.4.4. We have just proved the following theorem.

Theorem 3.4.1 *(Existence of Stream Function) If $u(x,y)$ and $v(x,y)$ have continuous first order partial derivatives and $u_x + v_y = 0$, then there is a stream function such that $(\Psi_x, \Psi_y) = (-v, u)$.*

3.4.7 Exercises

1. Consider the obstacle problem. Experiment with the different values of w and *eps*. Observe the number of iterations required for convergence.

2. Experiment with the mesh size and convince yourself the discrete problem has converged to the continuous problem.

3. Experiment with the physical parameters W, L and incoming velocity u_0.

4. Choose a different size and shape obstacle. Compare the velocities near the obstacle. What happens if there is more than one obstacle?

5. Prove the other part of Theorem 3.4.1 $\Psi_y = u$.

3.5   Deformed Membrane and Steepest Descent

3.5.1   Introduction

The objective of this and the next section is to introduce the conjugate gradient method for solving special algebraic systems where the coefficient matrix A is *symmetric* $(A = A^T)$ and *positive definite* $(x^T A x > 0$ for all nonzero real vectors $x)$. Properties of these matrices also will be carefully studied in Section 8.2. This method will be motivated by the applied problem to find the deformation of membrane if the position on the boundary and the pressure on the membrane are known. The model will initially be in terms of finding the deformation so that the potential energy of the membrane is a minimum, but it will be reformulated as a partial differential equation. Also, the method of steepest descent in a single direction will be introduced. In the next section this will be generalized from the steepest descent method from one to multiple directions, which will eventually give rise to the conjugate gradient method.

3.5.2   Applied Area

Consider a *membrane* whose edges are fixed, for example, a musical drum. If there is pressure (force per unit area) applied to the interior of the membrane, then the membrane will deform. The objective is to find the deformation for every location on the membrane. Here we will only focus on the time independent model, and also we will assume the deformation and its first order partial derivative are "relatively" small. These two assumptions will allow us to formulate a model, which is similar to the heat diffusion and fluid flow models in the previous sections.

3.5.3   Model

There will be three equivalent models. The formulation of the minimum potential energy model will yield the weak formulation and the partial differential equation model of a steady state membrane with small deformation. Let $u(x, y)$ be the deformation at the space location (x, y). The potential energy has two parts: one from the expanded surface area, and one from an applied pressure. Consider a small patch of the membrane above the rectangular region $\Delta x \Delta y$. The surface area above the region $\Delta x \Delta y$ is approximately, for a small patch,

$$\Delta S = (1 + u_x^2 + u_y^2)^{1/2} \Delta x \Delta y.$$

The potential energy of this patch from the expansion of the membrane will be proportional to the difference $\Delta S - \Delta x \Delta y$. Let the proportionality constant be given by the *tension* T. Then the potential energy of this patch from the expansion is

$$T(\Delta S - \Delta x \Delta y) = T((1 + u_x^2 + u_y^2)^{1/2} - 1)\Delta x \Delta y.$$

Now, apply the first order Taylor polynomial approximation to $f(p) = (1 + p)^{1/2} \approx f(0) + f'(0)(p - 0) \approx 1 + 1/2\ p$. Assume $p = u_x^2 + u_y^2$ is small to get an approximate potential energy

$$T(\Delta S - \Delta x \Delta y) \approx T/2\ (u_x^2 + u_y^2)\Delta x \Delta y. \qquad (3.5.1)$$

The potential energy from the applied pressure, $f(x, y)$, is the force times distance. Here force is pressure times area, $f(x, y)\Delta x \Delta y$. So, if the force is positive when it is applied upward, then the potential energy from the applied pressure is

$$-uf\Delta x \Delta y. \qquad (3.5.2)$$

Combine (3.5.1) and (3.5.2) so that the approximate potential energy for the patch is

$$T/2\ (u_x^2 + u_y^2)\Delta x \Delta y - uf\Delta x \Delta y. \qquad (3.5.3)$$

The potential energy for the entire membrane is found by adding the potential energy for each patch in (3.5.3) and letting $\Delta x, \Delta y$ go to zero

$$potential\ energy = P(u) \equiv \iint (T/2\ (u_x^2 + u_y^2) - uf)dxdy. \qquad (3.5.4)$$

The choice of suitable $u(x, y)$ should be a function such that this integral is finite and the given deformations at the edge of the membrane are satisfied. Denote this set of functions by S. The precise nature of S is a little complicated, but S should at least contain the continuous functions with piecewise continuous partial derivatives so that the double integral in (3.5.4) exists.

Definition. *The function u in S is called an energy solution of the steady state membrane problem if and only if $P(u) = \min_{v} P(v)$ where v is any function in S and $P(u)$ is from (3.5.4).*

The *weak formulation* is easily derived from the energy formulation. Let φ be any function that is zero on the boundary of the membrane and such that $u + \lambda \varphi$, for $-1 < \lambda < 1$, is also in the set of suitable functions, S. Define $F(\lambda) \equiv P(u + \lambda \varphi)$. If u is an energy solution, then F will be a minimum real valued function at $\lambda = 0$. Therefore, by expanding $P(u + \lambda \varphi)$, taking the derivative with respect to λ and setting $\lambda = 0$

$$0 = F'(0) = T \iint u_x \varphi_x + u_y \varphi_y dxdy - \iint \varphi f dxdy. \qquad (3.5.5)$$

Definition. *The function u in S is called a weak solution of the steady state membrane problem if and only if (3.5.5) holds for all φ that are zero on the boundary and $u + \lambda \varphi$ are in S.*

We just showed an energy solution must also be a weak solution. If there is a weak solution, then we can show there is only one such solution. Suppose

u and U are two weak solutions so that (3.5.5) must hold for both u and U. Subtract these two versions of (3.5.5) to get

$$
\begin{aligned}
0 &= T \iint u_x \varphi_x + u_y \varphi_y \, dx dy - T \iint U_x \varphi_x + U_y \varphi_y \, dx dy \\
&= T \iint (u - U)_x \varphi_x + (u - U)_y \varphi_y \, dx dy. \quad\quad (3.5.6)
\end{aligned}
$$

Now, let $w = u - U$ and note it is equal to zero on the boundary so that we may choose $\varphi = w$. Equation (3.5.6) becomes

$$
0 = \iint w_x^2 + w_y^2 \, dx dy.
$$

Then both partial derivatives of w must be zero so that w is a constant. But, w is zero on the boundary and so w must be zero giving u and U are equal.

A third formulation is the partial differential equation model. This is often called the classical model and requires second order partial derivatives of u. The first two models only require first order partial derivatives.

$$
-T(u_{xx} + u_{yy}) = f. \quad\quad (3.5.7)
$$

Definition.   *The classical solution of the steady state membrane problem requires u to satisfy (3.5.7) and the given values on the boundary.*

Any classical solution must be a weak solution. This follows from the conclusion of Green's theorem

$$
\iint Q_x - P_y \, dx dy = \oint P dx + Q dy.
$$

Let $Q = T \varphi u_x$ and $P = -T \varphi u_y$ and use $\varphi = 0$ on the boundary so that the line integral on the right side is zero. The left side is

$$
\iint (T \varphi u_x)_x - (-T \varphi u_y)_y \, dx dy = \iint T \varphi_x u_x + T \varphi_y u_y + T(u_{xx} + u_{yy}) \varphi \, dx dy.
$$

Because u is a classical solution and the right side is zero, the conclusion of Greens's theorem gives

$$
\iint T \varphi_x u_x + T \varphi_y u_y - \varphi f \, dx dy = 0.
$$

This is equivalent to (3.5.5) so that the classical solution must be a weak solution.

We have shown any energy or classical solution must be a weak solution. Since a weak solution must be unique, any energy or classical solution must be unique. In fact, the three formulations are equivalent under appropriate assumptions. In order to understand this, one must be more careful about the definition of a suitable set of functions, S. However, we do state this result even though this set has not been precisely defined. Note the energy and weak solutions do not directly require the existence of second order derivatives.

Theorem 3.5.1 *(Equivalence of Formulations) The energy, weak and classical formulations of the steady state membrane are equivalent.*

3.5.4 Method

The energy formulation can be discretized via the classical *Rayleigh-Ritz approximation* scheme where the solution is approximated by a linear combination of a finite number of suitable functions, $\varphi_j(x, y)$, where $j = 1, \cdots, n$

$$u(x, y) \approx \sum_{j=1}^{n} u_j \varphi_j(x, y). \tag{3.5.8}$$

These functions could be polynomials, trig functions or other likely candidates. The coefficients, u_j, in the linear combination are the unknowns and must be chosen so that the energy in (3.5.4) is a minimum

$$F(u_1, \cdots, u_n) \equiv P(\sum_{j=1}^{n} u_j \varphi_j(x, y)). \tag{3.5.9}$$

Definition. *The Rayleigh-Ritz approximation of the energy formulation is given by u in (3.5.8) where u_j are chosen such that $F : \mathbb{R}^n \to \mathbb{R}$ in (3.5.9) is a minimum.*

The u_j can be found from solving the algebraic system that comes from setting all the first order partial derivatives of F equal to zero.

$$
\begin{aligned}
0 &= F_{u_i} = \sum_{j=1}^{n} (T \iint \varphi_{ix}\varphi_{jx} + \varphi_{iy}\varphi_{jy} dxdy) u_j - \iint \varphi_i f dxdy \\
&= \sum_{j=1}^{n} a_{ij} u_j - d_i \\
&= \text{the } i \text{ component of } Au - d \text{ where} \qquad (3.5.10) \\
a_{ij} &\equiv T \iint \varphi_{ix}\varphi_{jx} + \varphi_{iy}\varphi_{jy} dxdy \text{ and} \\
d_i &\equiv \iint \varphi_i f dxdy.
\end{aligned}
$$

This algebraic system can also be found from the weak formulation by putting $u(x, y)$ in (3.5.8) and $\varphi = \varphi_i(x, y)$ into the weak equation (3.5.5). The matrix A has the following properties: (i) symmetric, (ii) positive definite and (iii) $F(u) = 1/2\ u^T Au - u^T d$. The symmetric property follows from the definition of a_{ij}. The positive definite property follows from

$$1/2\ u^T Au = T/2 \iint (\sum_{j=1}^{n} u_j \varphi_{jx})^2 + (\sum_{j=1}^{n} u_j \varphi_{jy})^2 dxdy > 0. \tag{3.5.11}$$

The third property follows from the definitions of F, A and d. The following important theorem shows that the algebraic problem $Au = d$ (3.5.10) is equivalent to the minimization problem given in (3.5.9). A partial proof is given at the end of this section, and this important result will be again used in the next section and in Chapter 9.

Theorem 3.5.2 *(Discrete Equivalence Formulations) Let A be any symmetric positive definite matrix. The following are equivalent: (i) $Ax = d$ and (ii) $J(x)$ $= \min_{y} J(y)$ where $J(x) \equiv \frac{1}{2} x^T Ax - x^T d$.*

The steepest descent method is based on minimizing the discrete energy integral, which we will now denote by $J(x)$. Suppose we make an initial guess, x, for the solution and desire to move in some direction p so that the new x, $x^+ = x + cp$, will make $J(x^+)$ a minimum. The direction, p, of steepest descent, where the directional derivative of $J(x)$ is largest, is given by $p = \nabla J(x) \equiv [J(x)_{x_i}] = -(d - Ax) \equiv -r$. Once this direction has been established we need to choose the c so that $f(c) = J(x + cr)$ is a minimum where

$$J(x + cr) = \frac{1}{2}(x + cr)^T A(x + cr) - (x + cr)^T d.$$

Because A is symmetric, $r^T Ax = x^T Ar$ and

$$
\begin{aligned}
J(x + cr) &= \frac{1}{2}x^T Ax + cr^T Ax + \frac{1}{2}c^2 r^T Ar - x^T d - cr^T d \\
&= J(x) - cr^T(d - Ax) + \frac{1}{2}c^2 r^T Ar \\
&= J(x) - cr^T r + \frac{1}{2}c^2 r^T Ar.
\end{aligned}
$$

Choose c so that $-cr^T r + \frac{1}{2}c^2 r^T Ar$ is a minimum. You can use derivatives or complete the square or you can use the discrete equivalence theorem. In the latter case x is replaced by c and the matrix A is replaced by the 1×1 matrix $r^T Ar$, which is positive for nonzero r because A is positive definite. Therefore, $c = r^T r/r^T Ar$.

Steepest Descent Method for $J(x) = \min_{y} J(y)$.

> Let x^0 be an initial guess
> $r^0 = d - Ax^0$
> for m = 0, maxm
> $\quad c = (r^m)^T r^m/(r^m)^T Ar^m$
> $\quad x^{m+1} = x^m + cr^m$
> $\quad r^{m+1} = r^m - cAr^m$
> $\quad$ test for convergence
> endloop.

In the above the next residual is computed in terms of the previous residual by $r^{m+1} = r^m - cAr^m$. This is valid because

$$r^{m+1} = d - A(x^m + cr^m) = d - Ax^m - A(cr^m) = r^m - cAr^m.$$

The test for convergence could be the norm of the new residual or the norm of the residual relative to the norm of d.

3.5.5   Implementation

MATLAB will be used to execute the steepest descent method as it is applied to the finite difference model for $-u_{xx} - u_{yy} = f$. The coefficient matrix is positive definite, and so, we can use this particular scheme. In the MATLAB code st.m the partial differential equation has right side equal to $200(1 + sin(\pi x)sin(\pi y))$, and the solution is required to be zero on the boundary of $(0, 1) \times (0, 1)$. The right side is computed and stored in lines 13-18. The vectors are represented as 2D arrays, and the sparse matrix A is not explicitly stored. Observe the use of array operations in lines 26, 32 and 35. The while loop is executed in lines 23-37. The matrix product Ar is stored in the 2D array q and is computed in lines 27-31 where we have used r is zero on the boundary nodes. The value for $c = r^T r / r^T Ar$ is alpha as computed in line 32. The output is given in lines 38 and 39 where the semilog plot for the norm of the error versus the iterations is generated by the MATLAB command semilogy(reserr).

MATLAB Code st.m

```
1.     clear;
2.     %
3.     % Solves -uxx -uyy = 200+200sin(pi x)sin(pi y)
4.     % Uses u = 0 on the boundary
5.     % Uses steepest descent
6.     % Uses 2D arrays for the column vectors
7.     % Does not explicitly store the matrix
8.     %
9.     n = 20;
10.    h = 1./n;
11.    u(1:n+1,1:n+1)= 0.0;
12.    r(1:n+1,1:n+1)= 0.0;
13.    r(2:n,2:n)= 1000.*h*h;
14.    for j= 2:n
15.         for i = 2:n
16.              r(i,j)= h*h*200*(1+sin(pi*(i-1)*h)*sin(pi*(j-1)*h));
17.         end
18.    end
19.    q(1:n+1,1:n+1)= 0.0;
20.    err = 1.0;
21.    m = 0;
```

```
22.     rho = 0.0;
23.     while ((err>.0001)*(m<200))
24.         m = m+1;
25.         oldrho = rho;
26.         rho = sum(sum(r(2:n,2:n).^2));      % dotproduct
27.         for j= 2:n     % sparse matrix product Ar
28.             for i = 2:n
29.                 q(i,j)=4.*r(i,j)-r(i-1,j)-r(i,j-1)-r(i+1,j)-r(i,j+1);
30.             end
31.         end
32.         alpha = rho/sum(sum(r.*q));      % dotproduct
33.         u = u + alpha*r;
34.         r = r - alpha*q;
35.         err = max(max(abs(r(2:n,2:n))));      % norm(r)
36.         reserr(m) = err;
37.     end
38.     m
39.     semilogy(reserr)
```

The steepest descent method appears to be converging, but after 200 iterations the norm of the residual is still only about .01. In the next section the conjugate gradient method will be described. A calculation with the conjugate gradient method shows that after only 26 iterations, the norm of the residual is about .0001. Generally, the steepest descent method is slow relative to the conjugate gradient method. This is because the minimization is only in one direction and not over higher dimensional sets.

3.5.6   Assessment

The Rayleigh-Ritz and steepest descent methods are classical methods, which serve as introductions to current numerical methods such as the finite element discretization method and the conjugate gradient iterative methods. MATLAB has a very nice partial differential equation toolbox that implements some of these. For more information on various conjugate gradient schemes use the MATLAB help command for pcg (preconditioned conjugate gradient).

The proof of the discrete equivalence theorem is based on the following matrix calculations. First, we will show if A is symmetric positive definite and if x satisfies $Ax = d$, then $J(x) \leq J(y)$ for all y. Let $y = x + (y - x)$ and use

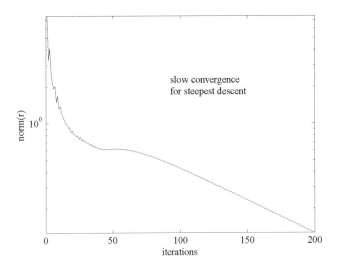

Figure 3.5.1: Steepest Descent norm(r)

$A = A^T$ and $r = d - Ax = 0$

$$
\begin{aligned}
J(y) &= \frac{1}{2}(x + (y - x))^T A(x + (y - x)) - (x + (y - x))^T d \\
&= \frac{1}{2}x^T Ax + (y - x)^T Ax \\
&\quad + \frac{1}{2}(y - x)^T A(y - x) - x^T d - (y - x)^T d \\
&= J(x) + \frac{1}{2}(y - x)^T A(y - x). \quad\quad (3.5.12)
\end{aligned}
$$

Because A is positive definite, $(y - x)^T A(y - x)$ is greater than or equal to zero. Thus, $J(y)$ is greater than or equal to $J(x)$.

Second, we show if $J(x) = \underset{y}{min}\ J(y)$,then $r = r(x) = 0$. Suppose r is not the zero vector so that $r^T r > 0$ and $r^T Ar > 0$. Choose y so that $y - x = rc$ and $0 \leq J(y) - J(x) = -cr^T r + \frac{1}{2} c^2 r^T Ar$. Let $c = \epsilon r^T r / r^T Ar$ to give a contradiction for small enough ϵ.

The weak formulation was used to show that any solution must be unique. It also can be used to formulate the *finite element method*, which is an alternative to the finite difference method. The finite difference method requires the domain to be a union of rectangles. One version of the finite element method uses the space domain as a finite union of triangles (elements) and the $\varphi_j(x, y)$ are piecewise linear functions on the triangles. For node j let $\varphi_j(x, y)$ be continuous, equal to 1.0 at node j, be a linear function for (x, y) in an adjacent triangles, and zero elsewhere. This allows for the numerical approximations on domains that are not rectangular. The interested reader should see the MATLAB code fem2d.m and R. E. White [27].

3.5.7 Exercises

1. Verify line (3.5.5).
2. Verify lines (3.5.10) and (3.5.11).
3. Why is the steepest descent direction equal to ∇J? Show $\nabla J = -r$, that is, $J(x)_{x_i} = -(d - Ax)_{x_i}$.
4. Show the formula for the $c = r^T r / r^T A r$ in the steepest descent method is correct via both derivative and completing the square.
5. Duplicate the MATLAB computations giving Figure 3.5.1. Experiment with the error tolerance $err = 2.0, 1.0, 0.5$ and 0.1.
6. In the MATLAB code st.m change the right side to $100xy + 40x^5$.
7. In the MATLAB code st.m change the boundary conditions to $u(x,1) = 10x(1-x)$ and zero elsewhere. Be careful in the matrix product $q = Ar$!
8. Fill in all the details leading to (3.5.12).

3.6 Conjugate Gradient Method

3.6.1 Introduction

The *conjugate gradient method* has three basic components: steepest descent in multiple directions, conjugate directions and preconditioning. The multiple direction version of steepest descent insures the largest possible decrease in the energy. The conjugate direction insures that solution of the reduced algebraic system is done with a minimum amount of computations. The preconditioning modifies the initial problem so that the convergence is more rapid.

3.6.2 Method

The steepest descent method hinges on the fact that for symmetric positive definite matrices the algebraic system $Ax = d$ is equivalent to minimizing a real valued function $J(x) = \frac{1}{2}x^T Ax - x^T d$, which for the membrane problem is a discrete approximation of the potential energy of the membrane. Make an initial guess, x, for the solution and move in some direction p so that the new x, $x^+ = x + cp$, will make $J(x^+)$ a minimum. The direction, p, of steepest descent, where the directional derivative of J is largest, is given by $p = -r$. Next choose the c so that $F(c) = J(x + cr)$ is a minimum, and this is $c = r^T r / r^T A r$. In the steepest descent method only the current residual is used. If a linear combination of all the previous residuals were to be used, then the "energy", $J(\hat{x}^+)$, would be smaller than the $J(x^+)$ for the steepest descent method.

For multiple directions the new x should be the old x plus a linear combination of the all the previous residuals

$$x^{m+1} = x^m + c_0 r^0 + c_1 r^1 + \cdots + c_m r^m. \tag{3.6.1}$$

This can be written in matrix form where R is $n \times (m+1)$, $m << n$, and is formed by the residual column vectors

$$R = \begin{bmatrix} r^0 & r^1 & \cdots & r^m \end{bmatrix}.$$

Then c is an $(m+1) \times 1$ column vector of the coefficients in the linear combination

$$x^{m+1} = x^m + Rc. \tag{3.6.2}$$

Choose c so that $J(x^m + Rc)$ is the smallest possible.

$$J(x^m + Rc) = \frac{1}{2}(x^m + Rc)^T A (x^m + Rc) - (x^m + Rc)^T d.$$

Because A is symmetric, $c^T R^T A x^m = (x^m)^T A R c$ so that

$$
\begin{aligned}
J(x^m + Rc) &= \frac{1}{2}(x^m)^T A x^m + c^T R^T A x^m + \frac{1}{2}c^T(R^T A R)c \\
&\quad -(x^m)^T d - c^T R^T d \\
&= J(x^m) - c^T R^T(d - A x^m) + \frac{1}{2}c^T(R^T A R)c \\
&= J(x^m) - c^T R^T r^m + \frac{1}{2}c^T(R^T A R)c. \tag{3.6.3}
\end{aligned}
$$

Now choose c so that $-c^T R^T r^m + \frac{1}{2}c^T(R^T A R)c$ is a minimum. If $R^T A R$ is symmetric positive definite, then use the discrete equivalence theorem. In this case x is replaced by c and the matrix A is replaced by the $(m+1) \times (m+1)$ matrix $R^T A R$. Since A is assumed to be symmetric positive definite, $R^T A R$ will be symmetric and positive definite if the columns of R are linearly independent ($Rc = 0$ implies $c = 0$). In this case c is $(m+1) \times 1$ and will be the solution of the reduced algebraic system

$$(R^T A R)c = R^T r^m. \tag{3.6.4}$$

The purpose of using the conjugate directions is to insure the matrix $R^T A R$ is easy to invert. The ij component of $R^T A R$ is $(r^i)^T A r^j$, and the i component of $R^T r^m$ is $(r^i)^T r^m$. $R^T A R$ would be easy to invert if it were a diagonal matrix, and in this case for i not equal to j $(r^i)^T A r^j = 0$. This means the column vectors would be "perpendicular" with respect to the inner product given by $x^T A y$ where A is symmetric positive definite.

Here we may apply the Gram-Schmidt process. For two directions r^0 and r^1 this has the form

$$p^0 = r^0 \text{ and } p^1 = r^1 + bp^0. \tag{3.6.5}$$

Now, b is chosen so that $(p^0)^T A p^1 = 0$

$$\begin{aligned}(p^0)^T A(r^1 + bp^0) &= 0 \\ (p^0)^T A r^1 + b(p^0)^T A p^0 &= \end{aligned}$$

and solve for

$$b = -(p^0)^T A r^1/(p^0)^T A p^0. \tag{3.6.6}$$

By the steepest descent step in the first direction

$$\begin{aligned} x^1 &= x^0 + cr^0 \text{ where} \\ c &= (r^0)^T r^0/(r^0)^T A r^0 \text{ and} \\ r^1 &= r^0 - cAr^0. \end{aligned} \tag{3.6.7}$$

The definitions of b in (3.6.6) and c in (3.6.7) yield the following additional equations

$$(p^0)^T r^1 = 0 \text{ and } (p^1)^T r^1 = (r^1)^T r^1. \tag{3.6.8}$$

Moreover, use $r^1 = r^0 - cAr^0$ in $b = -(p^0)^T A r^1/(p^0)^T A p^0$ and in $(r^1)^T r^1$ to show

$$b = (r^1)^T r^1/(p^0)^T p^0. \tag{3.6.9}$$

These equations allow for a simplification of (3.6.4) where R is now formed by the column vectors p^0 and p^1

$$\begin{bmatrix} (p^0)^T A p^0 & 0 \\ 0 & (p^1)^T A p^1 \end{bmatrix} \begin{bmatrix} c_0 \\ c_1 \end{bmatrix} = \begin{bmatrix} 0 \\ (r^1)^T r^1 \end{bmatrix}.$$

Thus, $c_0 = 0$ and $c_1 = (r^1)^T r^1/(p^1)^T A p^1$. From (3.6.1) with $m = 1$ and r^0 and r^1 replaced by p^0 and p^1

$$x^2 = x^1 + c_0 p^0 + c_1 p^1 = x^1 + 0p^0 + c_1 p^1. \tag{3.6.10}$$

For the three direction case we let $p^2 = r^2 + bp^1$ and choose this new b to be such that $(p^2)^T A p^1 = 0$ so that $b = -(p^1)^T A r^2/(p^1)^T A p^1$. Use this new b and the previous arguments to show $(p^0)^T r^2 = 0$, $(p^1)^T r^2 = 0$, $(p^2)^T r^2 = (r^2)^T r^2$ and $(p^0)^T A p^2 = 0$. Moreover, one can show $b = (r^2)^T r^2/(p^1)^T p^1$. The equations give a 3×3 simplification of (3.6.4)

$$\begin{bmatrix} (p^0)^T A p^0 & 0 & 0 \\ 0 & (p^1)^T A p^1 & 0 \\ 0 & 0 & (p^2)^T A p^2 \end{bmatrix} \begin{bmatrix} c_0 \\ c_1 \\ c_2 \end{bmatrix} = \begin{bmatrix} 0 \\ 0 \\ (r^2)^T r^2 \end{bmatrix}.$$

Thus, $c_0 = c_1 = 0$ and $c_2 = (r^2)^T r^2/(p^2)^T A p^2$. From (3.6.1) with $m = 2$ and r^0, r^1 and r^2 replaced by p^0, p^1, and p^2

$$x^3 = x^2 + c_0 p^0 + c_1 p^1 + c_2 p^2 = x^2 + 0p^0 + 0p^1 + c_2 p^3. \tag{3.6.11}$$

Fortunately, this process continues, and one can show by mathematical induction that the reduced matrix in (3.6.4) will always be a diagonal matrix and the right side will have only one nonzero component, namely, the last component. Thus, the use of conjugate directions substantially reduces the amount of computations, and the previous search direction vector does not need to be stored.

In the following description the conjugate gradient method corresponds to the case where the preconditioner is $M = I$. One common preconditioner is SSOR where the SOR scheme is executed in a forward and then a backward sweep. If $A = D - L - L^T$ where D is the diagonal part of A and $-L$ is the strictly lower triangular part of A, then M is

$$M = (D - wL)(1/((2 - w)w))D^{-1}(D - wL^T).$$

The solve step is relatively easy because there is a lower triangular solve, a diagonal product and an upper triangular solve. If the matrix is sparse, then these solves will also be sparse solves. Other preconditioners can be found via MATLAB help pcg and in Section 9.2.

Preconditioned Conjugate Gradient Method.

> Let x^0 be an initial guess
> $r^0 = d - Ax^0$
> solve $M\hat{r}^0 = r^0$ and set $p^0 = \hat{r}^0$
> for m = 0, maxm
> $c = (\hat{r}^m)^T r^m/(p^m)^T Ap^m$
> $x^{m+1} = x^m + cp^m$
> $r^{m+1} = r^m - cAp^m$
>   test for convergence
>   solve $M\hat{r}^{m+1} = r^{m+1}$
> $b = (\hat{r}^{m+1})^T r^{m+1}/(\hat{r}^m)^T r^m$
> $p^{m+1} = \hat{r}^{m+1} + bp^m$
> endloop.

3.6.3 Implementation

MATLAB will be used to execute the preconditioned conjugate gradient method with the SSOR preconditioner as it is applied to the finite difference model for $-u_{xx} - u_{yy} = f$. The coefficient matrix is symmetric positive definite, and so, one can use this particular scheme. Here the partial differential equation has right side equal to $200(1 + sin(\pi x)sin(\pi y))$ and the solution is required to be zero on the boundary of $(0, 1) \times (0, 1)$.

In the MATLAB code precg.m observe the use of array operations. The vectors are represented as 2D arrays, and the sparse matrix A is not explicitly stored. The preconditioning is done in lines 23 and 48 where a call to the user defined MATLAB function ssor.m is used. The conjugate gradient method is executed by the while loop in lines 29-52. In lines 33-37 the product Ap is computed and stored in the 2D array q; note how $p = 0$ on the boundary grid is used in the computation of Ap. The values for $c = alpha$ and $b = newrho/rho$ are computed in lines 40 and 50. The conjugate direction is defined in line 51.

MATLAB Codes precg.m and ssor.m

```
1.    clear;
2.    %
3.    % Solves -uxx -uyy = 200+200sin(pi x)sin(pi y)
4.    % Uses PCG with SSOR preconditioner
5.    % Uses 2D arrays for the column vectors
6.    % Does not explicitly store the matrix
7.    %
8.    w = 1.5;
9.    n = 20;
10.   h = 1./n;
11.   u(1:n+1,1:n+1)= 0.0;
12.   r(1:n+1,1:n+1)= 0.0;
13.   rhat(1:n+1,1:n+1) = 0.0;
14.   p(1:n+1,1:n+1)= 0.0;
15.   q(1:n+1,1:n+1)= 0.0;
16.   % Define right side of PDE
17.   for j= 2:n
18.        for i = 2:n
19.             r(i,j)= h*h*(200+200*sin(pi*(i-1)*h)*sin(pi*(j-1)*h));
20.        end
21.   end
22.   % Execute SSOR preconditioner
23.   rhat = ssor(r,n,w);
24.   p(2:n,2:n)= rhat(2:n,2:n);
25.   err = 1.0;
26.   m = 0;
27.   newrho = sum(sum(rhat.*r));
28.   % Begin PCG iterations
29.   while ((err>.0001)*(m<200))
30.        m = m+1;
31.        % Executes the matrix product q = Ap
32.        % Does without storage of A
33.        for j= 2:n
34.             for i = 2:n
35.                  q(i,j)=4.*p(i,j)-p(i-1,j)-p(i,j-1)-p(i+1,j)-p(i,j+1);
36.             end
37.        end
38.        % Executes the steepest descent segment
39.        rho = newrho;
```

```
40.          alpha = rho/sum(sum(p.*q));
41.          u = u + alpha*p;
42.          r = r - alpha*q;
43.          % Test for convergence
44.          % Use the infinity norm of the residual
45.          err = max(max(abs(r(2:n,2:n))));
46.          reserr(m) = err;
47.          % Execute SSOR preconditioner
48.          rhat = ssor(r,n,w);
49.          % Find new conjugate direction
50.          newrho = sum(sum(rhat.*r));
51.          p = rhat + (newrho/rho)*p;
52.      end
53.      m
54.      semilogy(reserr)
```

```
1.    function rhat=ssor(r,n,w)
2.    rhat = zeros(n+1);
3.    for j= 2:n
4.        for i = 2:n
5.            rhat(i,j)=w*(r(i,j)+rhat(i-1,j)+rhat(i,j-1))/4.;
6.        end
7.    end
8.    rhat(2:n,2:n) = ((2.-w)/w)*(4.)*rhat(2:n,2:n);
9.    for j= n:-1:2
10.       for i = n:-1:2
11.           rhat(i,j)=w*(rhat(i,j)+rhat(i+1,j)+rhat(i,j+1))/4.;
12.       end
13.   end
```

Generally, the steepest descent method is slow relative to the conjugate gradient method. For this problem, the steepest descent method did not converge in 200 iterations; the conjugate gradient method did converge in 26 iterations, and the SSOR preconditioned conjugate gradient method converged in 11 iterations. The overall convergence of both methods is recorded in Figure 3.6.1.

3.6.4   Assessment

The conjugate gradient method that we have described is for a symmetric positive definite coefficient matrix. There are a number of variations when the matrix is not symmetric positive definite. The choice of preconditioners is important, but in practice this choice is often done somewhat experimentally or is based on similar computations. The preconditioner can account for about 40% of the computation for a single iteration, but it can substantially reduce the number of iterations that are required for convergence. Another expensive

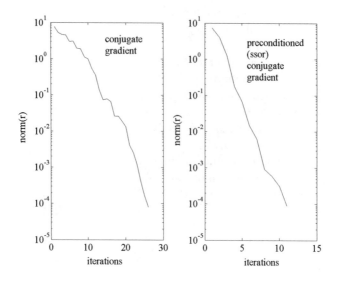

Figure 3.6.1: Convergence for CG and PCG

component of the conjugate gradient method is the matrix-vector product, and so one should pay particular attention to the implementation of this.

3.6.5   Exercises

1.   Show if A is symmetric positive definite and the columns of R are linearly independent, then $R^T AR$ is also symmetric positive definite.
2.   Verify line (3.6.8).
3.   Verify line (3.6.9).
4.   Duplicate the MATLAB computations that give Figure 3.6.1. Experiment with the SSOR parameter in the preconditioner.
5.   In the MATLAB code precg.m change the right side to $100xy + 40x^5$.
6.   In the MATLAB code precg.m change the boundary conditions to $u(x,1) = 10x(1 - x)$ and zero elsewhere. Be careful in the matrix product $q = Ap$!
7.   Read the MATLAB help pcg file and Section 9.3. Try some other preconditioners and some of the MATLAB conjugate gradient codes.

Chapter 4

Nonlinear and 3D Models

In the previous chapter linear iterative methods were used to approximate the solution to two dimensional steady state space problems. This often results in three nested loops where the outermost loop is the iteration of the method and the two innermost loops are for the two space directions. If the two dimensional problem is nonlinear or if the problem is linear and in three directions, then there must be one additional loop. In the first three sections, nonlinear problems, the Picard and Newton methods are introduced. The last three sections are devoted to three space dimension problems, and these often require the use of high performance computing. Applications will include linear and nonlinear heat transfer, and in the next chapter space dependent population models, image restoration and value of option contracts. A basic introduction to nonlinear methods can be found in Burden and Faires [4]. A more current description of nonlinear methods can be found in C. T. Kelley [11].

4.1 Nonlinear Problems in One Variable

4.1.1 Introduction

Nonlinear problems can be formulated as a *fixed point* of a function $x = g(x)$, or equivalently, as a *root* of $f(x) \equiv x - g(x) = 0$. This is a common problem that arises in computations, and a more general problem is to find N unknowns when N equations are given. The bisection algorithm does not generalize very well to these more complicated problems. In this section we will present two algorithms, Picard and Newton, which do generalize to problems with more than one unknown.

Newton's algorithm is one of the most important numerical schemes because, under appropriate conditions, it has local and quadratic convergence properties. Local convergence means that if the initial guess is sufficiently close to the root, then the algorithm will converge to the root. Quadratic convergence means that the error at the next step will be proportional to the square of the error at the current step. In general, the Picard algorithm only has first order convergence where the error at the next step is proportional to the error at the present step. But, the Picard algorithm may converge to a fixed point regardless of the initial guess.

4.1.2   Applied Area and Model

Consider the rapid cooling of an object, which has uniform temperature with respect to the space variable. Heat loss by transfer from the object to the surrounding region may be governed by equations that are different from Newton's law. Suppose a thin wire is glowing hot so that the main heat loss is via radiation. Then Newton's law of cooling may not be an accurate model. A more accurate model is the *Stefan radiation law*

$$
\begin{aligned}
u_t &= c(u_{sur}^4 - u^4) = F(u) \text{ and } u(0) = 973 \text{ where } c = A\varepsilon\sigma \\
A &= 1 \text{ is the area,} \\
\varepsilon &= .022 \text{ is the emissivity,} \\
\sigma &= 5.68 \ 10^{-8} \text{ is the Stefan-Boltzmann constant and} \\
u_{sur} &= 273 \text{ is the surrounding temperature.}
\end{aligned}
$$

The derivative of F is $-4cu^3$ and is large and negative for temperature near the initial temperature, $F'(973) = -4.6043$. Problems of this nature are called *stiff* differential equations. Since the right side is very large, very small time steps are required in Euler's method where u^+ is the approximation of $u(t)$ at the next time step and h is the increment in time $u^+ = u + hF(u)$. An alternative is to evaluate $F(u(t))$ at the next time step so that an implicit variation on Euler's method is $u^+ = u + hF(u^+)$. So, at each time step one must solve a fixed point problem.

4.1.3   Method: Picard

We will need to use an algorithm that is suitable for stiff differential equations. The model is a fixed point problem

$$u = g(u) \equiv u_{old} + hF(u). \tag{4.1.1}$$

For small enough h this can be solved by the *Picard algorithm*

$$u^{m+1} = g(u^m) \tag{4.1.2}$$

where the m indicates an inner iteration and not the time step. The initial guess for this iteration can be taken from one step of the Euler algorithm.

Example 1. Consider the first time step for $u_t = f(t, u) = t^2 + u^2$ and $u(0) = 1$. A variation on equation (4.1.1) has the form

$$
\begin{aligned}
u &= g(u) = 1 + (h/2)(f(0, 1) + f(h, u)) \\
&= 1 + (h/2)((0 + 1) + (h^2 + u^2)).
\end{aligned}
$$

This can be solved using the quadratic formula, but for small enough h one can use several iterations of (4.1.2). Let $h = .1$ and let the first guess be $u^0 = 1$ $(m = 0)$. Then the calculations from (4.1.2) will be: 1.100500, 1.111055, 1.112222, 1.112351. If we are "satisfied" with the last calculation, then let it be the value of the next time set, u^k where $k = 1$ is the first time step so that this is an approximation of $u(1h)$.

Consider the general problem of finding the fixed point of $g(x)$

$$g(x) = x. \tag{4.1.3}$$

The Picard algorithm has the form of successive approximation as in (4.1.2), but for more general $g(x)$. In the algorithm we continue to iterate (4.1.2) until there is little difference in two successive calculations. Another possible stopping criteria is to examine the size of the nonlinear residual $f(x) = g(x) - x$.

Example 2. Find the square root of 2. This could also be written either as $0 = 2 - x^2$ for the root, or as $x = x + 2 - x^2 = g(x)$ for the fixed point of $g(x)$. Try an initial approximation of the fixed point, say, $x^0 = 1$. Then the subsequent iterations are $x^1 = g(1) = 2$, $x^2 = g(2) = 0$, $x^3 = g(0) = 2$, and so on $0, 2, 0, 2$......! So, the iteration does not converge. Try another initial $x^0 = 1.5$ and it still does not converge $x^1 = g(1.5) = 1.25$, $x^2 = g(1.25) = 1.6875$, $x^3 = g(1.6875) = .83984375$! Note, this last sequence of numbers is diverging from the solution. A good way to analyze this is to use the mean value theorem $g(x) - g(\sqrt{2}) = g'(c)(x - \sqrt{2})$ where c is somewhere between x and $\sqrt{2}$. Here $g'(c) = 1 - 2c$. So, regardless of how close x is to $\sqrt{2}$, $g'(c)$ will approach $1 - 2\sqrt{2}$, which is strictly less than -1. Hence for x near $\sqrt{2}$ we have $|g(x) - g(\sqrt{2})| > |x - \sqrt{2}|$!

In order to obtain convergence, it seems plausible to require $g(x)$ to move points closer together, which is in contrast to the above example where they are moved farther apart.

Definition. $g : [a, b] \to \mathbb{R}$ *is called* contractive *on [a,b] if and only if for all x and y in [a,b] and positive $r < 1$*

$$|g(x) - g(y)| \le r|x - y|. \tag{4.1.4}$$

Example 3. Consider $u_t = u/(1 + u)$ with $u(0) = 1$. The implicit Euler method has the form $u^{k+1} = u^k + hu^{k+1}/(1 + u^{k+1})$ where k is the time step. For the first time step with $x = u^{k+1}$ the resulting fixed point problem is $x = 1 + hx/(1 + x) = g(x)$. One can verify that the first 6 iterates of Picard's algorithm with $h = 1$ are 1.5, 1.6, 1.6153, 1.6176, 1.6180 and 1.6180. The algorithm has converged to within 10^{-4}, and we stop and set $u^1 = 1.610$. The function $g(x)$ is contractive, and this can be seen by direct calculation

$$g(x) - g(y) = h[1/((1 + x)(1 + y))](x - y). \tag{4.1.5}$$

The term in the square bracket is less then one if both x and y are positive.

Example 4. Let us return to the radiative cooling problem in example 1 where we must solve a stiff differential equation. If we use the above algorithm with a Picard solver, then we will have to find a fixed point of $g(x) = \hat{x} + (h/2)(f(\hat{x}) + f(x))$ where $f(x) = c(u_{sur}^4 - x^4)$. In order to show that $g(x)$ is contractive, use the mean value theorem so that for some c between x and y

$$g(x) - g(y) = g'(c)(x - y)$$
$$|g(x) - g(y)| \leq max|g'(c)||x - y|. \tag{4.1.6}$$

So, we must require $r = max|g'(c)| < 1$. In our case, $g'(x) = (h/2)f'(x) = (h/2)(-4cx^3)$. For temperatures between 273 and 973, this means

$$(h/2)4.6043 < 1., \text{ that is, } h < (2/4.6043).$$

4.1.4 Method: Newton

Consider the problem of finding the root of the equation

$$f(x) = 0. \tag{4.1.7}$$

The idea behind Newton's algorithm is to approximate the function $f(x)$ at a given point by a straight line. Then find the root of the equation associated with this straight line. One continues to repeat this until little change in approximated roots is observed. The equation for the straight line at the iteration m is $(y - f(x^m))/(x - x^m) = f'(x^m)$. Define x^{m+1} so that $y = 0$ where the straight line intersects the x axis $(0 - f(x^m))/(x^{m+1} - x^m) = f'(x^m)$. Solve for x^{m+1} to obtain *Newton algorithm*

$$x^{m+1} = x^m - f(x^m)/f'(x^m). \tag{4.1.8}$$

There are two common stopping criteria for Newton's algorithm. The first test requires two successive iterates to be close. The second test requires the function to be near zero.

Example 5. Consider $f(x) = 2 - x^2 = 0$. The derivative of $f(x)$ is $-2x$, and the iteration in (4.1.8) can be viewed as a special Picard algorithm where $g(x) = x - (2 - x^2)/(-2x)$. Note $g'(x) = -1/x^2 + 1/2$ so that $g(x)$ is contractive near the root. Let $x^0 = 1$. The iterates converge and did so *quadratically* as is indicated in Table 4.1.1.

Example 6. Consider $f(x) = x^{1/3} - 1$. The derivative of $f(x)$ is $(1/3)x^{-2/3}$. The corresponding Picard iterative function is $g(x) = -2x + 3x^{2/3}$. Here $g'(x) = -2 + 2x^{-1/3}$ so that it is contractive suitably close to the root $x = 1$. Table 4.1.2 illustrates the *local convergence* for a variety of initial guesses.

4.1.5 Implementation

The MATLAB file picard.m uses the Picard algorithm for solving the fixed point problem $x = 1 + h(x/(1 + x))$, which is defined in the function file gpic.m. The

Table 4.1.1: Quadratic Convergence

m	x^m	E^m	$E^m/(E^{m-1})^2$
0	1.0	0.414213	
1	1.5	0.085786	2.000005
2	1.4166666	0.002453	3.000097
3	1.4142156	0.000002	3.008604

Table 4.1.2: Local Convergence

x^0	m for conv.	x^0	m for conv.
10.0	no conv.	00.1	4
05.0	no conv.	-0.5	6
04.0	6	-0.8	8
03.0	5	-0.9	20
01.8	3	-1.0	no conv.

iterates are computed in the loop given by lines 4-9. The algorithm is stopped in lines 6-8 when the difference between two iterates is less than .0001.

MATLAB Codes picard.m and gpic.m

```
1.    clear;
2.    x(1) = 1.0;
3.    eps = .0001;
4.    for m=1:20
5.          x(m+1) = gpic(x(m));
6.          if abs(x(m+1)-x(m))<eps
7.                break;
8.          end
9.    end
10.   x'
11.   m
12.   fixed_point = x(m+1)

function gpic = gpic(x)
      gpic = 1. + 1.0*(x/(1. + x));
```

Output from picard.m:
```
      ans =
      1.0000
      1.5000
      1.6000
      1.6154
      1.6176
      1.6180
      1.6180
```

m =
6
fixed_point =
1.6180

The MATLAB file newton.m contains the Newton algorithm for solving the root problem $0 = 2 - x^2$, which is defined in the function file fnewt.m. The iterates are computed in the loop given by lines 4-9. The algorithm is stopped in lines 6-8 when the residual $f(x^m)$ is less than .0001. The numerical results are in Table 4.4.1 where convergence is obtained after three iterations on Newton's method. A similar code is newton.f90 written in Fortran 9x.

MATLAB Codes newton.m, fnewt.m and fnewtp.m

```
1.      clear;
2.      x(1) = 1.0;
3.      eps = .0001;
4.      for m=1:20
5.              x(m+1) = x(m) - fnewt(x(m))/fnewtp(x(m));
6.              if abs(fnewt(x(m+1)))<eps
7.                      break;
8.              end
9.      end
10.     x'
11.     m
12.     fixed_point = x(m+1)

function fnewt = fnewt(x)
        fnewt =2 - x^2;

function fnewtp = fnewtp(x)
        fnewtp = -2*x;
```

4.1.6   Assessment

In the radiative cooling model we have also ignored the good possibility that there will be differences in temperature according to the location in space. In such cases there will be diffusion of heat, and one must model this mode of heat transfer.

We indicated that the Picard algorithm may converge if the mapping $g(x)$ is contractive. The following theorem makes this more precise. Under some additional assumptions the new error is bounded by the old error.

Theorem 4.1.1 *(Picard Convergence) Let $g:[a,b] \to [a,b]$ and assume that x is a fixed point of g and x is in $[a,b]$. If g is contractive on $[a,b]$, then the Picard algorithm in (4.1.2) converges to the fixed point. Moreover, the fixed point is unique.*

Proof. Let $x^{m+1} = g(x^m)$ and $x = g(x)$. Repeatedly use the contraction property (4.1.4).

$$\begin{aligned}
|x^{m+1} - x| &= |g(x^m) - g(x)| \\
&\leq r|x^m - x| \\
&= r|g(x^{m-1}) - g(x)| \\
&\leq r^2|x^{m-2} - x| \\
&\quad \vdots \\
&\leq r^{m+1}|x^0 - x|.
\end{aligned} \tag{4.1.9}$$

Since $0 \leq r < 1, r^{m+1}$ must go to zero as m increases.

If there is a second fixed point y, then $|x - y| = |g(x) - g(y)| \leq r|y - x|$ where $r < 1$. So, if x and y are different, then $|y - x|$ is not zero. Divide both sides by $|y - x|$ to get $1 \leq r$, which is a contradiction to our assumption that $r < 1$. Evidently, $x = y$. ∎

In the above examples we noted that Newton's algorithm was a special case of the Picard algorithm with $g(x) = x - f(x)/f'(x)$. In order to show $g(x)$ is contractive, we need to have, as in (4.1.6), $|g'(x)| < 1$.

$$g'(x) = 1 - (f'(x)^2 - f(x)f''(x))/f'(x)^2 = f(x)f''(x)/f'(x)^2 \tag{4.1.10}$$

If $\hat{x}$ is a solution of $f(x) = 0$ and $f(x)$ is continuous, then we can make $f(x)$ as small as we wish by choosing x close to $\hat{x}$. So, if $f''(x)/f'(x)^2$ is bounded, then $g(x)$ will be contractive for x near $\hat{x}$. Under the conditions listed in the following theorem this establishes the *local convergence*.

Theorem 4.1.2 *(Newton's Convergence) Consider $f(x) = 0$ and assume $\hat{x}$ is a root. If $f'(\hat{x})$ is not zero and $f''(x)$ is continuous on an interval containing $\hat{x}$, then*

1. *Newton's algorithm converges locally to the $\hat{x}$, that is, for x^0 suitably close to $\hat{x}$ and*

2. *Newton's algorithm converges quadratically, that is,*

$$|x^{m+1} - \hat{x}| \leq [max|f''(x)|/(2\,min|f'(x)|]|x^m - \hat{x}|^2. \tag{4.1.11}$$

Proof. In order to prove the quadratic convergence, use the extended mean value theorem where $a = x^m$ and $x = \hat{x}$ to conclude that there is some c such that

$$0 = f(\hat{x}) = f(x^m) + f'(x^m)(\hat{x} - x^m) + (f''(c)/2)(\hat{x} - x^m)^2.$$

Divide the above by $f'(x^m)$ and use the definition of x^{m+1}

$$\begin{aligned}
0 &= -(x^{m+1} - x^m) + (\hat{x} - x^m) + (f''(c)/(2f'(x^m)))(\hat{x} - x^m)^2 \\
&= -(x^{m+1} - \hat{x}) + (f''(c)/(2f'(x^m)))(\hat{x} - x^m)^2.
\end{aligned}$$

Since $f'(x)$ for some interval about $\hat{x}$ must be bounded away from zero, and $f''(x)$ and $f'(x)$ are continuous, the inequality in (4.1.11) must hold. ∎

4.1.7 Exercises

1.   Consider the fixed point example 1 and verify those computations. Experiment with increased sizes of h. Notice the algorithm may not converge if $|g'(u)| > 1$.

2.   Verify the example 3 for $u_t = u/(1 + u)$. Also, find the exact solution and compare it with the two discretization methods: Euler and implicit Euler. Observe the order of the errors.

3.   Consider the applied problem with radiative cooling in example 4. Solve the fixed point problems $x = g(x)$, with $g(x)$ in example 4, by the Picard algorithm using a selection of step sizes. Observe how this affects the convergence of the Picard iterations.

4.   Solve for x such that $x = e^{-x}$.

5.   Use Newton's algorithm to solve $0 = 7 - x^3$. Observe quadratic convergence.

4.2   Nonlinear Heat Transfer in a Wire

4.2.1   Introduction

In the analysis for most of the heat transfer problems we assumed the temperature varied over a small range so that the thermal properties could be approximated by constants. This always resulted in a linear algebraic problem, which could be solved by a variety of methods. Two possible difficulties are nonlinear thermal properties or larger problems, which are a result of diffusion in two or three directions. In this section we consider the nonlinear problems.

4.2.2   Applied Area

The properties of density, specific heat and thermal conductivity can be nonlinear. The exact nature of the nonlinearity will depend on the material and the range of the temperature variation. Usually, data is collected that reflects these properties, and a least squares curve fit is done for a suitable approximating function. Other nonlinear terms can evolve from the heat source or sink terms in either the boundary conditions or the source term on the right side of the heat equation. We consider one such case.

Consider a cooling fin or plate, which is *glowing hot*, say at 900 degrees Kelvin. Here heat is being lost by radiation to the surrounding region. In this case the heat lost is *not* proportional, as in Newton's law of cooling, to the difference in the surrounding temperature, u_{sur}, and the temperature of the glowing mass, u. Observations indicate that the heat loss through a surface area, A, in a time interval, Δt, is equal to $\Delta t\, A\varepsilon\sigma(u_{sur}^4 - u^4)$ where ε is the emissivity of the surface and σ is the Stefan-Boltzmann constant. If the temperature is not uniform with respect to space, then couple this with the Fourier heat law to form various nonlinear differential equations or boundary conditions.

4.2.3 Model

Consider a thin wire of length L and radius r. Let the ends of the wire have a fixed temperature of 900 and let the surrounding region be $u_{sur} = 300$. Suppose the surface of the wire is being cooled via radiation. The lateral surface area of a small cylindrical portion of the wire has area $A = 2\pi rh$. Therefore, the heat leaving the lateral surface in Δt time is

$$\Delta t(2\pi rh)(\varepsilon\sigma(u_{sur}^4 - u^4)).$$

Assume steady state heat diffusion in one direction and apply the Fourier heat law to get

$$
\begin{aligned}
0 \approx{}& \Delta t(2\pi rh)(\varepsilon\sigma(u_{sur}^4 - u^4)) + \\
& \Delta tK(\pi r^2)u_x(x + h/2) - \Delta tK(\pi r^2)u_x(x - h/2).
\end{aligned}
$$

Divide by $\Delta t(\pi r^2)h$ and let h go to zero so that

$$0 = (2\varepsilon\sigma/r)(u_{sur}^4 - u^4) + (Ku_x)_x.$$

The *continuous model* for the heat transfer is

$$
\begin{aligned}
-(Ku_x)_x &= c(u_{sur}^4 - u^4) \text{ where } c = 2\varepsilon\sigma/r \text{ and} & (4.2.1)\\
u(0) &= 900 = u(L). & (4.2.2)
\end{aligned}
$$

The thermal conductivity will also be temperature dependent, but for simplicity assume K is a constant and will be incorporated into c.

Consider the nonlinear differential equation $-u_{xx} = f(u)$. The finite difference model is for $h = L/(n + 1)$ and $u_i \approx u(ih)$ with $u_0 = 900 = u_{n+1}$

$$-u_{i-1} + 2u_i - u_{i+1} = h^2 f(u_i) \text{ for } i = 1, ..., n.$$

This *discrete model* has n unknowns, u_i, and n equations

$$F_i(u) \equiv h^2 f(u_i) + u_{i-1} - 2u_i + u_{i+1} = 0. \qquad (4.2.3)$$

Nonlinear problems can have multiple solutions. For example, consider the intersection of the unit circle $x^2 + y^2 - 1 = 0$ and the hyperbola $x^2 - y^2 - 1/2 = 0$. Here $n = 2$ with $u_1 = x$ and $u_2 = y$, and there are four solutions. In applications this can present problems in choosing the solution that most often exists in nature.

4.2.4 Method

In order to derive Newton's method for n equations and n unknowns, it is instructive to review the one unknown and one equation case. The idea behind Newton's algorithm is to approximate the function $f(x)$ at a given point by a

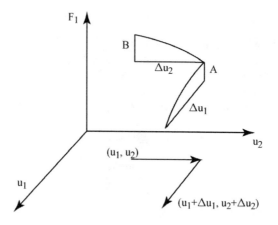

Figure 4.2.1: Change in F_1

straight line. Then find the root of the equation associated with this straight line. We make use of the approximation

$$\Delta f \approx f'(x)\Delta x. \tag{4.2.4}$$

The equation for the straight line at iteration m is

$$(y - f(x^m)) = f'(x^m)(x - x^m). \tag{4.2.5}$$

Define x^{m+1} so that $y = 0$ and solve for x^{m+1} to obtain Newton's algorithm

$$x^{m+1} = x^m - f(x^m)/f'(x^m). \tag{4.2.6}$$

The derivation of Newton's method for more than one equation and one unknown requires an analog of the approximation in (4.2.4). Consider $F_i(u)$ as a function of n variables u_j. If only the j component of u changes, then (4.2.4) will hold with x replaced by u_j and $f(x)$ replaced by $F_i(u)$. If all of the components change, then the net change in $F_i(u)$ can be approximated by sum of the partial derivatives of $F_i(u)$ with respect to u_j times the change in u_j:

$$
\begin{aligned}
\Delta F_i &= F_i(u_1 + \Delta u_1, \cdots, u_n + \Delta u_n) - F_i(u_1, \cdots, u_n) \\
&\approx F_{iu_1}(u)\Delta u_1 + \cdots + F_{iu_n}(u)\Delta u_n.
\end{aligned} \tag{4.2.7}
$$

For $n = 2$ this is depicted by Figure 4.2.1 with $i = 1$ and

$$
\begin{aligned}
\Delta F_i &= A + B \text{ where} \\
A &\approx F_{iu_1}(u)\Delta u_1 \text{ and } B \approx F_{iu_2}(u)\Delta u_2.
\end{aligned}
$$

The equation approximations in (4.2.7) can be put into matrix form

$$\Delta F \approx F'(u)\Delta u \tag{4.2.8}$$

where $\Delta F = [\ \Delta F_1 \ \ \cdots \ \ \Delta F_n \ ]^T$ and $\Delta u = [\ \Delta u_1 \ \ \cdots \ \ \Delta u_n \ ]^T$ are $n \times 1$ column vectors, and $F'(u)$ is defined as the $n \times n$ *derivative or Jacobian matrix*

$$F' \equiv \begin{bmatrix} F_{1u_1} & \cdots & F_{1u_n} \\ \vdots & \ddots & \vdots \\ F_{nu_1} & \cdots & F_{nu_n} \end{bmatrix}.$$

Newton's method is obtained by letting $u = u^m$, $\Delta u = u^{m+1} - u^m$ and $\Delta F = 0 - F(u^m)$. The vector approximation in (4.2.8) is replaced by an equality to get

$$0 - F(u^m) = F'(u^m)(u^{m+1} - u^m). \tag{4.2.9}$$

This vector equation can be solved for u^{m+1}, and we have the n variable Newton method

$$u^{m+1} = u^m - F'(u^{m+1})^{-1}F(u^m). \tag{4.2.10}$$

In practice the inverse of the Jacobian matrix is not used, but one must find the solution, Δu, of

$$0 - F(u^m) = F'(u^m)\Delta u. \tag{4.2.11}$$

Consequently, Newton's method consists of solving a sequence of linear problems. One usually stops when either F is "small", or Δu is "small "

Newton Algorithm.

> choose initial u^0
> for m = 1,maxit
> > compute $F(u^m)$ and $F'(u^m)$
> > solve $F'(u^m)\Delta u = -F(u^m)$
> > $u^{m+1} = u^m + \Delta u$
> > test for convergence
> endloop.

Example 1. Let $n = 2$, $F_1(u) = u_1^2 + u_2^2 - 1$ and $F_2(u) = u_1^2 - u_2^2 - 1/2$. The Jacobian matrix is 2×2, and it will be nonsingular if both variables are nonzero

$$F'(u) = \begin{bmatrix} 2u_1 & 2u_2 \\ 2u_1 & -2u_2 \end{bmatrix}. \tag{4.2.12}$$

If the initial guess is near a solution in a particular quadrant, then Newton's method may converge to the solution in that quadrant.

Example 2. Consider the nonlinear differential equation for the radiative heat transfer problem in (4.2.1)-(4.2.3) where

$$F_i(u) = h^2 f(u_i) + u_{i-1} - 2u_i + u_{i+1} = 0. \tag{4.2.13}$$

The Jacobian matrix is easily computed and must be tridiagonal because each $F_i(u)$ only depends on u_{i-1}, u_i and u_{i+1}

$$F'(u) = \begin{bmatrix} h^2 f'(u_1) - 2 & 1 & & \\ 1 & h^2 f'(u_2) - 2 & \ddots & \\ & \ddots & \ddots & 1 \\ & & 1 & h^2 f'(u_n) - 2 \end{bmatrix}.$$

For the Stefan cooling model where the absolute temperature is positive $f'(u) <$ 0. Thus, the Jacobian matrix is strictly diagonally dominant and must be non-singular so that the solve step can be done in Newton's method.

4.2.5   Implementation

The following is a MATLAB code, which uses Newton's method to solve the 1D diffusion problem with heat loss due to radiation. We have used the MATLAB command A\d to solve each linear subproblem. One could use an iterative method, and this might be the best way for larger problems where there is diffusion of heat in more than one direction.

In the MATLAB code nonlin.m the Newton iteration is done in the outer loop in lines 13-36. The inner loop in lines 14-29 recomputes the Jacobian matrix by rows $FP = F'(u)$ and updates the column vector $F = F(u)$. The solve step and the update to the approximate solution are done in lines 30 and 31. In lines 32-35 the Euclidean norm of the residual is used to test for convergence. The output is generated by lines 37-41.

MATLAB Codes nonlin.m, fnonl.m and fnonlp.m

```
1.    clear;
2.    % This code is for a nonlinear ODE.
3.    % Stefan radiative heat lose is modeled.
4.    % Newton's method is used.
5.    % The linear steps are solved by A\d.
6.    uo = 900.;
7.    n = 19;
8.    h = 1./(n+1);
9.    FP = zeros(n);
10.   F = zeros(n,1);
11.   u = ones(n,1)*uo;
12.   %     begin Newton iteration
13.   for m =1:20
14.       for i = 1:n       %compute Jacobian matrix
15.           if i==1
16.               F(i) = fnonl(u(i))*h*h + u(i+1) - 2*u(i) + uo;
17.               FP(i,i) = fnonlp(u(i))*h*h - 2;
18.               FP(i,i+1) = 1;
19.           elseif i<n
20.               F(i) = fnonl(u(i))*h*h + u(i+1) - 2*u(i) + u(i-1);
```

Table 4.2.1: Newton's Rapid Convergence

m	Norm of F
1	706.1416
2	197.4837
3	049.2847
4	008.2123
5	000.3967
6	000.0011
7	7.3703e-09

```
21.                  FP(i,i) = fnonlp(u(i))*h*h - 2;
22.                  FP(i,i-1) = 1;
23.                  FP(i,i+1) = 1;
24.          else
25.                  F(i) = fnonl(u(i))*h*h - 2*u(i) + u(i-1) + uo;
26.                  FP(i,i) = fnonlp(u(i))*h*h - 2;
27.                  FP(i,i-1) = 1;
28.          end
29.      end
30.      du = FP\F;       % solve linear system
31.      u = u - du;
32.      error = norm(F);
33.      if error<.0001
34.              break;
35.      end
36.  end
37.  m;
38.  error;
39.  uu = [900 u' 900];
40.  x = 0:h:1;
41.  plot(x,uu)

function fnonl = fnonl(u)
    fnonl = .00000005*(300^4 - u^4);

function fnonlp = fnonlp(u)
    fnonlp = .00000005*(-4)*u^3;
```

We have experimented with $c = 10^{-8}, 10^{-7}$ and 10^{-6}. The curves in Figure 4.2.2 indicate the larger the c the more the cooling, that is, the lower the temperature. Recall, from (4.2.1) $c = (2\varepsilon\sigma/r)/K$ so variable ε, K or r can change c.

The next calculations were done to illustrate the very rapid convergence of Newton's method. The second column in Table 4.2.1 has norms of the residual as a function of the Newton iterations m.

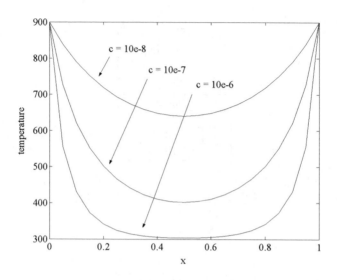

Figure 4.2.2: Temperatures for Variable c

4.2.6   Assessment

Nonlinear problems are very common, but they are often linearized by using
linear Taylor polynomial approximations of the nonlinear terms. This is done
because it is easier to solve one linear problem than a nonlinear problem where
one must solve a sequence of linear subproblems. However, Newton's method
has, under some assumption on $F(u)$, the two very important properties of local
convergence and quadratic convergence. These two properties have contributed
to the wide use and many variations of Newton's method for solving nonlinear
algebraic systems.

Another nonlinear method is a *Picard* method in which the nonlinear terms
are evaluated at the previous iteration, and the resulting linear problem is
solved for the next iterate. For example, consider the problem $-u_{xx} = f(u)$
with u given on the boundary. Let u^m be given and solve the linear problem
$-u_{xx}^{m+1} = f(u^m)$ for the next iterate u^{m+1}. This method does not always
converge, and in general it does not converge quadratically.

4.2.7   Exercises

1.   Apply Newton's method to example 1 with $n = 2$. Experiment with
different initial guesses in each quadrant. Observe local and quadratic convergence.

2.   Apply Newton's method to the radiative heat transfer problem. Experiment with different n, eps, L and emissivities. Observe local and quadratic

convergence as well as the number of Newton iterations required for convergence.

3. In exercise 2 determine how much heat is lost through the wire per unit time.

4. Consider the linearized version of $-u_{xx} = c(u_{sur}^4 - u^4) = f(u)$ where $f(u)$ is replaced by its first order Taylor polynomial $f(u_{sur}) + f'(u_{sur})(u - u_{sur})$. Compare the nonlinear and linearized solutions.

5. Try the Picard method on $-u_{xx} = c(u_{sur}^4 - u^4)$.

6. Consider the 1D diffusion problem where

$$K(u) = .001(1 + .01u + .000002u^2).$$

Find the nonlinear algebraic system and solve it using Newton's method.

7. Consider a 2D cooling plate that satisfies

$$-(Ku_x)_x - (Ku_y)_y = c(u_{sur}^4 - u^4).$$

Use Newton's method coupled with a linear solver that uses SOR to solve this nonlinear problem.

4.3 Nonlinear Heat Transfer in 2D

4.3.1 Introduction

Assume the temperature varies over a large range so that the thermal properties cannot be approximated by constants. In this section we will consider the nonlinear 2D problem where the thermal conductivity is a function of the temperature. The Picard nonlinear algorithm with a preconditioned conjugate gradient method for each of the linear solves will be used.

4.3.2 Applied Area

Consider a cooling fin or plate, which is attached to a hot mass. Assume the *nonlinear thermal conductivity* does have a least squares fit to the data to find the three coefficients in a quadratic function for $K(u)$. For example, if the thermal conductivity is given by $c_0 = .001, c_1 = .01, c_2 = .00002$ and

$$K(u) = c_0(1. + c_1u + c_2u^2), \tag{4.3.1}$$

then at $u = 100$ $K(100) = .001(1. + 1. + .2)$ is over double what it is at $u = 0$ where $K(0) = .001$.

4.3.3 Model

Consider a thin 2D plate whose edges have a given temperature. Suppose the thermal conductivity $K = K(u)$ is a quadratic function of the temperature.

The *continuous model* for the heat transfer is

$$-(K(u)u_x)x - (K(u)u_y)_y = f \text{ and} \qquad (4.3.2)$$

$$u = g \text{ on the boundary.} \qquad (4.3.3)$$

If there is source or sink of heat on the surface of the plate, then f will be nonzero. The temperature, g, on the boundary will be given and independent of time. One could have more complicated boundary conditions that involve space derivatives of the temperature.

The finite difference approximation of $K(u)u_x$ requires approximations of the thermal conductivity at the left and right sides of the rectangular region $\Delta x \Delta y$. Here we will compute the average of K, and at the right side this is

$$K_{i+1/2,j} \equiv (K(u_{i+1,j}) + K(u_{ij}))/2.$$

Then the approximation is

$$(K(u)u_x)_x \approx [K_{i+1/2,j}(u_{i+1,j} - u_{ij})/\Delta x -$$
$$K_{i-1/2,j}(u_{i,j} - u_{i-1,j})/\Delta x]/\Delta x.$$

Repeat this for the y direction to get the *discrete finite difference* model

$$f_{ij} = -[K_{i+1/2,j}(u_{i+1,j} - u_{ij})/\Delta x -$$
$$K_{i-1/2,j}(u_{i,j} - u_{i-1,j})/\Delta x]/\Delta x$$
$$-[K_{i,j+1/2}(u_{i,j+1} - u_{ij})/\Delta y -$$
$$K_{i,j-1/2}(u_{i,j} - u_{i,j-1})/\Delta y]/\Delta y. \qquad (4.3.4)$$

One can think of this in matrix form where the nonlinear parts of the problem come from the components of the coefficient matrix and the evaluation of the thermal conductivity. The nonzero components, up to five nonzero components in each row, in the matrix will have the same pattern as in the linear problem, but the values of the components will change with the temperature.

Nonlinear Algebraic Problem.

$$A(u)u = f. \qquad (4.3.5)$$

This class of nonlinear problems could be solved using Newton's method. However, the computation of the derivative or Jacobian matrix could be costly if the n^2 partial derivatives of the component functions are hard to compute.

4.3.4   Method

Picard's method will be used. We simply make an initial guess, compute the thermal conductivity at each point in space, evaluate the matrix $A(u)$ and solve for the next possible temperature. The solve step may be done by any method we choose.

Picard Algorithm for (4.3.5).

> choose initial u^0
> for m = 1,maxit
> > compute $A(u^m)$
> > solve $A(u^m)u^{m+1} = f$
> > test for convergence
> endloop.

One can think of this as a fixed point method where the problem (4.3.5) has the form

$$u = A(u)^{-1}f \equiv G(u). \tag{4.3.6}$$

The iterative scheme then is

$$u^{m+1} = G(u^m). \tag{4.3.7}$$

The convergence of such schemes requires $G(u)$ to be "contractive". We will not try to verify this, but we will just try the Picard method and see if it works.

4.3.5 Implementation

The following is a MATLAB code, which executes the Picard method and does the linear solve step by the preconditioned conjugate gradient method with the SSOR preconditioner. An alternative to MATLAB is Fortran, which is a compiled code, and therefore it will run faster than noncompiled codes such as MATLAB versions before release 13. The corresponding Fortran code is picpcg.f90.

The picpcg.m code uses two MATLAB function files: cond.m for the nonlinear thermal conductivity and pcgssor.m for the SSOR preconditioned conjugate gradient linear solver. The main program is initialized in lines 1-19 where the initial guess is zero. The for loop in line 21-46 executes the Picard algorithm. The nonlinear coefficient matrix is not stored as a full matrix, but only the five nonzero components per row are, in lines 22-31, computed as given in (4.3.4) and stored in the arrays *an*, *as*, *ae*, *aw* and *ac* for the coefficients of $u_{i,j+1}$, $u_{i,j-1}$, $u_{i+1,j}$, $u_{i-1,j}$ and $u_{i,j}$, respectively. The call to pcgssor is done in line 35, and here one should examine how the above arrays are used in this version of the preconditioned conjugate gradient method. Also, note the pcgssor is an iterative scheme, and so the linear solve is not done exactly and will depend on the error tolerance within this subroutine, see lines 20 and 60 in pcgssor. In line 37 of piccg.m the test for convergence of the Picard outer iteration is done.

MATLAB Codes picpcg.m, pcgssor.m and cond.m

```
1.    clear;
2.    % This progran solves -(K(u)ux)x - (K(u)uy)y = f.
3.    % K(u) is defined in the function cond(u).
4.    % The Picard nonlinear method is used.
```

```
5.      % The solve step is done in the subroutine pcgssor.
6.      % It uses the PCG method with SSOR preconditioner.
7.      maxmpic = 50;
8.      tol = .001;
9.      n = 20;
10.     up = zeros(n+1);
11.     rhs = zeros(n+1);
12.     up = zeros(n+1);
13.     h = 1./n;
14.     % Defines the right side of PDE.
15.     for j = 2:n
16.         for i = 2:n
17.             rhs(i,j) = h*h*200.*sin(3.14*(i-1)*h)*sin(3.14*(j-1)*h);
18.         end
19.     end
20.     % Start the Picard iteration.
21.     for mpic=1:maxmpic
22.         % Defines the five nonzero row components in the matrix.
23.         for j = 2:n
24.             for i = 2:n
25.                 an(i,j) = -(cond(up(i,j))+cond(up(i,j+1)))*.5;
26.                 as(i,j) = -(cond(up(i,j))+cond(up(i,j-1)))*.5;
27.                 ae(i,j) = -(cond(up(i,j))+cond(up(i+1,j)))*.5;
28.                 aw(i,j) = -(cond(up(i,j))+cond(up(i-1,j)))*.5;
29.                 ac(i,j) = -(an(i,j)+as(i,j)+ae(i,j)+aw(i,j));
30.             end
31.         end
32.         %
33.         % The solve step is done by PCG with SSOR.
34.         %
35.         [u , mpcg] = pcgssor(an,as,aw,ae,ac,up,rhs,n);
36.         %
37.         errpic = max(max(abs(up(2:n,2:n)-u(2:n,2:n))));
38.         fprintf('Picard iteration = %6.0f\n',mpic)
39.         fprintf('Number of PCG iterations = %6.0f\n',mpcg)
40.         fprintf('Picard error = %6.4e\n',errpic)
41.         fprintf('Max u = %6.4f\n', max(max(u)))
42.         up = u;
43.         if (errpic<tol)
44.             break;
45.         end
46.     end


1.      % PCG subroutine with SSOR preconditioner
2.      function [u , mpcg]= pcgssor(an,as,aw,ae,ac,up,rhs,n)
```

```
3.      w = 1.5;
4.      u = up;
5.      r = zeros(n+1);
6.      rhat = zeros(n+1);
7.      q = zeros(n+1);
8.      p = zeros(n+1);
9.      % Use the previous Picard iterate as an initial guess for PCG.
10.     for j = 2:n
11.          for i = 2:n
12.              r(i,j) = rhs(i,j)-(ac(i,j)*up(i,j) ...
13.                       +aw(i,j)*up(i-1,j)+ae(i,j)*up(i+1,j) ...
14.                       +as(i,j)*up(i,j-1)+an(i,j)*up(i,j+1));
15.          end
16.     end
17.     error = 1. ;
18.     m = 0;
19.     rho = 0.0;
20.     while ((error>.0001)&(m<200))
21.          m = m+1;
22.          oldrho = rho;
23.          % Execute SSOR preconditioner.
24.          for j= 2:n
25.               for i = 2:n
26.                   rhat(i,j) = w*(r(i,j)-aw(i,j)*rhat(i-1,j) ...
27.                               -as(i,j)*rhat(i,j-1))/ac(i,j);
28.               end
29.          end
30.          for j= 2:n
31.               for i = 2:n
32.                   rhat(i,j) = ((2.-w)/w)*ac(i,j)*rhat(i,j);
33.               end
34.          end
35.          for j= n:-1:2
36.               for i = n:-1:2
37.                   rhat(i,j) = w*(rhat(i,j)-ae(i,j)*rhat(i+1,j) ...
38.                               -an(i,j)*rhat(i,j+1))/ac(i,j);
39.               end
40.          end
41.          % Find conjugate direction.
42.          rho = sum(sum(r(2:n,2:n).*rhat(2:n,2:n)));
43.          if (m==1)
44.               p = rhat;
45.          else
46.               p = rhat + (rho/oldrho)*p ;
47.          end
```

```
48.          % Execute matrix product q = Ap.
49.          for j = 2:n
50.               for i = 2:n
51.                    q(i,j)=ac(i,j)*p(i,j)+aw(i,j)*p(i-1,j) ...
52.                         +ae(i,j)*p(i+1,j)+as(i,j)*p(i,j-1) ...
53.                         +an(i,j)*p(i,j+1);
54.               end
55.          end
56.          % Find steepest descent.
57.          alpha = rho/sum(sum(p.*q));
58.          u = u + alpha*p;
59.          r = r - alpha*q;
60.          error = max(max(abs(r(2:n,2:n))));
61.     end
62.     mpcg = m;
```

```
1.     % Function for thermal conductivity
2.     function cond = cond(x)
3.     c0 = 1.;
4.     c1 = .10;
5.     c2 = .02;
6.     cond = c0*(1.+ c1*x + c2*x*x);
```

The nonlinear term is in the thermal conductivity where $K(u) = 1.(1.+.1u+ .02u^2)$. If one considers the linear problem where the coefficients of u and u^2 are set equal to zero, then the solution is the first iterate in the Picard method where the maximum value of the linear solution is 10.15. In our nonlinear problem the maximum value of the solution is 6.37. This smaller value is attributed to the larger thermal conductivity, and this allows for greater heat flow to the boundary where the solution must be zero.

The Picard scheme converged in seven iterations when the absolute error equaled .001. The inner iterations in the PCG converge within 11 iterations when the residual error equaled .0001. The initial guess for the PCG method was the previous Picard iterate, and consequently, the number of PCG iterates required for convergence decreased as the Picard iterates increased. The following is the output at each stage of the Picard algorithm:

```
Picard iteration = 1
     Number of PCG iterations = 10
     Picard error = 10.1568
     Max u = 10.1568
Picard iteration = 2
     Number of PCG iterations = 11
     Picard error = 4.68381
     Max u = 5.47297
Picard iteration = 3
```

Number of PCG iterations = 11
Picard error = 1.13629
Max u = 6.60926
Picard iteration = 4
Number of PCG iterations = 9
Picard error = .276103
Max u = 6.33315
Picard iteration = 5
Number of PCG iterations = 7
Picard error = 5.238199E-02
Max u = 6.38553
Picard iteration = 6
Number of PCG iterations = 6
Picard error = 8.755684E-03
Max u = 6.37678
Picard iteration = 7
Number of PCG iterations = 2
Picard error = 9.822845E-04
Max u = 6.37776.

4.3.6   Assessment

For both Picard and Newton methods we must solve a sequence of linear problems. The matrix for the Picard's method is somewhat easier to compute than the matrix for Newton's method. However, Newton's method has, under some assumptions on $F(u)$, the two very important properties of local and quadratic convergence.

If the right side of $A(u)u = f$ depends on u so that $f = f(u)$, then one can still formulate a Picard iterative scheme by the following sequence $A(u^m)u^{m+1} = f(u^m)$ of linear solves. Of course, all this depends on whether or not $A(u^m)$ are nonsingular and on the convergence of the Picard algorithm.

4.3.7   Exercises

1.   Experiment with either the MATLAB picpcg.m or the Fortran picpcg.f90 codes. You may wish to print the output to a file so that MATLAB can graph the solution.
  (a).   Vary the convergence parameters.
  (b).   Vary the nonlinear parameters.
  (c).   Vary the right side of the PDE.
2.   Modify the code so that nonzero boundary conditions can be used. Pay careful attention to the implementation of the linear solver.
3.   Consider the linearized version of $-u_{xx} - u_{yy} = c(u_{sur}^4 - u^4) = f(u)$ where $f(u)$ is replaced by the first order Taylor polynomial $f(u_{sur}) + f'(u_{sur})(u - u_{sur})$. Compare the nonlinear and linearized solutions.

4.   Consider a 2D cooling plate whose model is $-(K(u)u_x)_x - (K(u)u_y)_y = c(u_{sur}^4 - u^4)$. Use Picard's method coupled with a linear solver of your choice.

4.4   Steady State 3D Heat Diffusion

4.4.1   Introduction

Consider the cooling fin where there is diffusion in all three directions. When each direction is discretized, say with N unknowns in each direction, then there will be N^3 total unknowns. So, if the N is doubled, then the total number of unknowns will increase by a factor of 8! Moreover, if one uses the full version of Gaussian elimination, the number of floating point operations will be of order $(N^3)^3/3$ so that a doubling of N will increase the floating point operations to execute the Gaussian elimination algorithm by a factor of 64! This is known as the curse of dimensionality, and it requires the use of faster computers and algorithms. Alternatives to full Gaussian elimination are block versions of Gaussian elimination as briefly described in Chapter 3 and iterative methods such as SOR and conjugate gradient algorithms. In this section a 3D version of SOR will be applied to a cooling fin with diffusion in all three directions.

4.4.2   Applied Area

Consider an electric transformer that is used on a power line. The electrical current flowing through the wires inside the transformer generates heat. In order to cool the transformer, fins that are not long or very thin in any direction are attached to the transformer. Thus, there will be significant temperature variations in each of the three directions, and consequently, there will be heat diffusion in all three directions. The problem is to find the *steady state heat distribution* in the 3D fin so that one can determine the fin's cooling effectiveness.

4.4.3   Model

In order to model the temperature, we will first assume temperature is given along the 3D boundary of the volume $(0, L) \times (0, W) \times (0, T)$. Consider a small mass within the fin whose volume is $\Delta x \Delta y \Delta z$. This volume will have heat sources or sinks via the two $\Delta x \Delta z$ surfaces, two $\Delta y \Delta z$ surfaces, and two $\Delta x \Delta y$ surfaces as well as any internal heat source given by $f(x, y, z)$ with units of heat/(vol. time). This is depicted in Figure 4.4.1 where the heat flowing through the right face $\Delta x \Delta z$ is given by the Fourier heat law $(\Delta x \Delta z) \Delta t K u_y(x, y + \Delta y, z)$.

The Fourier heat law applied to each of the three directions will give the

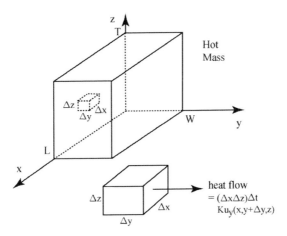

Figure 4.4.1: Heat Diffusion in 3D

heat flowing through these six surfaces. A steady state approximation is

$$
\begin{aligned}
0 \approx\ & f(x,y,z)(\Delta x \Delta y \Delta z)\Delta t \\
& +\Delta x \Delta y \Delta t (Ku_z(x,y,z+\Delta z/2) - Ku_z(x,y,z-\Delta z/2)) \\
& +\Delta x \Delta z \Delta t (Ku_y(x,y+\Delta y/2,z) - Ku_y(x,y-\Delta y/2,z)) \\
& +\Delta y \Delta z \Delta t (Ku_x(x+\Delta x/2,y,z) - Ku_x(x-\Delta x/2,y,z)). \quad (4.4.1)
\end{aligned}
$$

This approximation gets more accurate as Δx, Δy and Δz go to zero. So, divide by $(\Delta x \Delta y \Delta z)\Delta t$ and let Δx, Δy and Δz go to zero. This gives the *continuous model for the steady state 3D heat diffusion*

$$
\begin{aligned}
-(Ku_x)_x - (Ku_y)_y - (Ku_z)_z &= f &&(4.4.2) \\
u &= g \text{ on the boundary.} &&(4.4.3)
\end{aligned}
$$

Let u_{ijl} be the approximation of $u(i\Delta x, j\Delta y, l\Delta z)$ where $\Delta x = L/nx$, $\Delta y = W/ny$ and $\Delta z = T/nz$. Approximate the second order derivatives by the centered finite differences. There are $n = (nx-1)(ny-1)(nz-1)$ equations for n unknowns u_{ijl}. The *discrete finite difference 3D* model for $1 \le i \le nx - 1$, $1 \le j \le ny - 1$, $1 \le l \le nz - 1$ is

$$
\begin{aligned}
& -[K(u_{i+1,j,l} - u_{ijl})/\Delta x - K(u_{ijl} - u_{i-1,j,l})/\Delta x]/\Delta x \\
& -[K(u_{i,j+1,l} - u_{ijl})/\Delta y - K(u_{ijl} - u_{i,j-1,l})/\Delta y]/\Delta y \\
& -[K(u_{i,j,l+1} - u_{ijl})/\Delta z - K(u_{ijl} - u_{i,j,l-1})/\Delta z]/\Delta z \\
&= f(ih, jh, lh). \quad\quad (4.4.4)
\end{aligned}
$$

In order to keep the notation as simple as possible, we assume that the number of cells in each direction, nx, ny and nz, are such that $\Delta x = \Delta y = \Delta z = h$ and let $K = 1$. This equation simplifies to

$$
\begin{aligned}
6u_{ijl} \;=\; & f(ih, jh, lh)h^2 + u_{i,j,l-1} + u_{i,j-1,l} + u_{i-1,j,l} \\
& + u_{i,j,l+1} + u_{i,j+1,l} + u_{i+1,j,l}.
\end{aligned}
\tag{4.4.5}
$$

4.4.4   Method

Equation (4.4.5) suggests the use of the SOR algorithm where there are three nested loops within the SOR loop. The u_{ijl} are now stored in a 3D array, and either $f(ih, jh, lh)$ can be computed every SOR sweep, or $f(ih, jh, lh)$ can be computed once and stored in a 3D array. The classical order of ijl is to start with $l = 1$ (the bottom grid plane) and then use the classical order for ij starting with $j = 1$ (the bottom grid row in the grid plane l). This means the l loop is the outermost, j-loop is in the middle and the i-loop is the innermost loop.

Classical Order 3D SOR Algorithm for (4.4.5).

> choose nx, ny, nz such that h = L/nx = H/ny = T/nz
> for m = 1,maxit
>   for l = 1,nz
>   for j = 1,ny
>   for i = 1,nx
> $$
> \begin{aligned}
> utemp = \; & (f(ih, jh, lh) * h * h \\
> & + u(i-1, j, l) + u(i+1, j, l) \\
> & + u(i, j-1, l) + u(i, j+1, l) \\
> & + u(i, j, l-1) + u(i, j, l+1))/6 \\
> u(i, j, l) = \; & (1-w) * u(i, j, l) + w * utemp
> \end{aligned}
> $$
>   endloop
>   endloop
>   endloop
>   test for convergence
> endloop.

4.4.5   Implementation

The MATLAB code sor3d.m illustrates the 3D steady state cooling fin problem with the finite difference discrete model given in (4.4.5) where $f(x, y, z) = 0.0$. The following parameters were used: $L = W = T = 1$, $nx = ny = nz = 20$. There were $19^3 = 6859$ unknowns. In sor3d.m the initialization and boundary conditions are defined in lines 1-13. The SOR loop is in lines 14-33, where the lji-nested loop for all the interior nodes is executed in lines 16-29. The test for SOR convergence is in lines 22-26 and lines 30-32. Line 34 lists the SOR iterations needed for convergence, and line 35 has the MATLAB command $slice(u, [5\ 10\ 15\ 20], 10, 10)$, which generates a color coded 3D plot of the temperatures within the cooling fin.

MATLAB Code sor3d.m

```
1.    clear;
2.    % This is the SOR solution of a 3D problem.
3.    % Assume steady state heat diffusion.
4.    % Given temperature on the boundary.
5.    w = 1.8;
6.    eps = .001;
7.    maxit = 200;
8.    nx = 20;
9.    ny = 20;
10.   nz = 20;
11.   nunk = (nx-1)*(ny-1)*(nz-1);
12.   u = 70.*ones(nx+1,ny+1,nz+1); % initial guess
13.   u(1,:,:) = 200.;        % hot boundary at x = 0
14.   for iter = 1:maxit;      % begin SOR
15.        numi = 0;
16.        for l = 2:nz
17.             for j = 2:ny
18.                  for i = 2:nx
19.                       temp = u(i-1,j,l) + u(i,j-1,l) + u(i,j,l-1);
20.                       temp = (temp + u(i+1,j,l) + u(i,j+1,l)
                                              + u(i,j,l+1))/6.;
21.                       temp = (1. - w)*u(i,j,l) + w*temp;
22.                       error = abs(temp - u(i,j,l));
23.                       u(i,j,l) = temp;
24.                       if error<eps
25.                            numi = numi + 1;
26.                       end
27.                  end
28.             end
29.        end
30.        if numi==nunk
31.             break
32.        end
33.   end
34.   iter      % iterations for convergence
35.   slice(u, [5 10 15 20],10,10)      % creates color coded 3D plot
36.   colorbar
```

The SOR parameters $w = 1.6, 1.7$ and 1.8 were used with the convergence criteria $eps = 0.001$, and this resulted in convergence after 77, 50 and 62 iterations, respectively. In the Figure 4.4.2 the shades of gray indicate the varying temperatures inside the cooling fin. The lighter the shade of gray the warmer the temperature. So, this figure indicates the fin is very cool to the left, and so

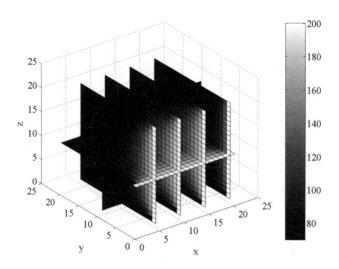

Figure 4.4.2: Temperatures Inside a 3D Fin

the fin for a hot boundary temperature equal to 200 is a little too long in the y direction.

4.4.6   Assessment

The output from the 3D code gives variable temperatures in all three directions. This indicates that a 1D or a 2D model is not applicable for this particular fin. A possible problem with the present 3D model is the given boundary condition for the portion of the surface, which is between the fin and the surrounding region. Here the alternative is a *derivative boundary condition*

$$K\frac{du}{dn} = c(u_{sur} - u) \text{ where } n \text{ is the unit outward normal.}$$

Both the surrounding temperature and the temperature of the transformer could vary with time. Thus, this really is a time dependent problem, and furthermore, one should consider the entire transformer and not just one fin.

4.4.7   Exercises

1.   Use the MATLAB code sor3d.m and experiment with the slice command.
2.   Experiment with different numbers of unknowns nx, ny and nz, and determine the best choices for the SOR parameter, w.
3.   Modify sor3d.m so that Δx, Δy and Δz do not have to be equal.

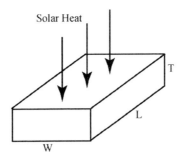

Figure 4.5.1: Passive Solar Storage

4.   Experiment with different L, W and T. Determine when a 1D or 2D model would be acceptable. Why is this important?

5.   Modify sor3d.m to include the derivative boundary condition. Experiment with the coefficient c as it ranges from 0 to infinity. For what values of c will the given boundary condition be acceptable?

4.5 Time Dependent 3D Diffusion

4.5.1 Introduction

We consider a time dependent 3D heat diffusion model of a passive solar energy storage unit similar to a concrete slab. The implicit time discretization method will be used so as to avoid the stability constraint on the time step. This generates a sequence of problems that are similar to the steady state 3D heat equation, and the preconditioned conjugate gradient (PCG) algorithm will be used to approximate the solutions of this sequence. Because of the size of the problem Fortran 90 will be used to generate the sequence of 3D arrays for the temperatures, and then MATLAB commands slice and mesh will be used to dynamically visualize all this data

4.5.2 Applied Area

Consider a *passive solar storage* unit, which collects energy by the day and gives it back at night. A simple example is a thick concrete floor in front of windows, which face the sun during the day. Figure 4.5.1 depicts a concrete slab with dimensions $(0, L) \times (0, W) \times (0, T)$ where top $z = T$ and the vertical sides and bottom have given temperature. Assume there is diffusion in all three directions. Since the surrounding temperature will vary with time, the amount of heat that diffuses in and out of the top will depend on time. The problem is to determine the effectiveness of the passive unit as a function of its geometric and thermal properties.

4.5.3 Model

The model has the form of a time and space dependent partial differential equation. The empirical Fourier heat law again governs the diffusion. For the nonsteady state problem we must consider the change in the amount of heat energy that a mass can retain as a function of temperature. If the temperature varies over a large range, or if there is a change in physical phase, then this relationship is nonlinear. However, for small variations in the temperature the change in heat for a small volume and small change in time is

$$\rho c_p (u(x, y, z, t + \Delta t) - u(x, y, z, t)) \, (volume)$$

where ρ is the density, c_p is the specific heat and u is the temperature. When this is coupled with an internal heat source $f(x, y, z, t)$ and diffusion in three directions for the $volume = \Delta x \Delta y \Delta z$, we get

$$
\begin{aligned}
\text{change in heat} \quad = \quad & \rho c_p(u(x,y,z,t+\Delta t) - u(x,y,z,t))\Delta x \Delta y \Delta z \\
\approx \quad & f(x,y,z,t+\Delta t)(\Delta x \Delta y \Delta z)\Delta t \\
& + (\Delta y \Delta z)\Delta t(Ku_x(x+\Delta x/2, y, z, t+\Delta t) \\
& \quad - Ku_x(x-\Delta x/2, y, z, t+\Delta t)) \\
& + (\Delta x \Delta z)\Delta t(Ku_y(x, y+\Delta y/2, z, t+\Delta t) \\
& \quad - Ku_y(x, y-\Delta y/2, z, t+\Delta t)) \\
& + (\Delta y \Delta x)\Delta t(Ku_z(x, y, z+\Delta z/2, t+\Delta t) \\
& \quad - Ku_z(x, y, z-\Delta z/2, t+\Delta t)). \quad (4.5.1)
\end{aligned}
$$

This approximation gets more accurate as Δx, Δy, Δz and Δt go to zero. So, divide by $(\Delta x \Delta y \Delta z)\Delta t$ and let Δx, Δy, Δz and Δt go to zero. Since $(u(x, y, z, t + \Delta t) - u(x, y, z, t))/\Delta t$ converges to the time derivative of u, u_t, as Δt goes to zero, (4.5.1) gives the partial differential equation in the 3D time dependent heat diffusion model.

Time Dependent 3D Heat Diffusion.

$$
\begin{aligned}
\rho c_p u_t \quad = \quad & f(x,y,z,t) + (Ku_x(x,y,z,t))_x + \\
& (Ku_y(x,y,z,t))_y + (Ku_z(x,y,z,t))_z \quad (4.5.2) \\
u \quad = \quad & 60 \text{ for } z = 0, \, x = 0, L, \, y = 0, W \quad (4.5.3) \\
u \quad = \quad & u_{sur}(t) = 60 + 30sin(\pi t/12) \text{ for } z = T \text{ and} \quad (4.5.4) \\
u \quad = \quad & 60 \text{ for } t = 0. \quad (4.5.5)
\end{aligned}
$$

4.5.4 Method

The derivation of (4.5.1) suggests the implicit time discretization method. Let k denote the time step with $u^k \approx u(x, y, z, k\Delta t)$. From the derivation of (4.5.2) one gets a sequence of steady state problems

$$
\begin{aligned}
\rho c_p(u^{k+1} - u^k)/\Delta t \quad = \quad & f^{k+1} + (Ku_x^{k+1})_x + \\
& (Ku_y^{k+1})_y + (Ku_x^{k+1})_y. \quad (4.5.6)
\end{aligned}
$$

The space variables can be discretized just as in the steady state heat diffusion problem. Thus, for each time step we must solve a linear algebraic system where the right side changes from one time step to the next and equals $f^{k+1}+\rho c_p u^k/\Delta t$ and the boundary condition (4.5.4) changes with time.

Implicit Time and Centered Finite Difference Algorithm.

$u^0 = u(x, y, z, 0)$ from (4.5.5)
for k = 1, maxk
      approximate the solution (4.5.6) by the finite difference method
      use the appropriate boundary conditions in (4.5.3) and (4.5.4)
      solve the resulting algebraic system such as in (4.5.8)
endloop.

This can be written as the seven point finite difference method, and here we let $h = \Delta x = \Delta y = \Delta z$ and $f(x, y, z, t) = 0$ so as to keep the code short. Use the notation $u_{ijl}^{k+1} \approx u(ih, jh, lh, (k+1)\Delta t)$ so that (4.5.6) is approximated by

$$
\begin{aligned}
\rho c_p(u_{ijl}^{k+1} - u_{ijl}^k)/\Delta t = \ & K/h^2(-6u_{ijl}^{k+1} \\
& +u_{i,j,l-1}^{k+1} + u_{i,j-1,l}^{k+1} + u_{i-1,j,l}^{k+1} \\
& +u_{i,j,l+1}^{k+1} + u_{i,j+1,l}^{k+1} + u_{i+1,j,l}^{k+1}).
\end{aligned} \tag{4.5.7}
$$

Let $\alpha = (\rho c_p/\Delta t)/(K/h^2)$ so that (4.5.7) simplifies to

$$
\begin{aligned}
(\alpha + 6)u_{ijl}^{k+1} = \ & \alpha u_{ijl}^k + u_{i,j,l-1}^{k+1} + u_{i,j-1,l}^{k+1} + u_{i-1,j,l}^{k+1} \\
& + u_{i,j,l+1}^{k+1} + u_{i,j+1,l}^{k+1} + u_{i+1,j,l}^{k+1}.
\end{aligned} \tag{4.5.8}
$$

4.5.5 Implementation

The Fortran 90 code solar3d.f90 is for time dependent heat transfer in a 3D volume. It uses the implicit time discretization as simplified in (4.5.8). The solve steps are done by the PCG with the SSOR preconditioner. The reader should note how the third dimension and the nonzero boundary temperatures are inserted into the code. The output is to the console with some information about the PCG iteration and temperatures inside the volume. Also, some of the temperatures are output to a file so that MATLAB's command slice can be used to produce a color coded 3D graph.

In solar3d.f90 the initialization is done in lines 1-22 where a 24 hour simulation is done in 48 time steps. The implicit time discretization is done in the do loop given by lines 24-35. The function subroutine usur(t) in lines 38-43 is for the top boundary whose temperature changes with time. The subroutine cgssor3d approximates the temperature at the next time step by using the preconditioned conjugate gradient method with SSOR. The output is to the console as well as to the file outsolar as indicated in lines 11 and 118-123. The file outsolar is a 2D table where each row in the table corresponds to a partial grid row of every third temperature. So every 121×11 segment in the table corresponds to the 3D temperatures for a single time.

Some of the other MATLAB codes also have Fortran versions so that the interested reader can gradually learn the rudiments of Fortran 90. These include heatl.f90, sor2d.f90, por2d.f90, newton.f90 and picpcg.f90.

Fortran Code solar3d.f90

```
1.     program solar3d
2.     ! This program solves
                density csp ut -(Kux)x-(Kuy)y-(Kuz)z = f.
3.     ! The thermal properties density, csp and K are constant.
4.     ! The implicit time discretization is used.
5.     ! The solve step is done in the subroutine cgssor3d.
6.     ! It uses the PCG method with the SSOR preconditioner.
7.     implicit none
8.     real,dimension(0:30,0:30,0:30):: u,up
9.     real :: dt,h,cond,density,csp,ac,time,ftime
10.    integer :: i,j,n,l,m,mtime,mpcg
11.    open(6,file='c:\MATLAB6p5\work\outsolar')
12.    mtime = 48
13.    ftime = 24.
14.    ! Define the initial condition.
15.    up = 60.0
16.    n = 30
17.    h = 1./n
18.    cond = 0.81
19.    density = 119.
20.    csp = .21
21.    dt = ftime/mtime
22.    ac = density*csp*h*h/(cond*dt)
23.    ! Start the time iteration.
24.    do m=1,mtime
25.        time = m*dt
26.    !
27.    ! The solve step is done by PCG with SSOR preconditioner.
28.    !
29.        call cgssor3d(ac,up,u,mpcg,n,time)
30.    !
31.        up =u
32.        print*,"Time = ",time
33.        print*," Number of PCG iterations = ",mpcg
34.        print*," Max u = ", maxval(u)
35.    end do
36.    close(6)
37.    end program
```

```
38.      ! Heat source function for top.
39.      function usur(t) result(fusur)
40.      implicit none
41.      real :: t,fusur
42.          fusur = 60. + 30.*sin(t*3.14/12.)
43.      end function

44.      ! PCG subroutine.
45.      subroutine cgssor3d(ac,up,u,mpcg,n,time)
46.      implicit none
47.      real,dimension(0:30,0:30,0:30):: p,q,r,rhat
48.      real,dimension(0:30,0:30,0:30),intent(in):: up
49.      real,dimension(0:30,0:30,0:30),intent(out):: u
50.      real :: oldrho, rho,alpha,error,w,ra,usur
51.      real ,intent(in):: ac,time
52.      integer :: i,j,l,m
53.      integer, intent(out):: mpcg
54.      integer, intent(in):: n
55.      w = 1.5
56.      ra = 1./(6.+ac)
57.      r = 0.0
58.      rhat = 0.0
59.      q = 0.0
60.      p = 0.0
61.      r = 0.0
62.      ! Uses previous temperature as an initial guess.
63.      u = up
64.      ! Updates the boundary condition on the top.
65.      do i = 0,n
66.          do j = 0,n
67.              u(i,j,n)=usur(time)
68.          end do
69.      end do
70.      r(1:n-1,1:n-1,1:n-1)=ac*up(1:n-1,1:n-1,1:n-1) &
71.                          -(6.0+ac)*u(1:n-1,1:n-1,1:n-1) &
72.                          +u(0:n-2,1:n-1,1:n-1)+u(2:n,1:n-1,1:n-1) &
73.                          +u(1:n-1,0:n-2,1:n-1)+u(1:n-1,2:n,1:n-1) &
74.                          +u(1:n-1,1:n-1,0:n-2)+u(1:n-1,1:n-1,2:n)
75.      error = 1.
76.      m = 0
77.      rho = 0.0
78.      do while ((error>.0001).and.(m<200))
79.          m = m+1
80.          oldrho = rho
81.      !      Execute SSOR preconditioner
```

```
82.          do l = 1,n-1
83.             do j= 1,n-1
84.                do i = 1,n-1
85.                   rhat(i,j,l) = w*(r(i,j,l)+rhat(i-1,j,l)&
86.                              +rhat(i,j-1,l) +rhat(i,j,l-1))*ra
87.                end do
88.             end do
89.          end do
90.          rhat(1:n-1,1:n-1,1:n-1) = ((2.-w)/w)*(6.+ac)
                              *rhat(1:n-1,1:n-1,1:n-1)
91.          do l = n-1,1,-1
92.             do j= n-1,1,-1
93.                do i = n-1,1,-1
94.                   rhat(i,j,l) = w*(rhat(i,j,l)+rhat(i+1,j,l) &
95.                              +rhat(i,j+1,l)+rhat(i,j,l+1))*ra
96.                end do
97.             end do
98.          end do
99.     !    Find conjugate direction
100.         rho = sum(r(1:n-1,1:n-1,1:n-1)*rhat(1:n-1,1:n-1,1:n-1))
101.         if (m.eq.1) then
102.            p = rhat
103.         else
104.            p = rhat + (rho/oldrho)*p
105.         endif
106.     ! Execute matrix product q = Ap
107.         q(1:n-1,1:n-1,1:n-1)=(6.0+ac)*p(1:n-1,1:n-1,1:n-1) &
108.                        -p(0:n-2,1:n-1,1:n-1)-p(2:n,1:n-1,1:n-1) &
109.                        -p(1:n-1,0:n-2,1:n-1)-p(1:n-1,2:n,1:n-1) &
110.                        -p(1:n-1,1:n-1,0:n-2)-p(1:n-1,1:n-1,2:n)
111.     !    Find steepest descent
112.         alpha = rho/sum(p*q)
113.         u = u + alpha*p
114.         r = r - alpha*q
115.         error = maxval(abs(r(1:n-1,1:n-1,1:n-1)))
116.      end do
117.      mpcg = m
118.      print*, m ,error,u(15,15,15),u(15,15,28)
119.      do l = 0,30,3
120.         do j = 0,30,3
121.            write(6,'(11f12.4)') (u(i,j,l),i=0,30,3)
122.         end do
123.      end do
124.      end subroutine
```

The MATLAB code movsolar3d is used to create a time sequence visualization of the temperatures inside the slab. In line 1 the MATLAB command load is used to import the table in the file outsolar, which was created by the Fortran 90 code solar3d.m, into a MATLAB array also called outsolar. You may need to adjust the directory in line one to fit your computer. This 2D array will have 48 segments of 121×11, that is, outsolar is a 5808×11 array. The nested loops in lines 6-12 store each 121×11 segment of outsolar into a 3D $11 \times 11 \times 11$ array A, whose components correspond to the temperatures within the slab. The visualization is done in line 13 by the MATLAB command slice, and this is illustrated in Figures 4.5.2 and 4.5.3. Also a cross section of the temperatures can be viewed using the MATLAB command mesh as is indicated in line 17.

MATLAB Code movsolar3d.m

```
1.      load c:\MATLAB6p5\work\outsolar;
2.      n = 11;
3.      mtime = 48;
4.      for k = 1:mtime
5.          start = (k-1)*n*n;
6.          for l = 1:n
7.              for j = 1:n
8.                  for i =1:n
9.                      A(i,j,l) = outsolar(n*(l-1)+i+start,j);
10.                 end
11.             end
12.         end
13.         slice(A,n,[10 6],[4])
14.         colorbar;
15.         section(:,:)=A(:,6,:);
16.         pause;
17.         % mesh(section);
18.         % pause;
19.     end
```

4.5.6 Assessment

The choice of step sizes in time or space variables is of considerable importance. The question concerning convergence of discrete solution to continuous solution is nontrivial. If the numerical solution does not vary much as the step sizes decrease and if the numerical solution seems "consistent" with the application, then one may be willing to accept the current step size as generating a "good" numerical model.

4.5.7 Exercises

1.   Experiment with different step sizes and observe convergence.

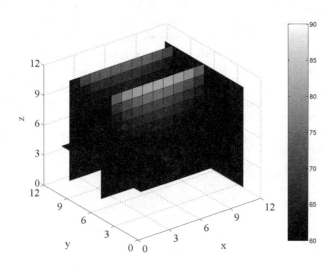

Figure 4.5.2: Slab is Gaining Heat

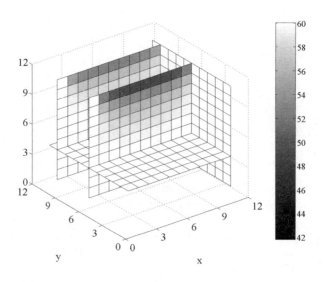

Figure 4.5.3: Slab is Cooling

2.   Modify solar3d.f90 to include the cases where Δx, Δy and Δz do not have to be equal.

3.   Experiment with the geometric parameters W, H and L.

4.   Experiment with the thermal parameters. What types of materials should be used and how does this affect the cost?

5.   Consider the derivative boundary condition on the top

$$\frac{du}{dz} = c(u_{sur}(t) - u) \text{ for } z = T.$$

Modify the above code to include this boundary condition. Experiment with the constant c.

6.   Calculate the change in heat content relative to the initial constant temperature of 60.

7.   Replace the cgssor3d() subroutine with a SOR subroutine and compare the computing times. Use (4.5.8) and be careful to distinguish between the time step index k and the SOR index m.

8.   Code the explicit method for the passive solar storage model, and observe the stability constraint on the change in time. Compare the explicit and implicit time discretizations for this problem.

4.6   High Performance Computations in 3D

4.6.1   Introduction

Many applications are not only 3D problems, but they often have more than one physical quantity associated with them. Two examples are aircraft modeling and weather prediction. In the case of an aircraft, the lift forces are determined by the velocity with three components, the pressure and in many cases the temperatures. So, there are at least five quantities, which all vary with 3D space and time. Weather forecasting models are much more complicated because there are more 3D quantities, often one does not precisely know the boundary conditions and there are chemical and physical changes in system. Such problems require very complicated models, and *faster algorithms and enhanced computing hardware* are essential to give realistic numerical simulations.

   In this section reordering schemes such as *coloring* the nodes and *domain decomposition* of the nodes will be introduced such that both direct and iterative methods will have some independent calculation. This will allow the use of high performance computers with vector pipelines (see Section 6.1) and multiprocessors (see Section 6.3). The implementation of these parallel methods can be challenging, and this will be more carefully studied in the last four chapters.

4.6.2   Methods via Red-Black Reordering

One can reorder nodes so that the vector pipelines or multiprocessors can be used to execute the SOR algorithm. First we do this for the 1D diffusion model with the unknown equal to zero at the boundary and

$$-(Ku_x)_x = f(x) \quad \text{(continuous model)} \quad (4.6.1)$$
$$K(-u_{i-1} + 2u_i - u_{i+1}) = h^2 f(ih) \quad \text{(discrete model)}. \quad (4.6.2)$$

The SOR method requires input of u_{i-1} and u_{i+1} in order to compute the new SOR value of u_i. Thus, if i is even, then only the u with odd subscripts are required as input. The vector version of each SOR iteration is to group all the even nodes and all the odd nodes: (i) use a vector pipe to do SOR over all the odd nodes, (ii) update all u for the odd nodes and (iii) use a vector pipe to do SOR over all the even nodes. This is sometimes called red-black ordering.

The matrix version also indicates that this could be useful for direct methods. Suppose there are seven unknowns so that the *classical* order is

$$\begin{bmatrix} u_1 & u_2 & u_3 & u_4 & u_5 & u_6 & u_7 \end{bmatrix}^T.$$

The corresponding algebraic system is

$$\begin{bmatrix} 2 & -1 & 0 & 0 & 0 & 0 & 0 \\ -1 & 2 & -1 & 0 & 0 & 0 & 0 \\ 0 & -1 & 2 & -1 & 0 & 0 & 0 \\ 0 & 0 & -1 & 2 & -1 & 0 & 0 \\ 0 & 0 & 0 & -1 & 2 & -1 & 0 \\ 0 & 0 & 0 & 0 & -1 & 2 & -1 \\ 0 & 0 & 0 & 0 & 0 & -1 & 2 \end{bmatrix} \begin{bmatrix} u_1 \\ u_2 \\ u_3 \\ u_4 \\ u_5 \\ u_6 \\ u_7 \end{bmatrix} = h^2 \begin{bmatrix} f_1 \\ f_2 \\ f_3 \\ f_4 \\ f_5 \\ f_6 \\ f_7 \end{bmatrix}.$$

The *red-black* order is

$$\begin{bmatrix} u_1 & u_3 & u_5 & u_7 & u_2 & u_4 & u_6 \end{bmatrix}^T.$$

The reordered algebraic system is

$$\begin{bmatrix} 2 & 0 & 0 & 0 & -1 & 0 & 0 \\ 0 & 2 & 0 & 0 & -1 & -1 & 0 \\ 0 & 0 & 2 & 0 & 0 & -1 & -1 \\ 0 & 0 & 0 & 2 & 0 & 0 & -1 \\ -1 & -1 & 0 & 0 & 2 & 0 & 0 \\ 0 & -1 & -1 & 0 & 0 & 2 & 0 \\ 0 & 0 & -1 & -1 & 0 & 0 & 2 \end{bmatrix} \begin{bmatrix} u_1 \\ u_3 \\ u_5 \\ u_7 \\ u_2 \\ u_4 \\ u_6 \end{bmatrix} = h^2 \begin{bmatrix} f_1 \\ f_3 \\ f_5 \\ f_7 \\ f_2 \\ f_4 \\ f_6 \end{bmatrix}.$$

The coefficient matrix for the red-black order is a block 2×2 matrix where the block diagonal matrices are pointwise diagonal. Therefore, the solution by block Gaussian elimination via the Schur complement, see Section 2.4, is easy to implement and has concurrent calculations.

Fortunately, the diffusion models for 2D and 3D will also have these desirable attributes. The simplified discrete models for 2D and 3D are, respectively,

$$K(-u_{i-1,j} - u_{i,j-1} + 4u_{ij} - u_{i+1,j} - u_{i,j+1}) = h^2 f(ih, jh) \text{ and} \quad (4.6.3)$$

$$K\left(-u_{i-1,j,l} - u_{i,j-1,l} - u_{i,j,l-1} + 6u_{ijl} - u_{i+1,j,l} - u_{i,j+1,l} - u_{i,j,l+1}\right)$$

$$= h^2 f(ih, jh, lh). \qquad (4.6.4)$$

In 2D diffusion the new values of u_{ij} are functions of $u_{i+1,j}$, $u_{i-1,j}$, $u_{i,j+1}$ and $u_{i,j-1}$ and so the SOR algorithm must be computed in a "checker board" order. In the first grid row start with the $j = 1$ and go in stride 2; for the second grid row start with $j = 2$ and go in stride 2. Repeat this for all pairs of grid rows. This will compute the newest u_{ij} for the same color, say, all the black nodes. In order to do all the red nodes, repeat this, but now start with $j = 2$ in the first grid row and then with $j = 1$ in the second grid row. Because the computation of the newest u_{ij} requires input from the nodes of a different color, all the calculations for the same color are independent. Therefore, the vector pipelines or multiprocessors can be used.

Red-Black Order 2D SOR for (4.6.3).

  choose nx, ny such that h = L/nx = W/ny
  for m = 1,maxit
   for j = 1,ny
    index = mod(j,2)
    for i = 2-index,nx,2
     $utemp = (f(i * h, j * h) * h * h$
      $+u(i - 1, j) + u(i + 1, j)$
      $+u(i, j - 1) + u(i, j + 1)) * .25$
     $u(i, j) = (1 - w) * u(i, j) + w * utemp$
    endloop
   endloop
   for j = 1,ny
    index = mod(j,2)
    for i = 1+index,nx,2
     $utemp = (f(i * h, j * h) * h * h$
      $+u(i - 1, j) + u(i + 1, j)$
      $+u(i, j - 1) + u(i, j + 1)) * .25$
     $u(i, j) = (1 - w) * u(i, j) + w * utemp$
    endloop
   endloop
   test for convergence
  endloop.

For 3D diffusion the new values of u_{ijl} are functions of $u_{i+1,j,l}$, $u_{i-1,j,l}$, $u_{i,j+1,l}$, $u_{i,j-1,l}$, $u_{i,j,l+1}$ and $u_{i,j,l-1}$ and so the SOR algorithm must be computed in a "3D checker board" order. The first grid plane should have a 2D checker board order, and then the next grid plane should have the interchanged color 2D checker board order. Because the computation of the newest u_{ijl} requires input from the nodes of a different color, all the calculations for the same color are independent. Therefore, the vector pipelines or multiprocessors can be used.

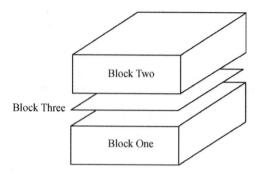

Figure 4.6.1: Domain Decompostion in 3D

4.6.3   Methods via Domain Decomposition Reordering

In order to introduce the domain decompostion order, again consider the 1D problem in (4.6.2) and use seven unknowns. Here the *domain decomposition order* is

$$\begin{bmatrix} u_1 & u_2 & u_3 & u_5 & u_6 & u_7 & u_4 \end{bmatrix}^T$$

where the center node, u_4, is listed last and the left and right blocks are listed first and second. The algebraic system with this order is

$$\begin{bmatrix} 2 & -1 & 0 & 0 & 0 & 0 & 0 \\ -1 & 2 & -1 & 0 & 0 & 0 & 0 \\ 0 & -1 & 2 & 0 & 0 & 0 & -1 \\ 0 & 0 & 0 & 2 & -1 & 0 & -1 \\ 0 & 0 & 0 & -1 & 2 & -1 & 0 \\ 0 & 0 & 0 & 0 & -1 & 2 & 0 \\ 0 & 0 & -1 & -1 & 0 & 0 & 2 \end{bmatrix} \begin{bmatrix} u_1 \\ u_2 \\ u_3 \\ u_5 \\ u_6 \\ u_7 \\ u_4 \end{bmatrix} = h^2 \begin{bmatrix} f_1 \\ f_2 \\ f_3 \\ f_5 \\ f_6 \\ f_7 \\ f_4 \end{bmatrix} .$$

Domain decomposition ordering can also be used in 2D and 3D applications. Consider the 3D case as depicted in Figure 4.6.1 where the nodes are partitioned into two large blocks and a smaller third block separating the two large blocks of nodes. Thus, if ijl is in block 1 (or 2), then only input from block 1 (or 2) and block 3 will be required to do the SOR computation. This suggests that one can reorder the nodes so that disjoint blocks of nodes, which are separated by a plane of nodes, can be computed concurrently in the SOR algorithm.

Domain Decomposition and 3D SOR Algorithm for (4.6.4).

> define blocks 1, 2 and 3
> for m = 1,maxit
>   concurrently do SOR on blocks 1 and 2
>   update u
>   do SOR on block 3
>   test for convergence
> endloop.

Domain decomposition order can also be used to directly solve for the unknowns. This was initially described in Section 2.4 where the Schur complement was studied. If the interface block 3 for the Poisson problem is listed last, then the algebraic system has the form

$$\begin{bmatrix} A_{11} & 0 & A_{13} \\ 0 & A_{22} & A_{23} \\ A_{31} & A_{32} & A_{33} \end{bmatrix} \begin{bmatrix} U_1 \\ U_2 \\ U_3 \end{bmatrix} = \begin{bmatrix} F_1 \\ F_2 \\ F_3 \end{bmatrix}. \tag{4.6.5}$$

In the Schur complement study in Section 2.4 B is the 2×2 block given by A_{11} and A_{22}, and C is A_{33}. Therefore, all the solves with B can be done concurrently, in this case with two processors. By partitioning the space domain into more blocks one can take advantage of additional processors. In the 3D case the big block solves will be smaller 3D subproblems and here one may need to use iterative methods. Note the conjugate gradient algorithm has a number of vector updates, dot products and matrix-vector products, and all these steps have independent parts.

In order to be more precise about the above, write the above 3×3 block matrix equation in block component form

$$A_{11}U_1 + A_{13}U_3 = F_1, \tag{4.6.6}$$
$$A_{22}U_2 + A_{23}U_3 = F_2 \text{ and} \tag{4.6.7}$$
$$A_{31}U_1 + A_{32}U_2 + A_{33}U_3 = F_3. \tag{4.6.8}$$

Now solve (4.6.6) and (4.6.7) for U_1 and U_2, and note the computations for $A_{11}^{-1}A_{13}$, $A_{11}^{-1}F_1$, $A_{22}^{-1}A_{23}$, and $A_{22}^{-1}F_2$ can be done concurrently. Put U_1 and U_2 into (4.6.8) and solve for U_3

$$\widehat{A}_{33}U_3 = \widehat{F}_3 \text{ where}$$
$$\widehat{A}_{33} = A_{33} - A_{31}A_{11}^{-1}A_{13} - A_{32}A_{22}^{-1}A_{23} \text{ and}$$
$$\widehat{F}_3 = F_3 - A_{31}A_{11}^{-1}F_1 - A_{32}A_{22}^{-1}F_2.$$

Then concurrently solve for $U_1 = A_{11}^{-1}F_1 - A_{11}^{-1}A_{13}U_3$ and $U_2 = A_{22}^{-1}F_2 - A_{22}^{-1}A_{23}U_3$.

4.6.4   Implementation of Gaussian Elimination via Domain Decomposition

Consider the 2D steady state heat diffusion problem. The MATLAB code gedd.m is block Gaussian elimination where the B matrix, in the 2×2 block matrix of the Schur complement formulation, is a block diagonal matrix with four blocks on its diagonal. The $C = A_{55}$ matrix is for the coefficients of the three interface grid rows between the four big blocks

$$
\begin{bmatrix}
A_{11} & 0 & 0 & 0 & A_{15} \\
0 & A_{22} & 0 & 0 & A_{25} \\
0 & 0 & A_{33} & 0 & A_{35} \\
0 & 0 & 0 & A_{44} & A_{45} \\
A_{15}^T & A_{12}^T & A_{13}^T & A_{14}^T & A_{55}
\end{bmatrix}
\begin{bmatrix}
U_1 \\ U_2 \\ U_3 \\ U_4 \\ U_5
\end{bmatrix}
=
\begin{bmatrix}
F_1 \\ F_2 \\ F_3 \\ F_4 \\ F_5
\end{bmatrix}.
\qquad (4.6.9)
$$

In the MATLAB code gedd.m the first 53 lines define the coefficient matrix that is associated with the 2D Poisson equation. The derivations of the steps for the Schur complement calculations are similar to those with two big blocks. The forward sweep to find the Schur complement matrix and right side is given in lines 54-64 where parallel computations with four processors can be done. The solution of the Schur complement reduced system is done in lines 66-69. The parallel computations for the other four blocks of unknowns are done in lines 70-74.

MATLAB Code gedd.m

```
1.     clear;
2.     % Solves a block tridiagonal SPD algebraic system.
3.     % Uses domain-decomposition and Schur complement.
4.     % Define the block 5x5 matrix AAA
5.     n = 5;
6.     A = zeros(n);
7.     for i = 1:n
8.         A(i,i) = 4;
9.         if (i>1)
10.             A(i,i-1)=-1;
11.         end
12.         if (i<n)
13.             A(i,i+1)=-1;
14.         end
15.     end
16.     I = eye(n);
17.     AA= zeros(n*n);
18.     for i =1:n
19.         newi = (i-1)*n +1;
20.         lasti = i*n;
21.         AA(newi:lasti,newi:lasti) = A;
```

```
22.          if (i>1)
23.                AA(newi:lasti,newi-n:lasti-n) = -I;
24.          end
25.          if (i<n)
26.                AA(newi:lasti,newi+n:lasti+n) = -I;
27.          end
28.     end
29.     Z = zeros(n);
30.     A0 = [A Z Z;Z A Z;Z Z A];
31.     A1 = zeros(n^2,3*n);
32.     A1(n^2-n+1:n^2,1:n)=-I;
33.     A2 = zeros(n^2,3*n);
34.     A2(1:n,1:n) = -I;
35.     A2(n^2-n+1:n^2,n+1:2*n) = -I;
36.     A3 = zeros(n^2,3*n);
37.     A3(1:n,n+1:2*n) = -I;
38.     A3(n^2-n+1:n^2,2*n+1:3*n) = -I;
39.     A4 = zeros(n^2,3*n);
40.     A4(1:n,2*n+1:3*n) = -I;
41.     ZZ =zeros(n^2);
42.     AAA = [AA ZZ ZZ ZZ A1;
43.     ZZ AA ZZ ZZ A2;
44.     ZZ ZZ AA ZZ A3;
45.     ZZ ZZ ZZ AA A4;
46.     A1' A2' A3' A4' A0];
47.     % Define the right side
48.     d1 =ones(n*n,1)*10*(1/(n+1)^2);
49.     d2 =ones(n*n,1)*10*(1/(n+1)^2);
50.     d3 =ones(n*n,1)*10*(1/(n+1)^2);
51.     d4 =ones(n*n,1)*10*(1/(n+1)^2);
52.     d0 =ones(3*n,1)*10*(1/(n+1)^2);
53.     d = [d1' d2' d3' d4' d0']';
54.     % Start the Schur complement method
55.     % Parallel computation with four processors
56.     Z1 = AA\[A1 d1];
57.     Z2 = AA\[A2 d2];
58.     Z3 = AA\[A3 d3];
59.     Z4 = AA\[A4 d4];
60.     % Parallel computation with four processors
61.     W1 = A1'*Z1;
62.     W2 = A2'*Z2;
63.     W3 = A3'*Z3;
64.     W4 = A4'*Z4;
65.     % Define the Schur complement system.
66.     Ahat = A0 -W1(1:3*n,1:3*n) - W2(1:3*n,1:3*n)
```

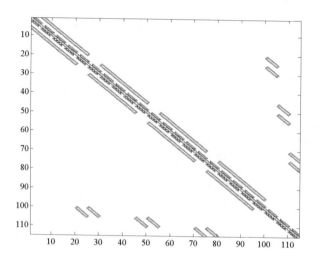

Figure 4.6.2: Domain Decomposition Matrix

                           - W3(1:3*n,1:3*n) -W4(1:3*n,1:3*n);

67.   dhat = d0 -W1(1:3*n,1+3*n) -W2(1:3*n,1+3*n)

                      -W3(1:3*n,1+3*n) -W4(1:3*n,1+3*n);

68.   % Solve the Schur complement system.

69.   x0 = Ahat\dhat;

70.   % Parallel computation with four processors

71.   x1 = AA\(d1 - A1*x0);

72.   x2 = AA\(d2 - A2*x0);

73.   x3 = AA\(d3 - A3*x0);

74.   x4 = AA\(d4 - A4*x0);

75.   % Compare with the full Gauss elimination method.

76.   norm(AAA\d - [x1;x2;x3;x4;x0])

Figure 4.6.2 is the coefficient matrix with the domain decomposition ordering. It was generated by the MATLAB file gedd.m and using the MATLAB command contour(AAA).

4.6.5   Exercises

1.   In (4.6.5)-(4.6.8) list the interface block first and not last. Find the solution with this new order.

2.   Discuss the benefits of listing the interface block first or last.

3.   Consider the parallel solution of (4.6.9). Use the Schur complement as in (4.6.5)-(4.6.8) to find the block matrix formula for the solution.

4. Use the results of the previous exercise to justify the lines 54-74 in the MATLAB code gedd.m.

Chapter 5

Epidemics, Images and Money

This chapter contains nonlinear models of epidemics, image restoration and value of option contracts. All three applications have diffusion-like terms, and so mathematically they are similar to the models in the previous chapters. In the epidemic model the unknown concentrations of the infected populations will depend on both time and space. A good reference is the second edition of A. Okubo and S. A. Levin [19]. Image restoration has applications to satellite imaging, signal processing and fish finders. The models are based on minimization of suitable real valued functions whose gradients are similar to the quasi-linear heat diffusion models. An excellent text with a number of MATLAB codes has been written by C. R. Vogel [26]. The third application is to the value of option contracts, which are agreements to sell or to buy an item at a future date and given price. The option contract can itself be sold and purchased, and the value of the option contract can be modeled by a partial differential equation that is similar to the heat equation. The text by P. Wilmott, S. Howison and J. Dewynne [28] presents a complete derivation of this model as well as a self-contained discussion of the relevant mathematics, numerical algorithms and related financial models.

5.1 Epidemics and Dispersion

5.1.1 Introduction

In this section we study a variation of an epidemic model, which is similar to measles. The population can be partitioned into three disjoint groups of susceptible, infected and recovered. One would like to precisely determine what parameters control or prevent the epidemic. The classical time dependent model has three ordinary differential equations. The modified model where the populations depend on both time and space will generate a system of nonlinear equations that must be solved at each time step. In populations that move in one direction such as along a river or a beach, Newton's method can easily be implemented and the linear subproblems will be solved by the MATLAB command A\b. In the following section dispersion in two directions will be considered. Here the linear subproblems in Newton's method can be solved by a sparse implementation of the preconditioned conjugate gradient method.

5.1.2 Application

Populations move in space for a number of reasons including search of food, mating and herding instincts. So they may tend to disperse or to group together. Dispersion can have the form of a random walk. In this case, if the population size and time duration are suitably large, then this can be modeled by Fick's motion law, which is similar to Fourier's heat law. Let $C = C(x,t)$ be the concentration (amount per volume) of matter such as spores, pollutant, molecules or a population.

Fick Motion Law. Consider the concentration $C(x,t)$ as a function of space in a single direction whose cross-sectional area is A. The change in the matter through A is given by

(a).   moves from high concentrations to low concentrations

(b).   change is proportional to the

      change in time,

      the cross section area and

      the derivative of the concentration with respect to x.

Let D be the proportionality constant, which is called the *dispersion*, so that the change in the amount via A at $x + \Delta x/2$ is

$$D \, \Delta t \, AC_x(x + \Delta x/2, t + \Delta t).$$

The dispersion from both the left and right of the volume $A\Delta x$ gives the approximate change in the amount

$$
\begin{aligned}
(C(x, t + \Delta t) - C(x,t))A\Delta x \;\approx\;\; & D \, \Delta t \, AC_x(x + \Delta x/2, t + \Delta t) \\
& -D \, \Delta t \, AC_x(x - \Delta x/2, t + \Delta t).
\end{aligned}
$$

Divide by $A\Delta x\Delta t$ and let Δx and Δt go to zero to get

$$C_t = (DC_x)_x. \qquad (5.1.1)$$

This is analogous to the heat equation where concentration is replaced by temperature and dispersion is replaced by thermal conductivity divided by density and specific heat. Because of this similarity the term diffusion is often associated with Fick's motion law.

5.1.3 Model

The SIR model is for the amounts or sizes of the populations as functions only of time. Assume the population is a disjoint union of susceptible $S(t)$, infected $I(t)$ and recovered $R(t)$. So, the total population is $S(t) + I(t) + R(t)$. Assume all infected eventually recover and all recovered are not susceptible. Assume the increase in the infected is proportional to the product of the change in time, the number of infected and the number of susceptible. The change in the infected population will increase from the susceptible group and will decrease into the recovered group.

$$I(t + \Delta t) - I(t) = \Delta t \, aS(t)I(t) - \Delta t \, bI(t) \qquad (5.1.2)$$

where a reflects how *contagious or infectious* the epidemic is and b reflects the *rate of recovery*. Now divide by Δt and let it go to zero to get the differential equation for I, $I' = aSI - bI$. The differential equations for S and R are obtained in a similar way.

SIR Epidemic Model.

$$
\begin{aligned}
S' &= -aSI \text{ with } S(0) = S_0, & (5.1.3) \\
I' &= aSI - bI \text{ with } I(0) = I_0 \text{ and} & (5.1.4) \\
R' &= bI \text{ with } R(0) = R_0. & (5.1.5)
\end{aligned}
$$

Note, $(S + I + R)' = S' + I' + R' = 0$ so that $S + I + R = $ constant and $S(0) + I(0) + R(0) = S_0 + I_0 + R_0$.

Note, $I'(0) = (aS(0) - b)I(0) > 0$ if and only if $aS(0) - b > 0$ and $I(0) > 0$. The epidemic cannot get started unless $aS(0) - b > 0$ so that the initial number of susceptible must be suitably large.

The SIR model will be modified in two ways. First, assume the infected do not recover but eventually die at a rate $-bI$. Second, assume the infected population disperses in one direction according to Fick's motion law, and the susceptible population does not disperse. This might be the case for populations that become infected with rabies. The unknown populations for the susceptible and the infected now are functions of time and space, and $S(x, t)$ and $I(x, t)$ are concentrations of the susceptible and the infected populations, respectively.

SI with Dispersion Epidemic Model.

$$
\begin{aligned}
S_t &= -aSI \text{ with } S(x,0) = S_0 \text{ and } 0 \le x \le L, & (5.1.6) \\
I_t &= aSI - bI + DI_{xx} \text{ with } I(x,0) = I_0 \text{ and} & (5.1.7) \\
I_x(0,t) &= 0 = I_x(L,t). & (5.1.8)
\end{aligned}
$$

In order to solve for the infected population, which has two space derivatives in its differential equation, boundary conditions on the infected population must be imposed. Here we have simply required that no inflected can move in or out of the left and right boundaries, that is, $I_x(0, t) = 0 = I_x(L, t)$.

5.1.4 Method

Discretize (5.1.6) and (5.1.7) implicitly with respect to the time variable to obtain a sequence of ordinary differential equations

$$S^{k+1} = S^k - \Delta t \, a S^{k+1} I^{k+1} \tag{5.1.9}$$
$$I^{k+1} = I^k + \Delta t \, a S^{k+1} I^{k+1} - \Delta t \, b I^{k+1} + \Delta t \, D I_{xx}^{k+1}. \tag{5.1.10}$$

As in the heat equation with derivative boundary conditions, use half cells at the boundaries and centered finite differences with $h = \Delta x = L/n$ so that there are $n+1$ unknowns for both $S = S^{k+1}$ and $I = I^{k+1}$. So, at each time step one must solve a system of $2(n+1)$ nonlinear equations for S_i and I_i given $\overline{S} = S^k$ and $\overline{I} = I^{k+1}$. Let $\mathbf{F} : \mathbb{R}^{2(n+1)} \longrightarrow \mathbb{R}^{2(n+1)}$ be the function of S_i and I_i where the $(S, I) \in \mathbb{R}^{2(n+1)}$ are listed by all the S_i and then all the I_i. Let $1 \leq i \leq n+1$, $\alpha = D \Delta t / h^2$, $\widehat{i} = i - (n+1)$ for $i > n+1$ and so that

$$1 \leq i \leq n+1: \qquad F_i = S_i - \overline{S}_i + \Delta t \, a S_i I_i$$

$$i = n+2: \qquad \begin{aligned} F_i &= I_{\widehat{i}} - \overline{I}_{\widehat{i}} - \Delta t \, a S_{\widehat{i}} I_{\widehat{i}} + \\ & \quad \Delta t \, b I_{\widehat{i}} - \alpha(-2 I_{\widehat{i}} + 2 I_{\widehat{i}+1}) \end{aligned}$$

$$n+2 < i < 2(n+1): \qquad \begin{aligned} F_i &= I_{\widehat{i}} - \overline{I}_{\widehat{i}} - \Delta t \, a S_{\widehat{i}} I_{\widehat{i}} + \\ & \quad \Delta t \, b I_{\widehat{i}} - \alpha(I_{\widehat{i}-1} - 2 I_{\widehat{i}} + I_{\widehat{i}+1}) \end{aligned}$$

$$i = 2(n+1): \qquad \begin{aligned} F_i &= I_{\widehat{i}} - \overline{I}_{\widehat{i}} - \Delta t \, a S_{\widehat{i}} I_{\widehat{i}} + \\ & \quad \Delta t \, b I_{\widehat{i}} - \alpha(2 I_{\widehat{i}-1} - 2 I_{\widehat{i}}). \end{aligned}$$

Newton's method will be used to solve $\mathbf{F}(S, I) = \mathbf{0}$. The nonzero components of the Jacobian $2(n+1) \times 2(n+1)$ matrix $\mathbf{F}'$ are

$$1 \leq i \leq n+1: \qquad F_{i S_i} = 1 + \Delta t \, a I_i \text{ and } F_{i I_i} = \Delta t \, a S_i$$

$$i = n+2: \qquad \begin{aligned} F_{i I_{\widehat{i}}} &= 1 + b \Delta t + 2\alpha - \Delta t \, a S_{\widehat{i}}, \\ F_{i I_{\widehat{i}+1}} &= -2\alpha \text{ and } F_{i S_{\widehat{i}}} = -\Delta t \, a I_{\widehat{i}} \end{aligned}$$

$$n+2 < i < 2(n+1): \qquad \begin{aligned} F_{i I_{\widehat{i}}} &= 1 + b \Delta t + 2\alpha - \Delta t \, a S_{\widehat{i}}, \\ F_{i I_{\widehat{i}+1}} &= -\alpha, \; F_{i I_{\widehat{i}-1}} = -\alpha \text{ and } F_{i S_{\widehat{i}}} = -\Delta t \, a I_{\widehat{i}} \end{aligned}$$

$$i = 2(n+1): \qquad \begin{aligned} F_{i I_{\widehat{i}}} &= 1 + b \Delta t + 2\alpha - \Delta t \, a S_{\widehat{i}}, \\ F_{i I_{\widehat{i}-1}} &= -2\alpha \text{ and } F_{i S_{\widehat{i}}} = -\Delta t \, a I_{\widehat{i}}. \end{aligned}$$

The matrix $\mathbf{F}'$ can be written as a block 2×2 matrix where the four blocks are $(n+1) \times (n+1)$ matrices

$$\mathbf{F}' = \begin{bmatrix} A & E \\ \widetilde{F} & C \end{bmatrix}. \tag{5.1.11}$$

A, E and $\widetilde{F}$ are diagonal matrices whose components are F_{iS_i}, F_{iI_i} and F_{iS_i}, respectively. The matrix C is tridiagonal, and for $n = 4$ it is

$$C = \begin{bmatrix} F_{6I_1} & -2\alpha & & & \\ -\alpha & F_{7I_2} & -\alpha & & \\ & -\alpha & F_{8I_3} & -\alpha & \\ & & -\alpha & F_{9I_4} & -\alpha \\ & & & -2\alpha & F_{10I_5} \end{bmatrix}. \qquad (5.1.12)$$

Since $\mathbf{F}'$ is relatively small, one can easily use a direct solver. Alternatively, because of the simple structure of $\mathbf{F}'$, the Schur complement could be used to do this solve. In the model with dispersion in two directions C will be block tridiagonal, and the solve step will be done using the Schur complement and the sparse PCG method.

5.1.5 Implementation

The MATLAB code SIDiff1d.m solves the system (5.1.6)-(5.1.8) by the above implicit time discretization with the centered finite difference discretization of the space variable. The resulting nonlinear algebraic system is solved at each time step by using Newton's method. The initial guess for Newton's method is the previous time values for the susceptible and the infected. The initial data is given in lines 1-28 with the parameters of the differential equation model defined in lines 9-11 and initial populations defined in lines 23-28. The time loop is in lines 29-84. Newton's method for each time step is executed in lines 30-70 with the $\mathbf{F}$ and $\mathbf{F}'$ computed in lines 32-62 and the linear solve step done in line 63. The Newton update is done in line 64. The output of populations versus space for each time step is given in lines 74-83, and populations versus time is given in lines 86 and 87.

MATLAB Code SIDiff1d.m

```
1.   clear;
2.   % This code is for susceptible/infected population.
3.   % The infected may disperse in 1D via Fick's law.
4.   % Newton's method is used.
5.   % The full Jacobian matrix is defined.
6.   % The linear steps are solved by A\d.
7.   sus0 = 50.;
8.   inf0 = 0.;
9.   a =20/50;
10.  b = 1;
11.  D = 10000;
12.  n = 20;
13.  nn = 2*n+2;
14.  maxk = 80;
15.  L = 900;
```

```
16.     dx = L./n;
17.     x = dx*(0:n);
18.     T = 3;
19.     dt = T/maxk;
20.     alpha = D*dt/(dx*dx);
21.     FP = zeros(nn);
22.     F = zeros(nn,1);
23.     sus = ones(n+1,1)*sus0;        % define initial populations
24.     sus(1:3) = 2;
25.     susp = sus;
26.     inf = ones(n+1,1)*inf0;
27.     inf(1:3) = 48;
28.     infp = inf;
29.     for k = 1:maxk        % begin time steps
30.         u = [susp; infp];        % begin Newton iteration
31.         for m =1:20
32.             for i = 1:nn         %compute Jacobian matrix
33.                 if i>=1&i<=n
34.                     F(i) = sus(i) - susp(i) + dt*a*sus(i)*inf(i);
35.                     FP(i,i) = 1 + dt*a*inf(i);
36.                     FP(i,i+n+1) = dt*a*sus(i);
37.                 end
38.                 if i==n+2
39.                     F(i) = inf(1) - infp(1) + b*dt*inf(1) -...
40.                             alpha*2*(-inf(1) + inf(2)) -
                                a*dt*sus(1)*inf(1);
41.                     FP(i,i) = 1+b*dt + alpha*2 - a*dt*sus(1);
42.                     FP(i,i+1) = -2*alpha;
43.                     FP(i,1) = -a*dt*inf(1);
44.                 end
45.                 if i>n+2&i<nn
46.                     i_shift = i - (n+1);
47.                     F(i) = inf(i_shift) - infp(i_shift) +
                                b*dt*inf(i_shift) - ...
48.                             alpha*(inf(i_shift-1) - 2*inf(i_shift) +
                                inf(i_shift+1)) - ...
49.                             a*dt*sus(i_shift)*inf(i_shift);
50.                     FP(i,i) = 1+b*dt + alpha*2 - a*dt*sus(i_shift);
51.                     FP(i,i-1) = -alpha;
52.                     FP(i,i+1) = -alpha;
53.                     FP(i, i_shift) = - a*dt*inf(i_shift);
54.                 end
55.                 if i==nn
56.                     F(i) = inf(n+1) - infp(n+1) + b*dt*inf(n+1) - ...
57.                             alpha*2*(-inf(n+1) + inf(n)) -
```

```
                                a*dt*sus(n+1)*inf(n+1);
58.                             FP(i,i) = 1+b*dt + alpha*2 - a*dt*sus(n+1);
59.                             FP(i,i-1) = -2*alpha;
60.                             FP(i,n+1) = -a*dt*inf(n+1);
61.                     end
62.                 end
63.                 du = FP\F;      % solve linear system
64.                 u = u - du;
65.                 sus(1:n+1) = u(1:n+1);
66.                 inf(1:n+1) = u(n+2:nn);
67.                 error = norm(F);
68.                 if error<.00001
69.                     break;
70.                 end
71.             end %        Newton iterations
72.             time(k) = k*dt;
73.             time(k)
74.             m
75.             error
76.             susp = sus;
77.             infp = inf;
78.             sustime(:,k) = sus(:);
79.             inftime(:,k) = inf(:);
80.             axis([0 900 0 60]);
81.             hold on;
82.             plot(x,sus,x,inf)
83.             pause
84.         end %time step
85.         hold off
86.         figure(2);
87.         plot(time,sustime(10,:),time,inftime(10,:))
```

In Figure 5.1.1 five time plots of infected and susceptible versus space are given. As time increases the locations of the largest concentrations of infected move from left to right. The left side of the infected will decrease as time increases because the concentration of the susceptible population decreases. Eventually, the infected population will start to decrease for all locations in space.

5.1.6   Assessment

Populations may or may not move in space according to Fick's law, and they may even move from regions of low concentration to high concentration! Populations may be moved by the flow of air or water. If populations do disperse according to Fick's law, then one must be careful to estimate the dispersion coefficient D and to understand the consequences of using this estimate. In

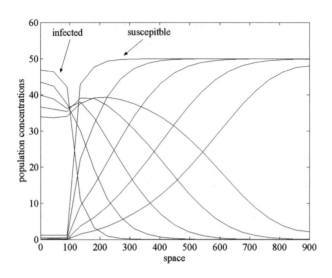

Figure 5.1.1: Infected and Susceptible versus Space

the epidemic model with dispersion in just one direction as given in (5.1.6)-(5.1.8) the coefficients a and b must also be estimated. Also, the population can disperse in more than one direction, and this will be studied in the next section.

5.1.7   Exercises

1.   Duplicate the calculations in Figure 5.1.1. Examine the solution as time increases.

2.   Find a steady state solution of (5.1.6)-(5.1.8). Does the solution in problem one converge to it?

3.   Experiment with the step sizes in the MATLAB code SIDiff1d.m: $n = 10, 20, 40$ and 80 and $kmax = 40, 80, 160$ and 320.

4.   Experiment with the contagious coefficient in the MATLAB code SIDiff1d.m: $a = 1/10, 2/10, 4/10$ and $8/10$.

5.   Experiment with the death coefficient in the MATLAB code SIDiff1d.m: $b = 1/2, 1, 2$ and 4.

6.   Experiment with the dispersion coefficient in the MATLAB code SIDiff1d.m: $D = 5000, 10000, 20000$ and 40000.

7.   Let an epidemic be dispersed by Fick's law as well as by the flow of a stream whose velocity is $v > 0$. Modify (5.1.7) to take this into account

$$I_t = aSI - bI + DI_{xx} - vI_x.$$

Formulate a numerical model and modify the MATLAB code SIDiff1d.m. Study the effect of variable stream velocities.

5.2 Epidemic Dispersion in 2D

5.2.1 Introduction

Consider populations that will depend on time, two space variables and will disperse according to Fick's law. The numerical model will also follow from an implicit time discretization and from centered finite differences in both space directions. This will generate a sequence of nonlinear equations, which will also be solved by Newton's method. The linear solve step in Newton's method will be done by a sparse version of the conjugate gradient method.

5.2.2 Application

The dispersion of a population can have the form of a random walk. In this case, if the population size and time duration are suitably large, then this can be modeled by Fick's motion law, which is similar to Fourier's heat law. Let $C = C(x, y, t)$ be the concentration (amount per volume) of matter such as a population. Consider the concentration $C(x, y, t)$ as a function of space in two directions whose volume is $H\Delta x\Delta y$ where H is the small distance in the z direction. The change in the matter through an area A is given by

(a).   matter moves from high concentrations to low concentrations

(b).   change is proportional to the

   change in time,

   the cross section area and

   the derivative of the concentration normal to A.

Let D be the proportionality constant, which is called the *dispersion*. Next consider dispersion from the left and right where $A = H\Delta y$, and the front and back where $A = H\Delta x$.

$$
\begin{aligned}
(C(x, y, t + \Delta t) - C(x, y, t))H\Delta x\Delta y \approx\ & D\,\Delta t\, H\Delta y C_x(x + \Delta x/2, y, t + \Delta t) \\
& -D\,\Delta t\, H\Delta y C_x(x - \Delta x/2, y, t + \Delta t) \\
& +D\,\Delta t\, H\Delta x C_y(x, y + \Delta y/2, t + \Delta t) \\
& -D\,\Delta t\, H\Delta x C_y(x, y - \Delta y/2, t + \Delta t).
\end{aligned}
$$

Divide by $H\Delta x\Delta y\Delta t$ and let $\Delta x, \Delta y$ and Δt go to zero to get

$$
C_t = (DC_x)_x + (DC_y)_y. \tag{5.2.1}
$$

This is analogous to the heat equation with diffusion of heat in two directions.

5.2.3   Model

The SIR model will be modified in two ways. First, assume the infected do not recover but eventually die at a rate $-bI$. Second, assume the infected population disperses in two directions according to Fick's motion law, and the susceptible population does not disperse. The unknown populations for the susceptible and the infected will be functions of time and space in two directions, and the $S(x, y, t)$ and $I(x, y, t)$ will now be concentrations of the susceptible and the infected populations.

SI with Dispersion in 2D Epidemic Model.

$$
\begin{array}{rcll}
S_t & = & -aSI \text{ with } S(x, y, 0) = S_0 \text{ and } 0 \leq x, y \leq L, & (5.2.2) \\
I_t & = & aSI - bI + DI_{xx} + DI_{yy} \text{ with } I(x, y, 0) = I_0, & (5.2.3) \\
I_x(0, y, t) & = & 0 = I_x(L, y, t) \text{ and} & (5.2.4) \\
I_y(x, 0, t) & = & 0 = I_y(x, L, t). & (5.2.5)
\end{array}
$$

In order to solve for the infected population, which has two space derivatives in its differential equation, boundary conditions on the infected population must be imposed. Here we have simply required that no inflected can move in or out of the left and right boundaries (5.2.4), and the front and back boundaries (5.2.5).

5.2.4   Method

Discretize (5.2.2) and (5.2.3) implicitly with respect to the time variable to obtain a sequence of partial differential equations with respect to x and y

$$
\begin{array}{rcll}
S^{k+1} & = & S^k - \Delta t\, aS^{k+1}I^{k+1} & (5.2.6) \\
I^{k+1} & = & I^k + \Delta t\, aS^{k+1}I^{k+1} - \Delta t\, bI^{k+1} & \\
& & +\Delta t\, DI_{xx}^{k+1} + \Delta t\, DI_{yy}^{k+1}. & (5.2.7)
\end{array}
$$

The space variables will be discretized by using centered differences, and the space grid will be slightly different from using half cells at the boundary. Here we will use $\Delta x = L/(n-1) = \Delta y = h$ and not $\Delta x = L/n$, and will use artificial nodes outside the domain as indicated in Figure 5.2.1 where $n = 4$ with a total of $(n-1)^2 = 9$ interior grid points and $4n = 16$ artificial grid points.

At each time step we must solve (5.2.6) and (5.2.7). Let $S = S^{k+1}$ and $I = I^{k+1}$ be approximated by $S_{i,j}$ and $I_{i,j}$ where $1 \leq i, j \leq n+1$ so that there are $(n+1)^2$ unknowns. The equations for the artificial nodes are derived from the derivative boundary conditions (5.2.4) and (5.2.5):

$$
I_{1,j} = I_{2,j}, \ I_{n+1,j} = I_{n,j}, \ I_{i,1} = I_{i,2} \text{ and } I_{i,n+1} = I_{i,n}. \qquad (5.2.8)
$$

The equations for $S_{i,j}$ with $2 \leq i, j \leq n$ follow from (5.2.6):

$$
0 = G_{i,j} \equiv S_{i,j} - \overline{S}_{i,j} + \Delta t\, aS_{i,j}I_{i,j}. \qquad (5.2.9)
$$

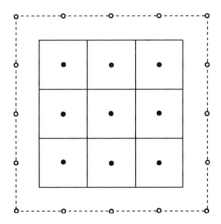

Figure 5.2.1: Grid with Artificial Grid Points

Equations for $I_{i,j}$ with $2 \leq i, j \leq n$ follow from (5.2.7) with $\alpha = \Delta t\, D/h^2$:

$$
\begin{aligned}
0 &= H_{i,j} \equiv I_{i,j} - \overline{I}_{i,j} - \Delta t\, a S_{i,j} I_{i,j} + \Delta t\, b I_{i,j} \\
&\quad -\alpha(I_{i-1,j} + I_{i,j-1} - 4I_{i,j} + I_{i+1,j} + I_{i,j+1}).
\end{aligned} \tag{5.2.10}
$$

Next use (5.2.8) to modify (5.2.10) for the nodes on the grid boundary. For example, if $i = j = 2$, then

$$
\begin{aligned}
H_{i,j} &\equiv I_{i,j} - \overline{I}_{i,j} - \Delta t\, a S_{i,j} I_{i,j} + \Delta t\, b I_{i,j} \\
&\quad -\alpha(-2I_{i,j} + I_{i+1,j} + I_{i,j+1}).
\end{aligned} \tag{5.2.11}
$$

Do this for all four corners and four sides in the grid boundary to get the final version of $H_{i,j}$. The nonlinear system of equations that must be solved at each time step has the form $\mathbf{F}(S, I) = \mathbf{0}$ where $\mathbf{F} : \mathbb{R}^{2(n-1)^2} \longrightarrow \mathbb{R}^{2(n-1)^2}$ and $\mathbf{F}(S, I) = (G, H)$.

Newton's method is used to solve for S and I. The Jacobian matrix is

$$
\mathbf{F}' = \begin{bmatrix} A & E \\ \widetilde{F} & C \end{bmatrix} = \begin{bmatrix} G_S & G_I \\ H_S & H_I \end{bmatrix}. \tag{5.2.12}
$$

$G_S = \frac{\partial G}{\partial S}$, $G_I = \frac{\partial G}{\partial I}$ and $H_S = \frac{\partial H}{\partial S}$ are diagonal matrices whose components are $1 + \Delta t\, a I_{i,j}$, $\Delta t\, a S_{i,j}$ and $-\Delta t\, a I_{i,j}$, respectively. $H_I = \frac{\partial H}{\partial I}$ is block tridiagonal with the off diagonal blocks being diagonal and the diagonal blocks being tridiagonal. For example, for $n = 4$

$$
H_I = \frac{\partial H}{\partial I} = \begin{bmatrix} C_{11} & C_{12} & 0 \\ C_{21} & C_{22} & C_{23} \\ 0 & C_{32} & C_{33} \end{bmatrix} \text{ where} \tag{5.2.13}
$$

$$C_{12} = C_{21} = C_{23} = C_{32} = \begin{bmatrix} -\alpha & 0 & 0 \\ 0 & -\alpha & 0 \\ 0 & 0 & -\alpha \end{bmatrix},$$

$$\beta_{i,j} = 1 - \Delta t \, a S_{i,j} + \Delta t \, b,$$

$$C_{11} = \begin{bmatrix} \beta_{2,2} + \alpha 2 & -\alpha & 0 \\ -\alpha & \beta_{3,2} + \alpha 3 & -\alpha \\ 0 & -\alpha & \beta_{4,2} + \alpha 2 \end{bmatrix},$$

$$C_{22} = \begin{bmatrix} \beta_{2,3} + \alpha 3 & -\alpha & 0 \\ -\alpha & \beta_{3,3} + \alpha 4 & -\alpha \\ 0 & -\alpha & \beta_{4,3} + \alpha 3 \end{bmatrix} \quad \text{and}$$

$$C_{33} = \begin{bmatrix} \beta_{2,4} + \alpha 2 & -\alpha & 0 \\ -\alpha & \beta_{3,4} + \alpha 3 & -\alpha \\ 0 & -\alpha & \beta_{4,4} + \alpha 2 \end{bmatrix}.$$

Newton's method for this problem has the form, with m denoting the Newton iteration and not the time step,

$$\begin{bmatrix} S^{m+1} \\ I^{m+1} \end{bmatrix} = \begin{bmatrix} S^m \\ I^m \end{bmatrix} - \begin{bmatrix} \Delta S \\ \Delta I \end{bmatrix} \quad \text{where} \tag{5.2.14}$$

$$\begin{bmatrix} G_S & G_I \\ H_S & H_I \end{bmatrix} \begin{bmatrix} \Delta S \\ \Delta I \end{bmatrix} = \begin{bmatrix} G(S^m, I^m) \\ H(S^m, I^m) \end{bmatrix}. \tag{5.2.15}$$

The solution of (5.2.15) can be easily found using the Schur complement because G_S, G_I and H_S are diagonal matrices. Use an elementary block matrix to zero the block (2,1) position in the 2×2 matrix in (5.2.15) so that in the following $\mathbf{I}$ is an $(n-1)^2 \times (n-1)^2$ identity matrix

$$\begin{bmatrix} \mathbf{I} & 0 \\ -H_S(G_S)^{-1} & \mathbf{I} \end{bmatrix} \begin{bmatrix} G_S & G_I \\ H_S & H_I \end{bmatrix} \begin{bmatrix} \Delta S \\ \Delta I \end{bmatrix} = \begin{bmatrix} \mathbf{I} & 0 \\ -H_S(G_S)^{-1} & \mathbf{I} \end{bmatrix} \begin{bmatrix} G \\ H \end{bmatrix}$$

$$\begin{bmatrix} G_S & G_I \\ 0 & H_I - H_S(G_S)^{-1}G_I \end{bmatrix} \begin{bmatrix} \Delta S \\ \Delta I \end{bmatrix} = \begin{bmatrix} G \\ H - H_S(G_S)^{-1}G \end{bmatrix}.$$

The matrix $H_I - H_S(G_S)^{-1}G_I$ is a pentadiagonal matrix with the same nonzero pattern as associated with the Laplace operator. So, the solution for ΔI can be done by a sparse conjugate gradient method

$$(H_I - H_S(G_S)^{-1}G_I) \Delta I = H - H_S(G_S)^{-1}G. \tag{5.2.16}$$

Once ΔI is known, then solve for ΔS

$$(G_S) \Delta S = G - G_I \Delta I. \tag{5.2.17}$$

5.2.5   Implementation

The MATLAB code SIDiff2d.m for (5.2.2)-(5.2.5) uses an implicit time discretization and finite difference in the space variables. This results in a sequence

of nonlinear problems $G(S, I) = 0$ and $H(S, I) = 0$ as indicated in equations
(5.2.8)-(5.2.11). In the code lines 1-30 initialize the data. Line 7 indicates three
m-files that are used, but are not listed. Line 29 defines the initial infected
concentration to be 48 near the origin. The time loop is in lines 31-72. Newton's method is executed in lines 32-58. The Jacobian is computed in lines
33-48. The coefficient matrix, the Schur complement, in equation (5.2.16) is
computed in line 45, and the right side of equation (5.2.16) is computed in
line 46. The linear system is solved in line 49 by the preconditioned conjugate
gradient method, which is implemented in the pcgssor.m function file and was
used in Section 4.3. Equation (5.2.17) is solved in line 50. The solution is in
line 51 and extended to the artificial grid points using (5.2.8) and the MATLAB
code update_bc.m. Lines 52 and 53 are the Newton updates for the solution
at a fixed time step. The output is given in graphical form for each time step
in lines 62-71.

MATLAB Code SIDiff2d.m

```
1.    clear;
2.    % This code is for susceptible/infected population.
3.    % The infected may disperse in 2D via Fick's law.
4.    % Newton's method is used.
5.    % The Schur complement is used on the Jacobian matrix.
6.    % The linear solve steps use a sparse pcg.
7.    % Uses m-files coeff_in_laplace.m, update_bc.m and pcgssor.m
8.    sus0 = 50;
9.    inf0 = 0;
10.   a = 20/50;
11.   b = 1;
12.   D = 10000;
13.   n = 21;
14.   maxk = 80;
15.   dx = 900./(n-1);
16.   x =dx*(0:n);
17.   dy = dx;
18.   y = x;
19.   T = 3;
20.   dt = T/maxk;
21.   alpha = D*dt/(dx*dx);
22.   coeff_in_laplace;     % define the coefficients in cs
23.   G = zeros(n+1);       % equation for sus (susceptible)
24.   H = zeros(n+1);       % equation for inf (infected)
25.   sus = ones(n+1)*sus0;    % define initial populations
26.   sus(1:3,1:3) = 2;
27.   susp = sus;
28.   inf = ones(n+1)*inf0;
29.   inf(1:3,1:3) = 48;
```

```
30.    infp = inf;
31.    for k = 1:maxk       % begin time steps
32.         for m =1:20        % begin Newton iteration
33.              for j = 2:n        % compute sparse Jacobian matrix
34.                   for i = 2:n
35.                        G(i,j) = sus(i,j) - susp(i,j) +
                                  dt*a*sus(i,j)*inf(i,j);
36.                        H(i,j) = inf(i,j) - infp(i,j) + b*dt*inf(i,j) - ...
37.                                  alpha*(cw(i,j)*inf(i-1,j)
                                  +ce(i,j)* inf(i+1,j)+ ...
38.                                  cs(i,j)*inf(i,j-1)+ cn(i,j)* inf(i,j+1)
                                  -cc(i,j)*inf(i,j))- ...
39.                                  a*dt*sus(i,j)*inf(i,j);
40.                        ac(i,j) = 1 + dt*b+alpha*cc(i,j)-dt*a*sus(i,j);
41.                        ae(i,j) = -alpha*ce(i,j);
42.                        aw(i,j) = -alpha*cw(i,j);
43.                        an(i,j) = -alpha*cn(i,j);
44.                        as(i,j) = -alpha*cs(i,j);
45.                        ac(i,j) = ac(i,j)-(dt*a*sus(i,j))*
                                  (-dt*a*inf(i,j))/(1+dt*a*inf(i,j));
46.                        rhs(i,j) = H(i,j) - (-dt*a*inf(i,j))*G(i,j)/
                                  (1+dt*a*inf(i,j));
47.                   end
48.              end
49.              [dinf , mpcg]= pcgssor(an,as,aw,ae,ac,inf,rhs,n);
50.              dsus(2:n,2:n) = G(2:n,2:n)-
                                  (dt*a*sus(2:n,2:n)).*dinf(2:n,2:n);
51.              update_bc;        % update the boundary values
52.              sus = sus - dsus;
53.              inf = inf - dinf;
54.              error = norm(H(2:n,2:n));
55.              if error<.0001
56.                   break;
57.              end
58.         end        % Newton iterations
59.         susp = sus;
60.         infp = inf;
61.         time(k) = k*dt;
62.         current_time = time(k)
63.         mpcg
64.         error
65.         subplot(1,2,1)
66.         mesh(x,y,inf)
67.         title('infected')
68.         subplot(1,2,2)
```

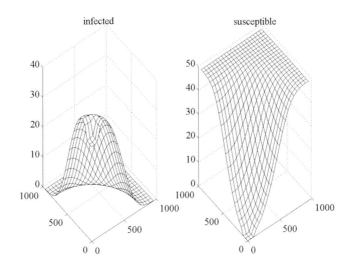

Figure 5.2.2: Infected and Susceptible at Time $= 0.3$

69.		mesh(x,y,sus)
70.		title('susceptible')
71.		pause
72.	end	%time step

Figure 5.2.2 indicates the population versus space for time equal to 0.3. The infected population had an initial concentration of 48 near $x = y = 0$. The left graph is for the infected population, and the peak is moving in the positive x and y directions. The concentration of the infected population is decreasing for small values of x and y, because the concentration of the susceptible population is nearly zero, as indicated in the right graph. This is similar to the one dimensional model of the previous section, see Figure 5.1.1.

5.2.6 Assessment

If populations do disperse according to Fick's law, then one must be careful to estimate the dispersion coefficient D and to understand the consequences of using this estimate. This is also true for the coefficients a and b. Populations can disperse in all three directions, and the above coefficients may not be constants. Furthermore, dispersion can also be a result of the population being carried in a fluid such as water or air, see exercise 7.

5.2.7 Exercises

1.   Duplicate the calculations in Figure 5.2.2. Examine the solution as time increases.

2.   Experiment with the step sizes in the MATLAB code SIDiff2d.m: $n = 11, 21, 41$ and 81 and $kmax = 40, 80, 160$ and 320.

3.   Experiment with the contagious coefficient in the MATLAB code SIDiff2d.m: $a = 1/10, 2/10, 4/10$ and $8/10$.

4.   Experiment with the death coefficient in the MATLAB code SIDiff2d.m: $b = 1/2, 1, 2$ and 4.

5.   Experiment with the dispersion coefficient in the MATLAB code SIDiff2d.m: $D = 5000, 10000, 20000$ and 40000.

6.   Let an epidemic be dispersed by Fick's law as well as by a flow in a lake whose velocity is $v = (v_1, v_2)$. The following is a modification of (5.2.3) to take this into account

$$I_t = aSI - bI + DI_{xx} + DI_{xx} - v_1 I_x - v_2 I_y.$$

Formulate a numerical model and modify the MATLAB code SIDiff2d.m. Study the effect of variable lake velocities.

5.3   Image Restoration

5.3.1   Introduction

In the next two sections images, which have been blurred and have noise, will be reconstructed so as to reduce the effects of this type of distortion. Applications may be from space exploration, security cameras, medical imaging and fish finders. There are a number of models for this, but one model reduces to the solution of a quasilinear elliptic partial differential equation, which is similar to the steady state heat conduction model that was considered in Section 4.3. In both sections the Picard iterative scheme will also be used. The linear solves for the 1D problems will be done directly, and for the 2D problem the conjugate gradient method will be used.

5.3.2   Application

On a rainy night the images that are seen by a person driving a car or airplane are distorted. One type of distortion is blurring where the lines of sight are altered. Another distortion is from equipment noise where additional random sources are introduced. Blurring can be modeled by a matrix product, and noise can be represented by a random number generator. For uncomplicated images the true image can be modeled by a one dimensional array so that the distorted image will be a matrix times the true image plus a random column vector

$$d \equiv K f_{true} + \eta. \tag{5.3.1}$$

The goal is to approximate the true image given the distorted image so that the residual

$$r(f) = d - Kf \tag{5.3.2}$$

is small and the approximate image given by f has a minimum number of erroneous curves.

5.3.3 Model

The blurring of a point i from point j will be modeled by a *Gaussian distribution* so that the blurring is

$$k_{ij}f_j \text{ where } k_{ij} = hC\exp((-(i-j)h)^2)/2\gamma^2). \tag{5.3.3}$$

Here $h = \Delta x$ is the step size and the number of points is suitably large. The cumulative effect at point i from all other points is given by summing with respect to j

$$[Kf]_i = \sum_j k_{ij}f_j. \tag{5.3.4}$$

At first glance the goal is to solve

$$Kf = d = Kf_{true} + \eta. \tag{5.3.5}$$

Since η is random with some bounded norm and K is often ill-conditioned, any variations in the right side of (5.3.5) may result in large variations in the solution of (5.3.5).

One possible resolution of this is to place an extra condition on the would-be solution so that unwanted oscillations in the restored image will not occur. One measure of this is the total variation of a discrete image. Consider a one dimensional image given by a column vector f whose components are function evaluations with respect to a partition $0 = x_0 < x_1 \cdots < x_n = L$ of an interval $[0\ L]$ with $\Delta x = x_i - x_{i-1}$. For this partition the *total variation* is

$$TV(f) = \sum_{j=1}^{n} \left| \frac{f_j - f_{j-1}}{\Delta x} \right| \Delta x. \tag{5.3.6}$$

As indicated in the following simple example the total variation can be viewed as a measure of the vertical portion of the curve. The total variation does depend on the choice of the partition, but for partitions with large n this can be a realistic estimate.

Example. Consider the three images in Figure 6.3.1 given by $n = 4, L = 4$ and $h = 1$ and

$$f = [1\ 3\ 2\ 3\ 1]^T$$
$$g = [1\ 3\ 4\ 3\ 1]^T$$
$$h = [1\ 3\ 3\ 3\ 1]^T.$$

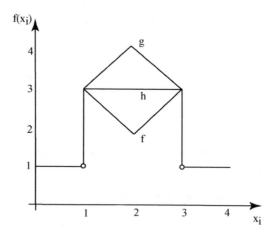

Figure 5.3.1: Three Curves with Jumps

Then the total variations are $TV(f) = 6, TV(g) = 6$ and $TV(h) = 4$ so that h, which is "flatter" than f or g, has the smallest total variation.

The following model attempts to minimize both the residual (5.3.2) and the total variation (5.3.6).

Tikhonov-TV Model for Image Restoration.

Let $r(f)$ be from (5.3.2), $TV(f)$ be from (5.3.6) and α be a given positive real number. Find $f \in \mathbb{R}^{n+1}$ so that the following real valued function is a minimum

$$T(f) = \frac{1}{2}\, r(f)^T r(f) + \alpha TV(f). \qquad (5.3.7)$$

The solution of this minimization problem can be attempted by setting all the partial derivatives of $T(f)$ with respect to f_i equal to zero and solving the resulting nonlinear system. However, the total variation term has an absolute value function in the summation, and so it does not have a derivative! A "fix" for this is to approximate the absolute value function by another function that has a continuous derivative. An example of such a function is

$$|t| \approx (t^2 + \beta^2)^{1/2}.$$

So, an approximation of the total variation uses $\Psi(t) = 2(t + \beta^2)^{1/2}$ and is

$$TV(f) \approx J_\beta(f) \equiv \frac{1}{2}\sum_{j=1}^{n} \Psi((\frac{f_j - f_{j-1}}{\Delta x})^2)\Delta x. \qquad (5.3.8)$$

The choice of the positive real numbers α and β can have significant impact on the model.

Modified Tikhonov-TV Model for Image Restoration.

Let α and β be given positive real numbers. Find $f \in \mathbb{R}^{n+1}$ so that the following real valued function is a minimum

$$T_{\alpha,\beta}(f) = \frac{1}{2} r(f)^T r(f) + \alpha J_\beta(f). \tag{5.3.9}$$

5.3.4 Method

In order to find the minimum of $T_{\alpha,\beta}(f)$, set the partial derivatives with respect to the components of f_i equal to zero. In the one dimensional case assume at the left and right boundary

$$f_0 = f_1 \text{ and } f_n = f_{n+1}. \tag{5.3.10}$$

Then there will be n unknowns and n nonlinear equations

$$\frac{\partial}{\partial f_i} T_{\alpha,\beta}(f) = 0. \tag{5.3.11}$$

Theorem 5.3.1 *Let (5.3.10) hold and use the gradient notation so that $grad(T_{\alpha,\beta}(f))$ as an $n \times 1$ column vector whose i components are $\frac{\partial}{\partial f_i}T_{\alpha,\beta}(f)$. Then*

$$
\begin{aligned}
grad(T_{\alpha,\beta}(f)) &= -K^T (d - Kf) + \alpha L(f)f \text{ where} \tag{5.3.12}\\
L(f) &\equiv D^T diag(\Psi' (D_i f))D \ \Delta x \ , \ i = 2, \cdots, n\\
D &\equiv (n-1) \times n \text{ with } -1/\Delta x \text{ on the diagonal}\\
&\quad \text{and } 1/\Delta x \text{ on the superdiagonal}\\
&\quad \text{and zero else where and}\\
D_i f &\equiv \frac{f_i - f_{i-1}}{\Delta x}.
\end{aligned}
$$

Proof. $grad(T_{\alpha,\beta}(f)) = grad(\frac{1}{2} r(f)^T r(f)) + \alpha grad(J_\beta(f))$.
First, we show $grad(\frac{1}{2} r(f)^T r(f)) = -K^T r(f)$.

$$
\begin{aligned}
\frac{\partial}{\partial f_i} \frac{1}{2} r(f)^T r(f) &= \frac{\partial}{\partial f_i} \frac{1}{2} \sum_l (d_l - \sum_j k_{lj} f_j)^2\\
&= \sum_l (d_l - \sum_j k_{lj} f_j)^{2-1} \frac{\partial}{\partial f_i}(d_l - \sum_j k_{lj} f_j)\\
&= \sum_l (d_l - \sum_j k_{lj} f_j)(0 - \sum_j k_{lj} \frac{\partial}{\partial f_i} f_j)\\
&= \sum_l (d_l - \sum_j k_{lj} f_j)(0 - k_{li})\\
&= -[K^T r(f)]_i.
\end{aligned}
$$

Second, the identity $grad(J_\beta(f)) = L(f)f$ is established for $n = 4$.

$$
\begin{aligned}
\frac{\partial}{\partial f_1} J_\beta(f) &= \frac{\partial}{\partial f_1} \frac{1}{2} \sum_{j=1}^{n} \Psi(((f_j - f_{j-1})/\Delta x)^2) \Delta x \\
&= \frac{1}{2} \Psi'((\frac{f_1 - f_0}{\Delta x})^2)) \frac{\partial}{\partial f_1} ((\frac{f_1 - f_0}{\Delta x})^2) \Delta x + \\
&\quad \frac{1}{2} \Psi'((\frac{f_2 - f_1}{\Delta x})^2)) \frac{\partial}{\partial f_1} ((\frac{f_2 - f_1}{\Delta x})^2) \Delta x \\
&= \Psi'((\frac{f_1 - f_0}{\Delta x})^2))(\frac{f_1 - f_0}{\Delta x}) \frac{1}{\Delta x} \Delta x + \\
&\quad \Psi'((\frac{f_2 - f_1}{\Delta x})^2))(\frac{f_2 - f_1}{\Delta x}) \frac{-1}{\Delta x} \Delta x \\
&= 0 + \Psi'((D_2 f)^2)) D_2 f \frac{-1}{\Delta x} \Delta x \qquad (5.3.13)
\end{aligned}
$$

$$
\begin{aligned}
\frac{\partial}{\partial f_2} J_\beta(f) &= \frac{\partial}{\partial f_2} \frac{1}{2} \sum_{j=1}^{n} \Psi((\frac{f_j - f_{j-1}}{\Delta x})^2) \Delta x \\
&= \frac{1}{2} \Psi'((\frac{f_2 - f_2}{\Delta x})^2)) \frac{\partial}{\partial f_2} ((\frac{f_2 - f_1}{\Delta x})^2) \Delta x + \\
&\quad \frac{1}{2} \Psi'((\frac{f_3 - f_2}{\Delta x})^2)) \frac{\partial}{\partial f_2} ((\frac{f_3 - f_2}{\Delta x})^2) \Delta x \\
&= \Psi'((D_2 f)^2)) D_2 f \frac{1}{\Delta x} \Delta x + \\
&\quad \Psi'((D_3 f)^2)) D_3 f \frac{-1}{\Delta x} \Delta x \qquad (5.3.14)
\end{aligned}
$$

$$
\begin{aligned}
\frac{\partial}{\partial f_3} J_\beta(f) &= \Psi'((D_3 f)^2)) D_3 f \frac{1}{\Delta x} \Delta x + \\
&\quad \Psi'((D_4 f)^2)) D_4 f \frac{-1}{\Delta x} \Delta x \qquad (5.3.15)
\end{aligned}
$$

$$
\frac{\partial}{\partial f_4} J_\beta(f) = \Psi'((D_4 f)^2)) D_4 f \frac{1}{\Delta x} \Delta x + 0. \qquad (5.3.16)
$$

The matrix form of the above four lines (5.2.13)-(5.3.16) with $\Psi'_i \equiv \Psi'((D_i f)^2))$ is

$$grad(J_\beta(f)) = \frac{1}{\Delta x} \begin{bmatrix} \Psi'_2 & -\Psi'_2 & & \\ -\Psi'_2 & \Psi'_2 + \Psi'_3 & -\Psi'_3 & \\ & -\Psi'_3 & \Psi'_3 + \Psi'_4 & -\Psi'_4 \\ & & -\Psi'_4 & \Psi'_4 \end{bmatrix} \begin{bmatrix} f_1 \\ f_2 \\ f_3 \\ f_4 \end{bmatrix}$$

$$= D^T \begin{bmatrix} \Psi'_2 & & \\ & \Psi'_3 & \\ & & \Psi'_4 \end{bmatrix} \Delta x D \begin{bmatrix} f_1 \\ f_2 \\ f_3 \\ f_4 \end{bmatrix} \quad \text{where} \quad (5.3.17)$$

$$D \equiv \begin{bmatrix} -1/\Delta x & 1/\Delta x & & \\ & -1/\Delta x & 1/\Delta x & \\ & & -1/\Delta x & 1/\Delta x \end{bmatrix}.$$

■

The identity in (5.3.12) suggests the use of the Picard algorithm to solve

$$grad(T_{\alpha,\beta}(f)) = -K^T (d - Kf) + \alpha L(f)f = 0. \qquad (5.3.18)$$

Simply evaluate $L(f)$ at the previous approximation of f and solve for the next approximation

$$\begin{aligned} -K^T \left(d - Kf^{m+1}\right) + \alpha L(f^m) f^{m+1} &= 0. \qquad (5.3.19) \\ (K^T K + \alpha L(f^m)) f^{m+1} &= K^T d \\ (K^T K + \alpha L(f^m))(f^{m+1} - f^m + f^m) &= K^T d \\ (K^T K + \alpha L(f^m))(\Delta f + f^m) &= K^T d \\ (K^T K + \alpha L(f^m)) \Delta f &= K^T d - (K^T K + \alpha L(f^m)) f^m. \end{aligned}$$

Picard Algorithm for the Solution of $-K^T (d - Kf) + \alpha L(f)f = 0.$

Let f^0 be the initial approximation of the solution
for $m = 0$ to $\max k$
    evaluate $L(f^m)$
    solve $(K^T K + \alpha L(f^m)) \Delta f = K^T d - (K^T K + \alpha L(f^m)) f^m$
    $f^{m+1} = \Delta f + f^m$
    test for convergence
endloop.

The solve step can be done directly if the algebraic system is not too large, and this is what is done for the following implementation of the one space dimension problem. Often the Picard's method tends to "converge" very slowly. Newton's method is an alternative scheme, which has many advantages.

5.3.5   Implementation

The MATLAB code image_1d.m makes use of the MATLAB Setup1d.m and function psi_prime.m files. Lines 14-26 initialize the Picard method. The call to Setup1d.m in line 14 defines the true image and distorts it by blurring and noise. The Picard method is executed in lines 27-43. The matrix H is defined in line 30, and the right hand side g is defined in line 31. The solve step is done in line 32 by $H \backslash g$, and the Picard update is done in line 33. The output to the second position in figure(1) is generated at each Picard step in lines 36-46.

MATLAB Codes image_1d.m and Setup1d.m

```
1.     % Variation on MATLAB code written by Curt Vogel,
2.     % Dept of Mathematical Sciences,
3.     % Montana State University,
4.     % for Chapter 8 of the SIAM Textbook,
5.     % "Computational Methods for Inverse Problems".
6.     %
7.     % Use Picard fixed point iteration to solve
8.     % grad(T(u)) = K'*(K*u-d) + alpha*L(u)*u = 0.
9.     % At each iteration solve for newu = u+du
10.    % (K'*K + alpha*L(u)) * newu = K'*d,
11.    % where
12.    % L(u) = grad(J(u)) =( D'*
13.    % diag(psi'(|[D*u]_i|^2,beta) * D * dx
14.    Setup1d       % Defines true image and distorts it
15.    alpha = .030      % Regularization parameter alpha
16.    beta = .01       %TV smoothing parameter beta
17.    fp_iter = 30;      % Number of fixed point iterations
18.    % Set up discretization of first derivative operator.
19.    D = zeros(n-1,n);
20.    for i =1:n-1
21.    D(i,i) = -1./h;
22.    D(i,i+1) = 1./h;
23.    end;
24.    % Initialization.
25.    dx = 1 / n;
26.    u_fp = zeros(n,1);
27.    for k = 1:fp_iter
28.        Du_sq = (D*u_fp).^2;
29.        L = D' * diag(psi_prime(Du_sq,beta)) * D * dx;
30.        H = K'*K + alpha*L;
31.        g = -H*u_fp + K'*d;
32.        du = H \ g;
33.        u_fp = u_fp + du;
34.        du_norm = norm(du)
```

```
35.            % Plot solution at each Picard step
36.            figure(1)
37.            subplot(1,2,2)
38.            plot( x,u_fp,'-')
39.            xlabel('x axis')
40.            title('TV Regularized Solution (-)')
41.            pause;
42.            drawnow
43.      end       % for fp_iter
44.      plot(x,f_true,'--', x,u_fp,'-')
45.      xlabel('x axis')
46.      title('TV Regularized Solution (-)')

1.       % MATLAB code Setup1d.m
2.       % Variation on MATLAB code written by Curt Vogel,
3.       % Dept of Mathematical Sciences,
4.       % Montana State University,
5.       % for Chapter 1 of the SIAM Textbook,
6.       % "Computational Methods for Inverse Problems".
7.       %
8.       % Set up a discretization of a convolution
9.       % integral operator K with a Gaussian kernel.
10.      % Generate a true solution and convolve it with the
11.      % kernel. Then add random error to the resulting data.
12.      % Set up parameters.
13.      clear;
14.      n = 100;        % nunber of grid points ;
15.      sig = .05;        % kernel width sigma
16.      err_lev = 10;        % input Percent error in data
17.      % Set up grid.
18.      h = 1/n;
19.      x = [h/2:h:1-h/2]';
20.      % Compute matrix K corresponding to convolution
                with Gaussian kernel.
21.      C=1/sqrt(pi)/sig
22.      for i = 1:n
23.          for j = 1:n
24.              K(i,j) = h*C* exp(-((i-j)*h)^2/(sig^2));
25.          end
26.      end
27.      % Set up true solution f_true and data d = K*f_true + error.
28.      f_true = .75*(.1<x&x<=.25) +.5*(.25<x&x<=.35)...
                +0.7*(.35<x&x<=.45) + .10*(.45<x&x<=.6)...
                +1.2*(.6<x&x<=.66)+1.6*(.66<x&x<=.70)
                +1.2*(.70<x&x<=.80)...
```

$$+1.6*(.80<x\&x<=.84)+1.2*(.84<x\&x<=.90)...$$
$$+0.3*(.90<x\&x<=1.0);$$

```
29.     Kf = K*f_true;
30.     % Define random error
31.     randn('state',0);
32.     eta = err_lev/100 * norm(Kf) * randn(n,1)/sqrt(n);
33.     d = Kf + eta;
34.     % Display the data.
35.     figure(1)
36.     subplot(1,2,1)
37.     %plot(x,f_true,'-', x,d,'o',x,Kf,'--')
38.     plot(x,d,'o')
39.     xlabel('x axis')
40.     title('Noisy and Blurred Data')
41.     pause

function s = psi_prime(t,beta)
    s = 1 ./ sqrt(t + beta^2);
```

Figure 5.3.2 has the output from image_1d.m. The left graph is generated by Setup1d.m where the parameter in line 16 of Setup1d.m controls the noise level. Line 28 in Setup1d.m defines the true image, which is depicted by the dashed line in the right graph of Figure 5.3.2. The solid line in the right graph is the restored image. The reader may find it interesting to experiment with the choice of α, β and n so as to better approximate the true image.

5.3.6   Assessment

Even for the cases where we know the true image, the "best" choice for the parameters in the modified Tikhonov-TV model is not clear. The convergence criteria range from a judgmental visual inspection of the "restored" image to monitoring the step error such as in line 34 of image_1d.m. The Picard scheme converges slowly, and other alternatives include variations on Newton's method. The absolute value function in the total variation may be approximated in other ways than using the square root function. Furthermore, total variation is not the only way to eliminate unwanted effects in the "restored" image. The interested reader should consult Curt Vogel's book [26] for a more complete discussion of these topics.

5.3.7   Exercises

1.   Duplicate the computations in Figure 5.3.2 and use different numbers of Picard iterations.
2.   Experiment with $n = 20, 40, 60$ and 80.
3.   Experiment with $\beta = 0.001, 0.010, 0.050$ and 0.10.
4.   Experiment with $\alpha = 0.01, 0.03, 0.08$ and 0.80.

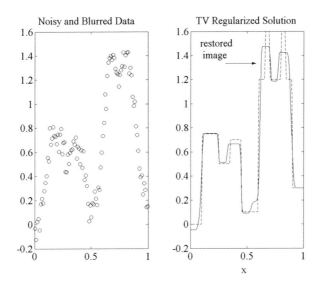

Figure 5.3.2: Restored 1D Image

5.   Experiment with different noise levels as given in line 16 in Setup1d.m.
6.   Experiment with different images as defined in line 28 in Setup1d.m.
7.   Verify lines (5.3.15) and (5.3.16).

5.4   Restoration in 2D

5.4.1   Introduction

In the previous section images that were piecewise functions of a single variable were represented by one dimensional arrays. If the image is more complicated so that the curves within the image are no longer a function of one variable, then one must use two dimensional arrays. For example, if the image is a solid figure, let the array have values equal to 100 if the array indices are inside the figure and value equal to zero if the array indices are outside the figure. Of course images have a number of attributes at each point or pixel, but for the purpose of this section we assume the arrays have nonnegative real values.

5.4.2   Application

The goal is to consider a distorted image, given by blurring and noise, and to reconstruct the two dimensional array so as to minimize the distortion and to preserve the essential attributes of the true image. The outline of the procedure will follow the previous section. This will be possible once the two dimensional array for the true image is converted to a single dimensional column vector. This is done by stacking the columns of the two dimensional arrays. For example, if the two dimensional image is a 3×3 array $f = [f_1 \ f_2 \ f_3]$ where f_j are now 3×1 column vectors, then let the bold version of f be a 9×1 column vector $\mathbf{f} = [f_1^T \ f_2^T \ f_3^T]^T$.

Let f_{true} be the $n \times n$ true image so that $\mathbf{f}_{true}$ is a $n^2 \times 1$ column vector. The the distorted image is a $n^2 \times n^2$ matrix times the true image plus a random $n^2 \times 1$ column vector

$$d \equiv K\mathbf{f}_{true} + \eta. \tag{5.4.1}$$

The goal is to approximate the true image given the distorted image so that the residual

$$r(\mathbf{f}) = d - K\mathbf{f} \tag{5.4.2}$$

is small and the approximate image given by $\mathbf{f}$ has a minimum number of erroneous surface oscillations.

5.4.3   Model

In order to minimize surface oscillations, a two dimensional version of the total variation is introduced. Consider a two dimensional image given by a matrix f whose components are function evaluations with respect to a partition of a square $[0 \ L] \times [0 \ L]$ $0 = x_0 < x_1 \cdots < x_n = L$ with $y_i = x_i$ and $h = \Delta y = \Delta x = x_i - x_{i-1}$. For this partition the *total variation* is

$$TV(f) = \sum_{i=1}^{n} \sum_{j=1}^{n} ((\frac{f_{i,j} - f_{i-1,j}}{\Delta x})^2 + (\frac{f_{i,j} - f_{i,j-1}}{\Delta y})^2)^{1/2} \Delta y \Delta x. \tag{5.4.3}$$

The total variation does depend on the choice of the partition, but for large partitions this can be a realistic estimate.

The total variation term has a square root function in the summation, and so it does not have a derivative at zero! Again a "fix" for this is to approximate the square root function by another function that has a continuous derivative such as

$$t^{1/2} \approx (t + \beta^2)^{1/2}.$$

So an approximation of the total variation uses $\Psi(t) = 2(t + \beta^2)^{1/2}$ and is

$$J_\beta(f) \equiv \frac{1}{2} \sum_{i=1}^{n} \sum_{j=1}^{n} \Psi((\frac{f_{i,j} - f_{i-1,j}}{\Delta x})^2 + (\frac{f_{i,j} - f_{i,j-1}}{\Delta y})^2) \Delta y \Delta x. \tag{5.4.4}$$

The choice of the positive real numbers α and β can have significant impact on the model.

Modified Tikhonov-TV Model for Image Restoration.

Let α and β be given positive real numbers. Find $f \in \mathbb{R}^{(n+1) \times (n+1)}$ so that the following real valued function is a minimum

$$T_{\alpha,\beta}(f) = \frac{1}{2} \ r(\mathbf{f})^T r(\mathbf{f}) + \alpha J_\beta(f). \tag{5.4.5}$$

5.4.4 Method

In order to find the minimum of $T_{\alpha,\beta}(f)$, set the partial derivatives with respect to the components of $f_{i,j}$ equal to zero. As in the one dimensional case assume at the boundary for $i,j = 1, \cdots, n$

$$f_{0,j} = f_{1,j}, \ f_{n,j} = f_{n+1,j}, \ f_{i,0} = f_{i,1} \text{ and } f_{i,n+1} = f_{i,n}. \tag{5.4.6}$$

Then there will be n^2 unknowns and n^2 nonlinear equations

$$\frac{\partial}{\partial f_{i,j}} T_{\alpha,\beta}(f) = 0. \tag{5.4.7}$$

The proof of the following theorem is similar to the one dimensional version, Theorem 5.3.1.

Theorem 5.4.1 *Let (5.4.6) hold and use the gradient notation* $grad(T_{\alpha,\beta}(f)$ *as a* $n^2 \times 1$ *column vector whose components are* $\frac{\partial}{\partial f_{i,j}} T_{\alpha,\beta}(f)$.

$$
\begin{aligned}
grad(T_{\alpha,\beta}(f)) &= -K^T (d - Kf) + \alpha L(f)f \text{ where} \qquad (5.4.8)\\
L(f) &\equiv (D^{xT} diag(\Psi'\,(D_i^x f))D^x \\
&\quad + D^{yT} diag(\Psi'\,(D_j^y f))D^y)\ \Delta x\ \Delta y\\
&\quad D^x \text{ and } D^y \text{ are } (n-1)^2 \times n^2 \text{ matrices via}\\
D_i^x f &\equiv \frac{f_{i,j} - f_{i-1,j}}{\Delta x} \text{ and } D_j^y f \equiv \frac{f_{i,j} - f_{i,j-1}}{\Delta y}\\
i,j &= 2, \cdots, n.
\end{aligned}
$$

Equations (5.4.7) and (5.4.8) require the solution of n^2 nonlinear equations for n^2 unknowns. As in the one dimensional case the Picard method is used.

Picard Algorithm for the Solution of $-K^T (d - Kf) + \alpha L(f)f = 0.$

Let f^0 be the initial approximation of the solution
for $m = 0$ to max k
    evaluate $L(f^m)$
    solve $(K^T K + \alpha L(f^m))\Delta f = K^T d - (K^T K + \alpha L(f^m))f^m$
    $f^{m+1} = \Delta f + f^m$
    test for convergence
endloop.

The solve step is attempted using the conjugate gradient iterative method. In the following implementation this inner iteration does not converge, but the outer iteration will still converge slowly!

5.4.5 Implementation

The following MATLAB code image_2d uses additional MATLAB files that are not listed: Setup2d.m, cgcrv.m, integral_op.m ,psi.m and psi_prime.m. Lines

1-38 initialize the data, the blurring matrix, and the true and distorted images, which are graphed in figure(1). The Picard iteration is done in lines 39-94, and the relative error is computed in line 95. The conjugate gradient method is used in lines 54 and 55 where an enhanced output of the "convergence" is given in figure(2), see lines 66-82. The Picard update is done in lines 56 and 57. Lines 83-89 complete figure(1) where the restored images are now graphed, see Figure 5.4.1. Lines 91-93 generate figure(4), which is a one dimensional plot of a cross-section.

MATLAB Code image_2d.m

```
1.     % Variation on MATLAB code written by Curt Vogel,
2.     % Dept of Mathematical Sciences,
3.     % Montana State University,
4.     % for Chapter 8 of the SIAM Textbook,
5.     % "Computational Methods for Inverse Problems".
6.     %
7.     % Use Picard fixed point iteration to solve
8.     % grad(T(u)) = K'*(K*u-d) + alpha*L(u)*u = 0.
9.     % At each iteration solve for newu = u+du
10.    % (K'*K + alpha*L(u)) * newu = K'*d where
11.    % L(u) =( D'* diag(psi'(|[D*u]_i|^2,beta) * D * dx
12.    Setup2d % Defines true2d image and distorts it
13.    max_fp_iter = input(' Max. no. of fixed point iterations = ');
14.    max_cg_iter = input(' Max. no. of CG iterations = ');
15.    cg_steptol = 1e-5;
16.    cg_residtol = 1e-5;
17.    cg_out_flag = 0; % If flag = 1, output CG convergence info.
18.    reset_flag = input(' Enter 1 to reset; else enter 0: ');
19.    if exist('f_alpha','var')
20.         e_fp = [];
21.    end
22.    alpha = input(' Regularization parameter alpha = ');
23.    beta = input(' TV smoothing parameter beta = ');
24.    % Set up discretization of first derivative operators.
25.    n = nfx;
26.    nsq = n^2;
27.    Delta_x = 1 / n;
28.    Delta_y = Delta_x;
29.    D = spdiags([-ones(n-1,1) ones(n-1,1)], [0 1], n-1,n) / Delta_x;
30.    I_trunc1 = spdiags(ones(n-1,1), 0, n-1,n);
31.    Dx1 = kron(D,I_trunc1);      % Forward differencing in x
32.    Dy1 = kron(I_trunc1,D);      % Forward differencing in y
33.    % Initialization.
34.    k_hat_sq = abs(k_hat).^2;
35.    Kstar_d = integral_op(dat,conj(k_hat),n,n); % Compute K'*dat.
```

```
36.     f_fp = zeros(n,n);
37.     fp_gradnorm = [];
38.     snorm_vec = [];
39.     for fp_iter = 1:max_fp_iter
40.         % Set up regularization operator L.
41.         fvec = f_fp(:);
42.         psi_prime1 = psi_prime((Dx1*fvec).^2
                                   + (Dy1*fvec).^2, beta);
43.         Dpsi_prime1 = spdiags(psi_prime1, 0, (n-1)^2,(n-1)^2);
44.         L1 = Dx1' * Dpsi_prime1 * Dx1
                         + Dy1' * Dpsi_prime1 * Dy1;
45.         L = L1 * Delta_x * Delta_y;
46.         KstarKf = integral_op(f_fp,k_hat_sq,n,n);
47.         Matf_fp =KstarKf(:)+ alpha*(L*f_fp(:));
48.         G = Matf_fp - Kstar_d(:);
49.         gradnorm = norm(G);
50.         fp_gradnorm = [fp_gradnorm; gradnorm];
51.         % Use CG iteration to solve linear system
52.         % (K'*K + alpha*L)*Delta_f = r
53.         fprintf(' ... solving linear system using cg iteration ... \n');
54.         [Delf,residnormvec,stepnormvec,cgiter] = ...
55.                 cgcrv(k_hat_sq,L,alpha,-G,max_cg_iter,
                                   cg_steptol,cg_residtol);
56.         Delta_f = reshape(Delf,n,n);
57.         f_fp = f_fp + Delta_f     % Update Picard iteration
58.         snorm = norm(Delf);
59.         snorm_vec = [snorm_vec; snorm];
60.         if exist('f_alpha','var')
61.             e_fp = [e_fp; norm(f_fp - f_alpha,'fro')
                               /norm(f_alpha,'fro')];
62.         end
63.         % Output fixed point convergence information.
64.         fprintf(' FP iter=%3.0f, ||grad||=%6.4e,
                           ||step||=%6.4e, nCG=%3.0f\n', ...
65.                         fp_iter, gradnorm, snorm, cgiter);
66.         figure(2)
67.         subplot(221)
68.             semilogy(residnormvec/residnormvec(1),'o')
69.             xlabel('CG iteration')
70.             title('CG Relative Residual Norm')
71.         subplot(222)
72.             semilogy(stepnormvec,'o')
73.             xlabel('CG iteration')
74.             title('CG Relative Step Norm')
75.         subplot(223)
```

```
76.                   semilogy([1:fp_iter],fp_gradnorm,'o-')
77.                   xlabel('Fixed Point Iteration')
78.                   title('Norm of FP Gradient')
79.               subplot(224)
80.                   semilogy([1:fp_iter],snorm_vec,'o-')
81.                   xlabel('Fixed Point Iteration')
82.                   title('Norm of FP Step')
83.               figure(1)
84.               subplot(223)
85.                   imagesc(f_fp), colorbar
86.                   title('Restoration')
87.               subplot(224)
88.                   mesh(f_fp), colorbar
89.                   title('Restoration')
90.               figure(4)
91.                   plot([1:nfx]',f_fp(ceil(nfx/2),:), [1:nfx]',
                                  f_true(ceil(nfx/2),:))
92.                   title('Cross Section of Reconstruction')
93.               drawnow
94.           end     % for fp_iter
95.           rel_soln_error = norm(f_fp(:)-f_true(:))/norm(f_true(:))
```

Figure 5.4.1 has the output from figure(1) in the above code. The upper left
graph is the true image, and the upper right is the distorted image. The lower
left is the restored image after 30 Picard iterations with 10 inner iterations of
conjugate gradient; $n = 100$, $\alpha = 1.0$ and $\beta = 0.1$ were used. The graph in the
lower right is a three dimensional mesh plot of the restored image.

5.4.6   Assessment

Like the one dimensional case, (i) the "best" choice for the parameters in the
modified Tikhonov-TV model is not clear, (ii) the convergence criteria range
from a judgmental visual inspection of the "restored" image to monitoring the
step error, (iii) the Picard scheme converges slowly and (iv) the total variation
is not the only way to eliminate unwanted effects in the "restored" image. In
the two dimensional case the conjugate gradient method was used because of
the increased size of the algebraic system. In the above calculations this did
not appear to converge, and here one should be using some preconditioner to
accelerate convergence. The interested reader should consult Curt Vogel's book
[26] for a more complete discussion of these topics. Also, other methods for
image restoration are discussed in M. Bertero and P. Boccacci [3].

5.4.7   Exercises

1.   Duplicate the computations in Figure 5.4.1. Use different numbers of
Picard and conjugate gradient iterations.

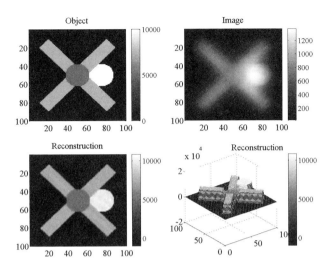

Figure 5.4.1: Restored 2D Image

2.   Experiment with $n = 20, 60, 100$ and 120.
3.   Experiment with $\beta = 0.05, 0.10, 0.50$ and 5.00.
4.   Experiment with $\alpha = 0.1, 1.0, 5.0$ and 10.0.
5.   Experiment with different noise levels as given in Setup2d.m.
6.   For the special case $n = 4$ prove the identity (5.4.8) in Theorem 5.4.1.
It might be helpful to execute image_2d with $n = 4$, and then to examine the
matrices in Theorem 5.4.1.

5.5   Option Contract Models

5.5.1   Introduction

Option contracts are agreements between two parties to buy or sell an underlying asset at a particular price on or before a given date. The underlying asset may be physical quantities or stocks or bonds or shares in a company. The value of the underlying asset may change with time. For example, if a farmer owns 100 tons of corn, and a food producer agrees to purchase this for a given price within six months, then the given price remains fixed but the price of corn in the open market may change during this six months! If the market price of corn goes down, then the option contract for the farmer has more value. If the market price of corn goes up, then the value of the option contract for the food producer goes up. Option contracts can also be sold and purchased, and here the overall goal is to estimate the value of an option contract.

5.5.2   Application

We will focus on a particular option contract called an *American put option*. This option contract obligates the writer of the contract to buy an underlying asset from the holder of the contract at a particular price, called the *exercise or strike price, E*. This must be done within a given time, called an *expiration date, T*. The holder of the option contract may or may not sell the *underlying asset* whose market value, S, will vary with time. The value of the American put option contract to the holder will vary with time, t, and S. If S gets large or if t gets close to T, then the value of the American put option contract, $P(S,t)$, will decrease. On the other hand, if S gets small, then the value of the American put option contract will increase towards the exercise price, that is, $P(S,t)$ will approach E as S goes to zero. If time exceeds expiration date, then the American put option will be worthless, that is, $P(S,t) = 0$ for $t > T$.

The objective is to determine the value of the American put option contract as a function of S, t and the other parameters E, T, r (the interest rate) and σ (the market volatility), which will be described later. In particular, the holder of the contract would like to know when is the "best" time to exercise the American put option contract. If the market value of the underlying contract is well above the exercise price, then the holder may want to sell on the open market and not to the writer of the American put option contract. If the market price of the underlying asset continues to fall below the exercise price, then at some "point" the holder will want to sell to the writer of the contract for the larger exercise price. Since the exercise price is fixed, the holder will sell as soon as this "point" is reached so that the money can be used for additional investment. This "point" refers to a particular market value $S = S_f(t)$, which is also unknown and is called the *optimal exercise price.*

The writers and the holders of American put option contracts are motivated to enter into such contracts by speculation of the future value of the underlying asset and by the need to minimize risk to a portfolio of investments. If an investor feels a particular asset has an under-valued market price, then entering into American put option contracts has a possible value if the speculation is that the market value of the underlying asset may increase. However, if an investor feels the underlying asset has an over-priced market value, then the investor may speculate that the underlying asset will decrease in value and may be tempted to become holder of an American put option contract.

The need to minimize risk in a portfolio is also very important. For example, suppose a portfolio has a number of investments in one sector of the economy. If this sector expands, then the portfolio increases in value. If this sector contracts, then this could cause some significant loss in value of the portfolio. If the investor becomes a holder of American put option contracts with some of the portfolio's assets as underlying assets, then as the market values of the assets decrease, the value of the American put option will increase. A proper distribution of underlying investments and option contracts can minimize risk to a portfolio.

5.5.3 Model

The value of the payoff from exercising the American put option is either the exercise price minus the market value of the underlying asset, or zero, that is, the *payoff* is $\max(E-S,0)$. The value of the American put option must almost always be greater than or equal to the payoff. This follows from the following risk free scheme: buy the underlying asset for S, buy the American put option contract for $P(S,t) < \max(E-S,0)$, and then immediately exercise this option contract for E, which would result in a profit $E-P-S > 0$. As this scheme is very attractive to investors, it does not exist for a very long time, and so one simply requires $P(S,t) \geq \max(E-S,0)$. In summary, the following conditions are placed on the value of the American put option contract for $t \leq T$. The boundary conditions and condition at time equal to T are

$$P(0,t) = E \tag{5.5.1}$$
$$P(L,t) = 0 \text{ for } L \gg E \tag{5.5.2}$$
$$P(S,T) = \max(E-S,0). \tag{5.5.3}$$

The inequality constraints are

$$P(S,t) \geq \max(E-S,0) \tag{5.5.4}$$
$$P_t(S,t) \leq 0 \text{ and } P_t(S_f+,t) = 0 \tag{5.5.5}$$
$$P(S_f(t),t) = E - S_f(t) \tag{5.5.6}$$
$$P(S,t) = E - S \text{ for } S < S_f(t) \tag{5.5.7}$$
$$\frac{d}{dt}S_f(t) > 0 \text{ exists.} \tag{5.5.8}$$

The graph of $P(S,t)$ for fixed time should have the form given in Figure 5.5.1.

The partial derivative of $P(S,t)$ with respect to S needs to be continuous at S_f so that the left and right derivatives must both be equal to -1. This needs some justification. From (5.5.7) $P_S(S_f-,t) = -1$ so we need to show $P_S(S_f+,t) = -1$. Since $P(S_f(t),t) = E - S_f(t)$,

$$\frac{d}{dt}P(S_f(t),t) = 0 - \frac{d}{dt}S_f(t)$$
$$P_S(S_f+,t)\frac{d}{dt}S_f(t) + P_t(S_f+,t) = -\frac{d}{dt}S_f(t)$$
$$P_t(S_f+,t) = -(1 + P_S(S_f+,t))\frac{d}{dt}S_f(t).$$

Since $P_t(S_f+,t) = 0$ and $\frac{d}{dt}S_f(t) > 0, 1 + P_S(S_f+,t) = 0$.

In the region in Figure 5.5.1 where $P(S,t) > \max(E-S,0)$, the value of the option contract must satisfy the celebrated Black-Scholes partial differential equation where r is the interest rate and σ is the volatility of the market for a particular underlying asset

$$P_t + \frac{\sigma^2}{2}S^2 P_{SS} + rSP_S - rP = 0. \tag{5.5.9}$$

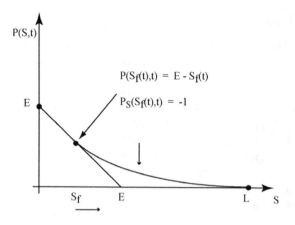

Figure 5.5.1: Value of American Put Option

The derivation of this equation is beyond the scope of this brief introduction to option contracts. The *Black-Scholes equation* differs from the partial differential equation for heat diffusion in three very important ways. First, it has variable coefficients. Second, it is a backward time problem where $P(S,t)$ is given at a future time $t = T$ as $P(S,T) = \max(E - S, 0)$. Third, the left boundary where $S = S_f(t)$ is unknown and varies with time.

Black-Scholes Model for the American Put Option Contract.

$$
\begin{aligned}
P(S,T) &= \max(E - S, 0) & (5.5.10)\\
P(0,t) &= E & (5.5.11)\\
P(L,t) &= 0 \text{ for } L \gg E & (5.5.12)\\
P(S_f(t),t) &= E - S_f(t) & (5.5.13)\\
P_S(S_f\pm,t) &= -1 & (5.5.14)\\
P(S,t) &\geq \max(E - S, 0) & (5.5.15)\\
P &= E - S \text{ for } S \leq S_f(t) & (5.5.16)\\
P_t + \frac{\sigma^2}{2}S^2 P_{SS} + rSP_S - rP &= 0 \text{ for } S > S_f(t). & (5.5.17)
\end{aligned}
$$

The volatility σ is an important parameter that can change with time and can be difficult to approximate. If the volatility is high, then there is more uncertainty in the market and the value of an American put option contract should increase. Generally, the market value of an asset will increase with time according to

$$
\frac{d}{dt}S = \mu S.
$$

The parameter μ can be approximated by using past data for $S_k = S(k\Delta t)$

$$\frac{S_{k+1} - S_k}{\Delta t} = \mu_k S_k.$$

The approximation for μ is given by an average of all the μ_k

$$\bar{\mu} = \frac{1}{K} \sum_{k=0}^{K-1} \mu_k = \frac{1}{K\Delta t} \sum_{k=0}^{K-1} \frac{S_{k+1} - S_k}{S_k}. \tag{5.5.18}$$

The *volatility* is the square root of the unbiased variance of the above data

$$\bar{\sigma}^2 = \frac{1}{(K-1)\Delta t} \sum_{k=0}^{K-1} \left(\frac{S_{k+1} - S_k}{S_k} - \bar{\mu}\Delta t \right)^2. \tag{5.5.19}$$

Thus, if $\bar{\sigma}^2$ is large, then one may expect in the future large variations in the market values of the underlying asset. Volatilities often range from near 0.05 for government bonds to near 0.40 for venture stocks.

5.5.4 Method

The numerical approximation to the Black-Scholes model in (5.5.10)-(5.5.17) is similar to the explicit method for heat diffusion in one space variable. Here we replace the space variable by the value of the underlying asset and the temperature by the value of the American put option contract. In order to obtain an initial condition, we replace the time variable by

$$\tau \equiv T - t. \tag{5.5.20}$$

Now abuse notation a little and write $P(S, \tau)$ in place of $P(S, t) = P(S, T - \tau)$ so that $-P_\tau$ replaces P_t in (5.5.17). Then the condition at the exercise date in (5.5.10) becomes the initial condition. With the boundary conditions in (5.5.11) and (5.5.12) one may apply the explicit finite difference method as used for the heat diffusion model to obtain P_i^{k+1} approximations for $P(i\Delta S, (k+1)\Delta\tau)$. But, the condition in (5.5.15) presents an obstacle to the value of the option. Here we simply choose the $P_i^{k+1} = \max(E - S_i, 0)$ if $P_i^{k+1} < \max(E - S_i, 0)$.

Explicit Method with Projection for (5.5.10)-(5.5.17).

Let $\alpha \equiv (\Delta\tau/(\Delta S)^2)(\sigma^2/2)$.

$$P_i^{k+1} = P_i^k + \alpha S_i^2(P_{i-1}^k - 2P_i^k + P_{i+1}^k)$$
$$+ (\Delta\tau/\Delta S)\,rS_i(P_{i+1}^k - P_i^k) - \Delta\tau\,rP_i^k$$

$$= \alpha S_i^2 P_{i-1}^k + (1 - 2\alpha S_i^2 - (\Delta\tau/\Delta S)\,rS_i - \Delta\tau\,r)P_i^k$$
$$+ (\alpha S_i^2 + (\Delta\tau/\Delta S)\,rS_i)P_{i+1}^k$$

$$P_i^{k+1} = \max(P_i^{k+1}, \max(E - S_i, 0)).$$

The conditions (5.5.13) and (5.5.14) at $S = S_f$ do not have to be explicitly implemented provided the time step $\Delta\tau$ is suitably small. This is another version of a stability condition.

Stability Condition.

$$\alpha \equiv (\Delta\tau/(\Delta S)^2)(\sigma^2/2)$$
$$1 - 2\alpha S_i^2 - (\Delta\tau/\Delta S)\,rS_i - \Delta\tau\,r > 0.$$

5.5.5   Implementation

The MATLAB code bs1d.m is an implementation of the explicit method for the American put option contract model. In the code the array x corresponds to the value of the underlying asset, and the array u corresponds to the value of the American put option contract. The time step, dt, is for the backward time step with initial condition corresponding to the exercise payoff of the option. Lines 1-14 define the parameters of the model with exercise price $E = 1.0$. The payoff obstacle is defined in lines 15-20, and the boundary conditions are given in lines 21 and 22. Lines 23 -37 are the implementation of the explicit scheme with projection to payoff obstacle given in lines 30-32. The approximate times when market prices correspond to the optimal exercise times are recorded in lines 33-35. These are the approximate points in asset space and time when (5.5.13) and (5.5.14) hold, and the output for time versus asset space is given in figure(1) by lines 38 and 39. Figure(2) generates the value of the American put option contract for four different times.

MATLAB Code bs1d.m

```
1.    % Black-Scholes Equation
2.    % One underlying asset
3.    % Explicit time with projection
4.    sig = .4
5.    r = .08;
6.    n = 100 ;
7.    maxk = 1000;
8.    f = 0.0;
9.    T = .5;
10.    dt = T/maxk;
```

```
11.    L = 2.;
12.    dx = L/n;
13.    alpha =.5*sig*sig*dt/(dx*dx);
14.    sur = zeros(maxk+1,1);
15.    % Define the payoff obstacle
16.    for i = 1:n+1
17.          x(i) = dx*(i-1);
18.          u(i,1) = max(1.0 - x(i),0.0);
19.          suro(i) = u(i,1);
20.    end
21.    u(1,1:maxk+1) = 1.0;      % left BC
22.    u(n+1,1:maxk+1) = 0.0;      % right BC
23.    % Use the explicit discretization
24.    for k = 1:maxk
25.          for i = 2:n
26.                u(i,k+1) = dt*f+...
27.                    x(i)*x(i)*alpha*(u(i-1,k)+ u(i+1,k)-2.*u(i,k))...
28.                    + u(i,k)*(1 -r*dt) ...
29.                    -r*x(i)*dt/dx*(u(i,k)-u(i+1,k));
30.                if (u(i,k+1)<suro(i))      % projection step
31.                    u(i,k+1) = suro(i);
32.                end
33.                if ((u(i,k+1)>suro(i)) & (u(i,k)==suro(i)))
34.                    sur(i) = (k+.5)*dt;
35.                end
36.          end
37.    end
38.    figure(1)
39.          plot(20*dx:dx:60*dx,sur(20:60))
40.    figure(2)
41.          %mesh(u)
42.          plot(x,u(:,201),x,u(:,401),x,u(:,601),x,u(:,maxk+1))
43.          xlabel('underlying asset')
44.          ylabel('value of option')
45.          title('American Put Option')
```

Figure 5.5.2 contains the output for the above code where the curves for the values of the American put option contracts are increasing with respect to τ (decreasing with respect to time t). Careful inspection of the curves will verify that the conditions at S_f in (5.5.13) and (5.5.14) are approximately satisfied.

Figure 5.5.3 has the curves for the value of the American put option contracts at time $t = 0.5$ and variable volatilities $\sigma = 0.4, 0.3, 0.2$ and 0.1. Note as the volatility decreases, the value of the option contract decreases towards the payoff value. This monotonicity property can be used to imply volatility parameters based on past market data.

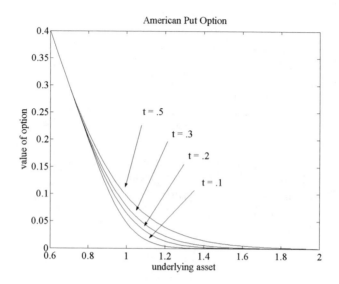

Figure 5.5.2: P(S,T-t) for Variable Times

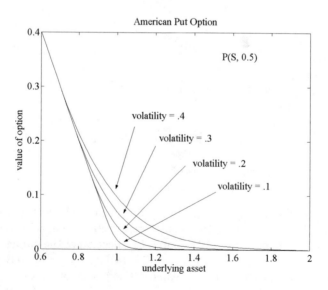

Figure 5.5.3: Option Values for Variable Volatilities

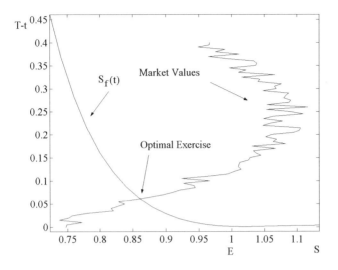

Figure 5.5.4: Optimal Exercise of an American Put

Figure 5.5.4 was generated in part by figure(1) in bs1d.m where the smooth curve represents the time when S equals the optimal exercise of the American put option contract. The vertical axis is $\tau = T - t$, and the horizontal axis is the value of the underlying asset. The non smooth curve is a simulation of the daily market values of the underlying asset. As long as the market values are above $S_f(t)$, the value of the American put option contract will be worth more than the value of the payoff, $\max(E - S, 0)$, of the American put option contract. As soon as the market value is equal to $S_f(t)$, then the American put option contract should be exercised This will generate revenue $E = 1.0$ where the t is about 0.06 before the expiration date and the market value is about $S_f(t) = 0.86$. Since the holder of the American put option contract will give up the underlying asset, the value of the payoff at this time is about $\max(E - S, 0) = \max(1.0 - 0.86, 0) = 0.14$.

5.5.6 Assessment

As usual the parameters in the Black-Scholes model may depend on time and may be difficult to estimate. The precise assumptions under which the Black-Scholes equation models option contracts should be carefully studied. There are a number of other option contracts, which are different from the American put option contract. Furthermore, there may be more than one underlying asset, and this is analogous to heat diffusion in more than one direction.

The question of convergence of the discrete model to the continuous model needs to be examined. These concepts are closely related to a well studied applied area on "free boundary value" problems, which have models in the form of variational inequalities and linear complementarity problems. Other applications include mechanical obstacle problems, heat transfer with a change in phase and fluid flow in earthen dams.

5.5.7 Exercises

1.   Experiment with variable time and asset space steps.
2.   Duplicate the calculations in Figures 5.5.2 and 5.5.3.
3.   Experiment with variable interest rates r.
4.   Experiment with variable exercise values E.
5.   Experiment with variable expiration time T, and examine figure(1) generated by bs1d.m.

5.6   Black-Scholes Model for Two Assets

5.6.1   Introduction

A portfolio of investments can have a number of assets as well as a variety of option contracts. Option contracts can have more than one underlying asset and different types of payoff functions such as illustrated in Figures 5.6.1-5.6.3. The Black-Scholes two assets model is similar to the heat diffusion model in two space variables, and the explicit time discretization will also be used to approximate the solution.

5.6.2   Application

Consider an American put option contract with two underlying assets and a payoff function $\max(E - S_1 - S_2, 0)$ where E is the exercise price and S_1 and S_2 are the values of the underlying assets. This is depicted in Figure 5.6.1 where the tilted plane is the positive part of the payoff function. The value of the put contract must be above or equal to the payoff function. The dotted curve indicates where the put value separates from the payoff function; this is analogous to the $S_f(t)$ in the one asset case. The dotted line will change with time so that as time approaches the expiration date the dotted line will move toward the line $E - S_1 - S_2 = 0 = P$. If the market values for the two underlying assets at a particular time are on the dotted line for this time, then the option should be exercised so as to optimize any profits.

5.6.3   Model

Along the two axes where one of the assets is zero, the model is just the Black-Scholes one asset model. So the boundary conditions for the two asset model must come from the solution of the one asset Black-Scholes model.

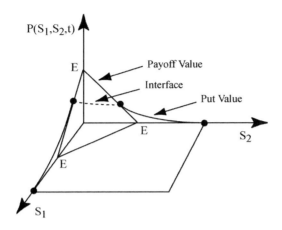

Figure 5.6.1: American Put with Two Assets

Let $P(S_1, S_2, t)$ be the value of the American put option contract. For positive values of the underlying assets the Black-Scholes equation is

$$P_t + \frac{\sigma_1^2}{2} S_1^2 P_{S_1 S_1} + \sigma_1 \sigma_1 \rho_{12} S_1 S_2 P_{S_1 S_2} + \frac{\sigma_2^2}{2} S_2^2 P_{S_2 S_2} + r S_1 P_{S_1} + r S_2 P_{S_2} - r P = 0.$$
$$(5.6.1)$$

The following initial and boundary conditions are required:

$$
\begin{aligned}
P(S_1, S_2, T) &= \max(E - S_1 - S_2, 0) & (5.6.2)\\
P(0, 0, t) &= E & (5.6.3)\\
P(L, S_2, t) &= 0 \text{ for } L \gg E & (5.6.4)\\
P(S_1, L, t) &= 0 & (5.6.5)\\
P(S_1, 0, t) &= \text{from the one asset model} & (5.6.6)\\
P(0, S_2, t) &= \text{from the one asset model.} & (5.6.7)
\end{aligned}
$$

The put contract value must be at least the value of the payoff function

$$P(S_1, S_2, t) \geq \max(E - S_1 - S_2, 0).$$
$$(5.6.8)$$

Other payoff functions can be used, and these will result in more complicated boundary conditions. For example, if the payoff function is $max(E_1 - S_1, 0.0) + max(E_2 - S_2, 0.0)$, then $P(L, S_2, t)$ and $P(S_1, L, t)$ will be nonzero solutions of two additional one asset models.

Black Scholes Model of an American Put with Two Assets.

Let the payoff function be $\max(E - S_1 - S_2, 0)$.
Require the inequality in (5.6.8) to hold.
The initial and boundary condition are in equations (5.6.2)-(5.6.7).
Either $P(S_1, S_2, t) = E - S_1 - S_2$ or $P(S_1, S_2, t)$ satisfies (5.6.1).

5.6.4   Method

The explicit time discretization with projection to the payoff obstacle will be used. Again replace the time variable by $\tau \equiv T - t$. Now abuse notation a little and write $P(S_1, S_2, \tau)$ in place of $P(S_1, S_2, t) = P(S_1, S_2, T - \tau)$ so that $-P_\tau$ replaces P_t. With the initial and boundary conditions given one may apply the explicit finite difference method as used for heat diffusion in two directions to obtain $P_{i,j}^{k+1}$ approximations for $P(i\Delta S_1, j\Delta S_2, (k+1)\Delta\tau)$. For the inequality condition simply choose the $P_{i,j}^{k+1} = \max(E - i\Delta S_1 - j\Delta S_2, 0)$ if $P_{i,j}^{k+1} < \max(E - i\Delta S_1 - j\Delta S_2, 0)$.

Explicit Method with Projection for (5.6.1)-(5.6.8).

Let $\alpha_1 \equiv (\Delta\tau/(\Delta S_1)^2)(\sigma_1^2/2)$,
$\quad \alpha_2 \equiv (\Delta\tau/(\Delta S_2)^2)(\sigma_2^2/2)$,
$\quad \alpha_{12} \equiv (\Delta\tau/(2\Delta S_1 2\Delta S_2))\rho_{12}(\sigma_1\sigma_2/2)$
$\quad S_{1i} \equiv i\Delta S_1$ and $S_{2j} \equiv j\Delta S_2$

$$P_{i,j}^{k+1} = P_{i,j}^k + \alpha_1 S_{1i}^2(P_{i-1,j}^k - 2P_{i,j}^k + P_{i+1,j}^k)$$
$$+2S_{1i}S_{2j}\alpha_{12}((P_{i+1,j+1} - P_{i-1,j+1}) - (P_{i+1,j-1} - P_{i-1,j-1}))$$
$$+\alpha_2 S_{2j}^2(P_{i,j-1}^k - 2P_{i,j}^k + P_{i,j+1}^k)$$
$$+ (\Delta\tau/\Delta S_1)\, rS_{1i}(P_{i+1,j}^k - P_{i,j}^k)$$
$$+ (\Delta\tau/\Delta S_2)\, rS_{2j}(P_{i,j+1}^k - P_{i,j}^k)$$
$$-\Delta\tau\, rP_{i,j}^k$$

$$P_{ij}^{k+1} = \max(P_{ij}^{k+1}, \max(E - i\Delta S_1 - j\Delta S_2, 0)).$$

5.6.5   Implementation

The two underlying assets model contains four one dimensional models along the boundary of the two underlying assets domain. In the interior of the domain the two dimensional Black-Scholes equation must be solved. In the MATLAB code bs2d.m the input is given in lines 1-35, and two possible payoff functions are defined in lines 26-35. The time loop is done in lines 36-124. The one dimensional Black-Scholes equation along the boundaries are solved in lines 44-57 for $y = 0$, 58-72 for $y = L$, 73-87 for $x = 0$ and 88-102 for $x = L$. For $y = 0$ the projection to the payoff obstacle is done in lines 50-52, and the times when the put value separates from the payoff obstacle is found in lines 53-56. The two dimensional Black-Scholes equation is solved for the interior nodes in lines 103-123. Lines 124-137 generate four graphs depicting the value of the put contract and the optimal exercise times, see Figures 5.6.2 and 5.6.3.

MATLAB Code bs2d.m

```
1.      % Program bs2d
2.      % Black-Scholes Equation
3.      % Two underlying assets
```

```
4.     % Explicit time with projection
5.     clear;
6.     n = 40;
7.     maxk = 1000;
8.     f = .00;
9.     T=.5;
10.    dt = T/maxk;
11.    L=4;
12.    dx = L/n;
13.    dy = L/n;
14.    sig1 = .4;
15.    sig2 = .4;
16.    rho12 = .3;
17.    E1 = 1;
18.    E2 = 1.5;
19.    total = E1 + E2;
20.    alpha1 = .5*sig1^2*dt/(dx*dx);
21.    alpha2 = .5*sig2^2*dt/(dy*dy);
22.    alpha12 = .5*sig1*sig2*rho12*dt/(2*dx*2*dy);
23.    r = .12;
24.    sur = zeros(n+1);
25.    % Insert Initial Condition
26.    for j = 1:n+1
27.          y(j) = dy*(j-1);
28.          for i = 1:n+1
29.                x(i) = dx*(i-1);
30.                %Define the payoff function
31.                u(i,j,1) = max(E1-x(i),0.0) + max(E2-y(j),0.0);
32.                % u(i,j,1) = max(total -(x(i) + y(j)),0.0);
33.                suro(i,j) = u(i,j,1);
34.          end;
35.    end;
36.    % Begin Time Steps
37.    for k = 1:maxk
38.          % Insert Boundary Conditions
39.          u(n+1,1,k+1) = E2;
40.          u(n+1,n+1,k+1) = 0.0;
41.          u(1,n+1,k+1) = E1 ;
42.          u(1,1,k+1) = total;
43.          % Do y = 0.
44.          j=1;
45.          for i = 2:n
46.                u(i,j,k+1) = dt*f+x(i)*x(i)*alpha1*...
47.                      (u(i-1,j,k) + u(i+1,j,k)-2.*u(i,j,k)) ...
48.                      +u(i,j,k)*(1 -r*dt)- ...
```

```
49.                    r*x(i)*dt/dx*(u(i,j,k)-u(i+1,j,k));
50.              if (u(i,j,k+1)<suro(i,j))
51.                   u(i,j,k+1) = suro(i,j);
52.              end
53.              if ((u(i,j,k+1)>suro(i,j))&...
54.                   (u(i,j,k)==suro(i,j)))
55.                   sur(i,j)= k+.5;
56.              end
57.          end
58.          % Do y = L.
59.          j=n+1;
60.          for i = 2:n
61.              u(i,j,k+1) = dt*f+x(i)*x(i)*alpha1*...
62.                   (u(i-1,j,k) + u(i+1,j,k)-2.*u(i,j,k)) ...
63.                   +u(i,j,k)*(1 -r*dt)- ...
64.                   r*x(i)*dt/dx*(u(i,j,k)-u(i+1,j,k));
65.              if (u(i,j,k+1)<suro(i,j))
66.                   u(i,j,k+1) = suro(i,j);
67.              end
68.              if ((u(i,j,k+1)>suro(i,j))&...
69.                   (u(i,j,k)==suro(i,j)))
70.                   sur(i,j)= k+.5;
71.              end
72.          end
73.          % Do x = 0.
74.          i=1;
75.          for j = 2:n
76.              u(i,j,k+1) = dt*f+y(j)*y(j)*alpha2*...
77.                   (u(i,j-1,k) + u(i,j+1,k)-2.*u(i,j,k))...
78.                   +u(i,j,k)*(1 -r*dt)-...
79.                   r*y(j)*dt/dy*(u(i,j,k)-u(i,j+1,k));
80.              if (u(i,j,k+1)<suro(i,j))
81.                   u(i,j,k+1) = suro(i,j);
82.              end
83.              if ((u(i,j,k+1)>suro(i,j)) &...
84.                   (u(i,j,k)==suro(i,j)))
85.                   sur(i,j)= k+.5;
86.              end
87.          end
88.          % Do x = L.
89.          i=n+1;
90.          for j = 2:n
91.              u(i,j,k+1) = dt*f+y(j)*y(j)*alpha2*...
92.                   (u(i,j-1,k) + u(i,j+1,k)-2.*u(i,j,k))...
93.                   +u(i,j,k)*(1 -r*dt)-...
```

```
94.                        r*y(j)*dt/dy*(u(i,j,k)-u(i,j+1,k));
95.                   if (u(i,j,k+1)<suro(i,j))
96.                        u(i,j,k+1) = suro(i,j);
97.                   end
98.                   if ((u(i,j,k+1)>suro(i,j))&...
99.                        (u(i,j,k)==suro(i,j)))
100.                       sur(i,j)= k+.5;
101.                  end
102.            end
103.            % Solve for Interior Nodes
104.            for j= 2:n
105.                for i = 2:n
106.                    u(i,j,k+1) = dt*f+x(i)*x(i)*alpha1*...
107.                         (u(i-1,j,k) + u(i+1,j,k)-2.*u(i,j,k))...
108.                         +u(i,j,k)*(1 -r*dt)...
109.                         -r*x(i)*dt/dx*(u(i,j,k)-u(i+1,j,k))...
110.                         +y(j)*y(j)*alpha2*...
111.                          (u(i,j-1,k) + u(i,j+1,k)-2.*u(i,j,k)) ...
112.                         -r*y(j)*dt/dy*(u(i,j,k)-u(i,j+1,k)) ...
113.                         +2.0*x(i)*y(j)*alpha12*...
114.                          (u(i+1,j+1,k)-u(i-1,j+1,k)
                              -u(i+1,j-1,k)+u(i-1,j-1,k));
115.                    if (u(i,j,k+1)<suro(i,j))
116.                         u(i,j,k+1) = suro(i,j);
117.                    end
118.                    if ((u(i,j,k+1)>suro(i,j)) &...
119.                         (u(i,j,k)==suro(i,j)))
120.                         sur(i,j)= k+.5;
121.                    end
122.                end
123.            end
124.       end
125.       figure(1)
126.       subplot(2,2,1)
127.       mesh(x,y,suro')
128.       title('Payoff Value')
129.       subplot(2,2,2)
130.       mesh(x,y,u(:,:,maxk+1)')
131.       title('Put Value')
132.       subplot(2,2,3)
133.       mesh(x,y,u(:,:,maxk+1)'-suro')
134.       title('Put Minus Payoff')
135.       subplot(2,2,4)
136.       mesh(x,y,sur'*dt)
137.       title('Optimal Exercise Times')
```

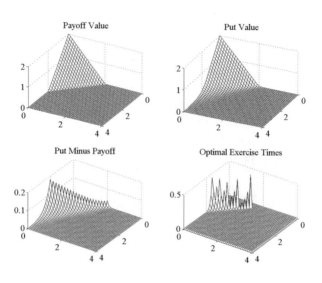

Figure 5.6.2: $\max(E_1 + E_2 - S_1 - S_2, 0)$

Figure 5.6.2 is for the *lumped payoff* $\max(E_1 + E_2 - S_1 - S_2, 0)$ with two assets with different volatilities $\sigma_1 = 0.4$ and $\sigma_2 = 0.1$. All four graphs have the underlying asset values on the x and y axes. The upper left graph is for the payoff function, and the upper right graph is for the value of the American put option contract at time equal to 0.5 before the expiration date, T. Note the difference in the graphs along the axes, which can be attributed to the larger volatility for the first underlying asset. The lower left graph depicts the difference in the put value at time equal to 0.5 and the value of the payoff. The lower right graph depicts time of optimal exercise versus the two underlying assets. Here the vertical axis has $\tau = T - t$, and the interface for the optimal exercise of the put moves towards the vertical axis as τ increases.

Figure 5.6.3 is for the *distributed payoff* $\max(E_1 - S_1, 0) + \max(E_2 - S_2, 0)$ with two assets and with equal volatilities $\sigma_1 = 0.4$ and $\sigma_2 = 0.4$. and with different exercise values $E_1 = 1.0$ and $E_2 = 1.5$. All four graphs have the underlying asset values on the x and y axes. The upper left graph is for the payoff function, and the upper right graph is for the value of the American put option contract at time equal to 0.5 before the expiration date. Note the difference in the graphs along the axes, which can be attributed to the different exercise values. The lower left graph depicts the difference in the put value at time equal to 0.5 and the value of the payoff. The lower right graph has time of optimal exercise versus the two underlying assets. Here the vertical axis has $\tau = T - t$ where T is the expiration date. There are three interface curves for possible optimal exercise times. One is inside the underlying asset region $[0 \ E_1] \times [0 \ E_2]$ and moves towards the z axis as τ increases. The other two are in

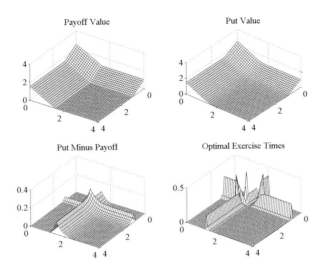

Figure 5.6.3: $\max(E_1 - S_1, 0) + \max(E_2 - S_2, 0)$

the regions $[E_1\ 4] \times [0\ E_2]$ and $[0\ E_1] \times [E_2\ 4]$ and move towards the vertical lines containing the points $(4, 0, 0)$ and $(0, 4, 0)$, respectively.

5.6.6 Assessment

The parameters in the Black-Scholes model may depend on time and may be difficult to estimate. The precise assumptions under which the Black-Scholes equation models option contracts should be carefully studied. The question of convergence of the discrete model to the continuous model needs to be examined.

The jump from one to multiple asset models presents some interesting differences from heat and mass transfer problems. The boundary conditions are implied from the solution of the one asset Black-Scholes equation. When going from one to two assets, there may be a variety of payoff or obstacle functions that will result in multiple interfaces for the possible optimal exercise opportunities. There may be many assets as contrasted to only three space directions for heat and mass transfer.

5.6.7 Exercises

1. Experiment with variable time and asset space steps.
2. Duplicate the calculations in Figures 5.6.2 and 5.6.3.
3. Experiment with other payoff functions.
4. Experiment with variable interest rates r.

5.   Experiment with variable exercise values E.
6.   Experiment with variable expiration time T.
7.   Experiment with variable volatility σ.

Chapter 6

High Performance Computing

Because many applications are very complicated such as weather prediction, there will often be a large number of unknowns. For heat diffusion in one direction the long thin wire was broken into n smaller cells and the temperature was approximated for all the time steps in each segment. Typical values for n could range from 50 to 100. If one needs to model temperature in a tropical storm, then there will be diffusion in three directions. So, if there are $n = 100$ segments in each direction, then there will be $n^3 = 100^3 = 10^6$ cells each with unknown temperatures, velocity components and pressures. Three dimensional problems present strong challenges in memory, computational units and storage of data. The first three sections are a very brief description of serial, vector and multiprocessing computer architectures. The last three sections illustrate the use of the IBM/SP and MPI for the parallel computation of matrix products and the two and three space variable models of heat diffusion and pollutant transfer. Chapter 7 contains a more detailed description of MPI and the essential subroutines for communication among the processors. Additional introductory materials on parallel computations can be found in P. S. Pacheco [21] and more advanced topics in Dongarra, Duff, Sorensen and van der Vorst [6].

6.1 Vector Computers and Matrix Products

6.1.1 Introduction

In this section we consider the components of a computer and the various ways they are connected. In particular, the idea behind a vector pipeline is introduced, and a model for speedup is presented. Applications to matrix-vector products and to heat and mass transfer in two directions will be presented. The sizes of matrix models substantially increase when the heat and mass transfer

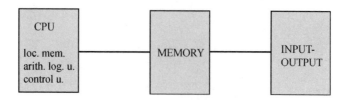

Figure 6.1.1: von Neumann Computer

in two or three directions are modeled. This is the reason for considering vector and multiprocessor computers.

The *von Neumann* definition of a computer contains three parts: main memory, input-output device and *central processing unit* (CPU). The CPU has three components: the arithmetic logic unit, the control unit and the local memory. The arithmetic logic unit does the floating point calculations while the control unit governs the instructions and data. Figure 6.1.1 illustrates a von Neumann computer with the three basic components. The local memory is small compared to the main memory, but moving data within the CPU is usually very fast. Hence, it is important to move data from the main memory to the local memory and do as much computation with this data as possible before moving it back to the main memory. Algorithms that have been optimized for a particular computer will take these facts into careful consideration.

Another way of describing a computer is the *hierarchical classification* of its components. There are three levels: the processor level with wide band communication paths, the register level with several bytes (8 bits per byte) pathways and the gate or logic level with several bits in its pathways. Figure 6.1.2 is a processor level depiction of a *multiprocessor* computer with four CPUs. The CPUs communicate with each other via the shared memory. The switch controls access to the shared memory, and here there is a potential for a bottleneck. The purpose of multiprocessors is to do more computation in less time. This is critical in many applications such as weather prediction.

Within the CPU is the arithmetic logic unit with many floating point adders, which are register level devices. A *floating point add* can be described in four distinct steps each requiring a distinct hardware segment. For example, use four digits to do a floating point add $100.1 + (-3.6)$:

CE: compare expressions $.1001 \cdot 10^3$ and $-.36 \cdot 10^1$
AE: mantissa alignment $.1001 \cdot 10^3$ and $-.0036 \cdot 10^3$
AD: mantissa add $1001 - 0036 = 0965$
NR: normalization $.9650 \cdot 10^2$.

This is depicted by Figure 6.1.3 where the lines indicate communication pathways with several bytes of data. The data moves from left to right in time intervals equal to the clock cycle time of the particular computer. If each step

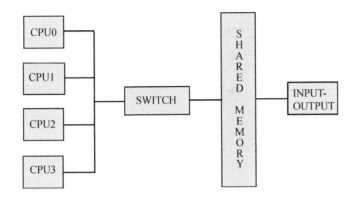

Figure 6.1.2: Shared Memory Multiprocessor

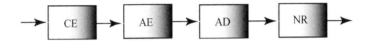

Figure 6.1.3: Floating Point Add

takes one clock cycle and the clock cycle time is 6 nanoseconds, then a floating point add will take 24 nanoseconds (10^{-9} sec.).

Within a floating point adder there are many devices that add integers. These devices typically deal with just a few bits of information and are examples of gate or logic level devices. Here the integer adds can be done by base two numbers. These devices are combinations of a small number of transistors designed to simulate truth tables that reflect basic binary operations. Table 6.1.1 indicates how one digit of a base two number with 64 digits (or bits) can be added. In Figure 6.1.4 the input is x, y and c_1 (the carry from the previous digit) and the output is z and c_0 (the carry to the next digit).

Table 6.1.1: Truth Table for Bit Adder

x	y	c_1	z	c_0
0	0	0	0	0
0	0	1	1	0
0	1	0	1	0
0	1	1	0	1
1	0	0	1	0
1	0	1	0	1
1	1	0	0	1
1	1	1	1	1

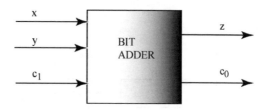

Figure 6.1.4: Bit Adder

6.1.2   Applied Area

Vector pipelines were introduced so as to make greater use of the register level hardware. We will focus on the operation of floating point addition, which requires four distinct steps for each addition. The segments of the device that execute these steps are only busy for one fourth of the time to perform a floating point add. The objective is to design computer hardware so that all of the segments will be busy most of the time. In the case of the four segment floating point adder this could give a speedup possibly close to four.

A *vector pipeline* is a register level device, which is usually in either the control unit or the arithmetic logic unit. It has a collection of distinct hardware modules or segments that execute the steps of an operation and each segment is required to be busy once the device is full. Figure 6.1.5 depicts a four segment *vector floating point adder* in the arithmetic logic unit. The first pair of floating point numbers is denoted by D1, and this pair enters the pipeline in the upper left in the figure. Segment CE on D1 is done during the first clock cycle. During the second clock cycle D1 moves to segment AE, and the second pair of floating point numbers D2 enters segment CE. Continue this process so that after three clock cycles the pipeline is full and a floating point add is produced every clock cycle. So, for large number of floating point adds with four segments the ideal speedup is four.

6.1.3   Model

A discrete model for the speedup of a particular pipeline is as follows. Such models are often used in the design phase of computers. Also they are used to determine how to best utilize vector pipelines on a selection of applications.

Vector Pipeline Timing Model.

$$
\begin{aligned}
\text{Let } K &= \text{time for one clock cycle,} \\
N &= \text{number of items of data,} \\
L &= \text{number of segments,} \\
S_v &= \text{startup time for the vector pipeline operation and} \\
S_s &= \text{startup time for a serial operation.} \\
\text{Tserial} &= \text{serial time} = S_s + (LK)N.
\end{aligned}
$$

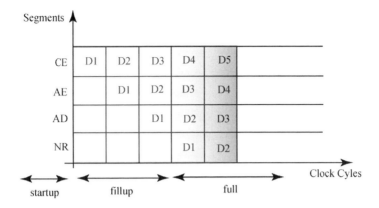

Figure 6.1.5: Vector Pipeline for Floating Point Add

$$\text{Tvector} \quad = \text{vector time} = [S_v + (L - 1)K] + KN.$$
$$\text{Speedup} = \text{Tserial/Tvector}.$$

The vector startup time is usually much larger than the serial startup time. So, for small amounts of data (small N), the serial time may be smaller than the vector time! The vector pipeline does not start to become effective until it is full, and this takes $S_v + (L - 1)K$ clock cycles. Note that the speedup approaches the number of segments L as N increases.

Another important consideration in the use of vector pipelines is the orderly and timely input of data. An example is matrix-vector multiplication and the Fortran programming language. Fortran stores arrays by listing the columns from left to right. So, if one wants to input data from the rows of a matrix, then this will be in stride equal to the length of the columns. On the other hand, if one inputs data from columns, then this will be in stride equal to one. For this reason, when vector pipelines are used to do matrix-vector products, the ji version performs much better than the ij version. This can have a very significant impact on the effective use of vector computers.

6.1.4 Implementation

In order to illustrate the benefits of vector pipelines, consider the basic matrix-vector product. The ij method uses products of rows times the column vector, and the ji method uses linear combinations of the columns. The advantage of the two methods is that it often allows one to make use of particular properties of a computer such as communication speeds and local versus main memory size. We shall use the following notation:

$x = [x_j]$ is a column vector where $j = 1, ..., n$ and
$A = [a_{i,j}]$ is a matrix where $i = 1, ..., m$ are the row numbers
and $j = 1, ..., n$ are the column numbers.

Matrix-vector Product (ij version) Ax = d.

for $i = 1, m$
$\qquad d_i = 0$
$\qquad$ for $j = 1, n$
$\qquad\qquad d_i = d_i + a_{i,j}x_j$
$\qquad$ endloop
endloop.

An alternate way to do matrix-vector products is based on the following reordering of the arithmetic operations. Consider the case where $n = m = 3$ and

$$
\begin{array}{rcl}
a_{1,1}x_1 + a_{1,2}x_2 + a_{1,3}x_3 & = & d_1 \\
a_{2,1}x_1 + a_{2,2}x_2 + a_{2,3}x_3 & = & d_2 \\
a_{3,1}x_1 + a_{3,2}x_2 + a_{3,3}x_3 & = & d_3.
\end{array}
$$

This can be written in vector form as

$$
\begin{bmatrix} a_{1,1} \\ a_{2,1} \\ a_{3,1} \end{bmatrix} x_1 +
\begin{bmatrix} a_{1,2} \\ a_{2,2} \\ a_{3,2} \end{bmatrix} x_2 +
\begin{bmatrix} a_{1,3} \\ a_{2,3} \\ a_{3,3} \end{bmatrix} x_3 =
\begin{bmatrix} d_1 \\ d_2 \\ d_3 \end{bmatrix}.
$$

In other words the product is a linear combination of the columns of the matrix. This amounts to reversing the order of the nested loop in the above algorithm. The matrix-matrix products are similar, but they will have three nested loops. Here there are 6 different orderings of the loops, and so, the analysis is a little more complicated.

Matrix-vector Product (ji version) Ax = d.

$d = 0$
for $j = 1, n$
$\qquad$ for $i = 1, m$
$\qquad\qquad d_i = d_i + a_{i,j}x_j$
$\qquad$ endloop
endloop.

The calculations in Table 6.1.2 were done on the Cray T916 at the North Carolina Supercomputing Center. All calculations were for a matrix-vector product where the matrix was 500×500 for $2(500)^2 = 5 \cdot 10^5$ floating point operations. In the ji Fortran code the inner loop was vectorized or not vectorized by using the Cray directives !dir$ vector and !dir$ novector just before the loop 30 in the following code segment:

Table 6.1.2: Matrix-vector Computation Times

Method	Time x 10^{-4} sec.
ji (vec. on, -O1)	008.86
ji (vec. off, -O1)	253.79
ij (vec. on, -O1)	029.89
ij (vec. off, -O1)	183.73
matmul (-O2)	033.45
ji (vec. on, -O2)	003.33

```
        do 20 j = 1,n
!dir$ vector
            do 30 i = 1,n
                prod(i) = prod(i) + a(i,j)*x(j)
30          continue
20      continue.
```

The first two calculations indicate a speedup of over 28 for the ji method. The next two calculations illustrate that the ij method is slower than the ji method. This is because Fortran stores numbers of a two dimensional array by columns. Since the ij method gets rows of the array, the input into the vector pipe will be in stride equal to n. The fifth calculation used the f90 intrinsic matmul for matrix-vector products. The last computation used full optimization, -O2. The loops 20 and 30 were recognized to be a matrix-vector product and an optimized BLAS2 subroutine was used. BLAS2 is a collection of basic linear algebra subroutines with order n^2 operations, see http://www.netlib.org or http://www.netlib.org/blas/sgemv.f. This gave an additional speedup over 2 for an overall speedup equal to about 76. The floating point operations per second for this last computation was $(5 \ 10^5)/ (3.33 \ 10^{-4})$ or about 1,500 megaflops.

6.1.5 Assessment

Vector pipes can be used to do a number of operations or combination of operations. The potential speedup depends on the number of segments in the pipe. If the computations are more complicated, then the speedups will decrease or the code may not be vectorized. There must be an orderly input of data into the pipe. Often there is a special memory called a register, which is used to input data into the pipe. Here it is important to make optimal use of the size and speed of these registers. This can be a time consuming effort, but many of the BLAS subroutines have been optimized for certain computers.

Not all computations have independent parts. For example, consider the following calculations $a(2) = c + a(1)$ and $a(3) = c + a(2)$. The order of the two computations will give different results! This is called a *data dependency*. Compilers will try to detect these, but they may or may not find them. Basic iterative methods such as Euler or Newton methods have data dependencies.

Some calculations can be reordered so that they have independent parts. For example, consider the traditional sum of four numbers $((a(1) + a(2)) + a(3))) + a(4)$. This can be reordered into partial sums $(a(1) + a(2)) + (a(3) + a(4))$ so that two processors can be used.

6.1.6 Exercises

1. Write a Fortran or C or MATLAB code for the ij and ji matrix vector products.

2. Write a Fortran or C or MATLAB code so that the BLAS subroutine sgemv() is used.

3. On your computing environment do calculations for a matrix-vector product similar to those in Table 6.1.2 that were done on the Cray T916. Compare the computing times.

4. Consider the explicit finite difference methods in Sections 1.2-1.5.

(a). Are the calculations in the outer time loops independent, vectorizable and why?

(b). Are the calculations in the inner space loops independent, vectorizable and why?

6.2 Vector Computations for Heat Diffusion

6.2.1 Introduction

Consider heat conduction in a thin plate, which is thermally insulated on its surface. The model of the temperature will have the form $u^{k+1} = Au^k + b$ for the time dependent case and $u = Au + b$ for the steady state case. In general, the matrix A can be extremely large, but it will also have a special structure with many more zeros than nonzero components. Here we will use vector pipelines to execute this computation, and we will also extend the model to heat diffusion in three directions.

6.2.2 Applied Area

Previously we considered the model of heat diffusion in a long thin wire and in a thin cooling fin. The temperature was a function of one or two space variables and time. A more realistic model of temperature requires it to be a function of three space variables and time. Consider a cooling fin that has diffusion in all three space directions as discussed in Section 4.4. The initial temperature of the fin will be given and one hot surface will be specified as well as the other five cooler surfaces. The objective is to predict the temperature in the interior of the fin in order to determine the effectiveness of the cooling fin.

6.2.3 Model

The model can be formulated as either a continuous model or as a discrete model. For appropriate choices of time and space steps the solutions should be close. In order to generate a 3D time dependent model for heat transfer diffusion, the Fourier heat law must be applied to the x, y and z directions. The continuous and discrete 3D models are very similar to the 2D versions. In the continuous 3D model the temperature u will depend on four variables, $u(x, y, z, t)$. In (6.2.1) $-(Ku_z)_z$ models the diffusion in the z direction where the heat is entering and leaving the top and bottom of the volume $\Delta x \Delta y \Delta z$.

Continuous 3D Model for $u = u(x, y, z, t)$.

$$\rho c u_t - (Ku_x)_x - (Ku_y)_y - (Ku_z)_z \quad = \quad f \tag{6.2.1}$$
$$u(x, y, z, 0) \quad = \quad \text{given and} \tag{6.2.2}$$
$$u(x, y, z, t) \quad = \quad \text{given on the boundary.} \tag{6.2.3}$$

Explicit Finite Difference 3D Model for $u_{i,j,l}^k \approx u(i\Delta x, j\Delta y, l\Delta z, k\Delta t)$.

$$
\begin{aligned}
u_{i,j,l}^{k+1} \quad &= \quad (\Delta t/\rho c) f_{i,j,l}^k + (1 - \alpha) u_{i,j,l}^k \\
&\quad + \Delta t/\Delta x^2 (u_{i+1,j,l}^k + u_{i-1,j,l}^k) \\
&\quad + \Delta t/\Delta y^2 (u_{i,j+1,l}^k + u_{i,j-1,l}^k) \\
&\quad + \Delta t/\Delta z^2 (u_{i,j,l+1}^k + u_{i,j,l-1}^k) \tag{6.2.4}
\end{aligned}
$$
$$\alpha \quad = \quad (K/\rho c)\Delta t(2/\Delta x^2 + 2/\Delta y^2 + 2/\Delta z^2),$$
$$i, j, l \quad = \quad 1, .., n-1 \text{ and } k = 0, .., maxk - 1,$$
$$u_{i,j,l}^0 \quad = \quad \text{given, } i, j, l = 1, .., n-1 \text{ and} \tag{6.2.5}$$
$$u_{i,j,l}^k \quad = \quad \text{given, } k = 1, ..., maxk, i, j, l \text{ on the boundary grid.} \tag{6.2.6}$$

Stability Condition.

$$1 - ((K/\rho c)\Delta t(2/\Delta x^2 + 2/\Delta y^2 + 2/\Delta z^2)) > 0.$$

6.2.4 Method

The computation of the above explicit model does not require the solution of a linear algebraic system at each time step, but it does require the time step to be suitably small so that the stability condition holds. Since there are three space directions, there are three indices associated with these directions. Therefore, there will be four nested loops with the time loop being the outer loop. The inner three loops can be in any order because the calculations for $u_{i,j,l}^{k+1}$ are independent with respect to the space indices. This means one could use vector or multiprocessing computers to do the inner loops.

Table 6.2.1: Heat Diffusion Vector Times

Loop Length	Serial Time	Vector Time	Speedup
30 (Alliant)	04.07	2.60	1.57
62 (Alliant)	06.21	2.88	2.16
126 (Alliant)	11.00	4.04	2.72
254 (Alliant)	20.70	6.21	3.33
256(Cray)	1.36	.0661	20.57
512(Cray)	2.73	.1184	23.06

6.2.5   Implementation

The following code segment for 2D diffusion was run on the Cray T916 computer. In code the Cray directive *!dir$ vector* before the beginning of loop 30 instructs the compiler to use the vector pipeline on loop 30. Note the index for loop 30 is i, and this is a row number in the array u. This ensures that the vector pipe can sequentially access the data in u.

```
        do 20 k = 1,maxit
    !dir$ vector
            do 30 i = 2,n
                u(i,k+1) = dt*f + alpha*(u(i-1,k) + u(i+1,k))
                         + (1 - 2*alpha)*u(i,k)
    30          continue
    20  continue.
```

Table 6.2.1 contains vector calculations for an older computer called the Alliant FX/40 and for the Cray T916. It indicates increased speedup as the length of loop 30 increases. This is because the startup time relative to the execution time is decreasing. The Alliant FX/40 has a vector pipe with four segments and has a limit of four for the speedup. The Cray T916 has more segments in its vector pipeline with speedups of about 20. Here the speedup for the Cray T916 is about 20 because the computations inside loop 30 are a little more involved than those in the matrix-vector product example.

The MATLAB code heat3d.m is for heat diffusion in a 3D cooling fin, which has initial temperature equal to 70, and with temperature at the boundary $x = 0$ equal to 370 for the first 50 time steps and then set equal to 70 after 50 time steps. The other temperatures on the boundary are always equal to 70. The code in heat3d.m generates a 4D array whose entries are the temperatures for 3D space and time. The input data is given in lines 1-28, the finite difference method is executed in the four nested loops in lines 37-47, and some of the output is graphed in the 3D plot, using the MATLAB command slice in line 50. The MATLAB commands slice and pause allow the user to view the heat moving from the hot mass towards the cooler sides of the fin. This is much more interesting than the single grayscale plot in Figure 6.2.1 at time 60. The coefficient, 1-alpha, in line 41 must be positive so that the stability condition holds.

MATLAB Code heat3d.m

```
1.    % Heat 3D Diffusion.
2.    % Uses the explicit method.
3.    % Given boundary conditions on all sides.
4.    clear;
5.    L = 2.0;
6.    W = 1.0;
7.    T = 1.0;
8.    Tend = 100.;
9.    maxk = 200;
10.   dt = Tend/maxk;
11.   nx = 10.;
12.   ny = 10;
13.   nz = 10;
14.   u(1:nx+1,1:ny+1,1:nz+1,1:maxk+1) = 70.; % Initial temperature.
15.   dx = L/nx;
16.   dy = W/ny;
17.   dz = T/nz;
18.   rdx2 = 1./(dx*dx);
19.   rdy2 = 1./(dy*dy);
20.   rdz2 = 1./(dz*dz);
21.   cond = .001;
22.   spheat = 1.0;
23.   rho = 1.;
24.   a = cond/(spheat*rho);
25.   alpha = dt*a*2*(rdx2+rdy2+rdz2);
26.   x = dx*(0:nx);
27.   y = dy*(0:ny);
28.   z = dz*(0:nz);
29.   for k=1:maxk+1        % Hot side of fin.
30.        time(k) = (k-1)*dt;
31.        for l=1:nz+1
32.             for i=1:nx+1
33.                  u(i,1,l,k) =300.*(time(k)<50)+ 70.;
34.             end
35.        end
36.   end
37.   for k=1:maxk        % Explicit method.
38.        for l=2:nz
39.             for j = 2:ny
40.                  for i = 2:nx
41.                       u(i,j,l,k+1) =(1-alpha)*u(i,j,l,k) ...
42.                            +dt*a*(rdx2*(u(i-1,j,l,k)+u(i+1,j,l,k))...
```

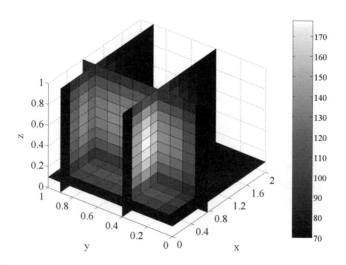

Figure 6.2.1: Temperature in Fin at t = 60

```
43.                                            +rdy2*(u(i,j-1,l,k)+u(i,j+1,l,k))...
44.                                            +rdz2*(u(i,j,l-1,k)+u(i,j,l+1,k)));
45.                        end
46.                    end
47.                end
48.            v=u(:,:,:,k);
49.            time(k)
50.            slice(x,y,z,v,.75,[.4 .9],.1)
51.            colorbar
52.            pause
53.        end
```

6.2.6   Assessment

The explicit time discretization is an example of a method that is ideally vector-izable. The computations in the inner space loops are independent so that the inner loops can be executed using a vector or multiprocessing computer. However, the stability condition on the step sizes can still be a serious constraint. An alternative is to use an implicit time discretization, but as indicated in Section 4.5 this generates a sequence of linear systems, which require some additional computations at each time step.

6.2.7 Exercises

1.   Duplicate the calculations in heat3d.m. Experiment with the slice parameters in the MATLAB command slice.

2.   In heat3d.m experiment with different time mesh sizes, $maxk = 150, 300$ and 450. Be sure to consider the stability constraint.

3.   In heat3d.m experiment with different space mesh sizes, nx or ny or $nz = 10, 20$ and 40. Be sure to consider the stability constraint.

4.   In heat3d.m experiment with different thermal conductivities $K = cond = .01, .02$ and $.04$. Be sure to make any adjustments to the time step so that the stability condition holds.

5.   Suppose heat is being generated at a rate of 3 units of heat per unit volume per unit time.

(a).   Modify heat3d.m to implement this source of heat.

(b).   Experiment with different values for this heat source $f = 0, 1, 2$ and 3.

6.3 Multiprocessors and Mass Transfer

6.3.1 Introduction

Since computations for 3D heat diffusion require four nested loops, the computational demands increase. In such cases the use of vector or multiprocessing computers could be very effective. Another similar application is the concentration of a pollutant as it is dispersed within a deep lake. Here the concentration is a function of time and three space variables. This problem, like heat diffusion in 3D, will also require more computing power. In this section we will describe and use a multiprocessing computer.

A *multiprocessing* computer is a computer with more than one "tightly" coupled CPU. Here "tightly" means that there is relatively fast communication among the CPUs. There are several classification schemes that are commonly used to describe various multiprocessors: memory, communication connections and data streams.

Two examples of the memory classification are shared and distributed. The *shared memory* multiprocessors communicate via the global shared memory, and Figure 6.1.2 is a depiction of a four processor shared memory multiprocessor. Shared memory multiprocessors often have in-code directives that indicate the code segments to be executed concurrently. The *distributed* memory multiprocessors communicate by explicit message passing, which must be part of the computer code. Figures 6.3.1 and 6.3.2 illustrate three types of distributed memory computers. In these depictions each node could have several CPUs, for example some for computation and one for communication. Another illustration is each node could be a shared memory computer, and the IBM/SP is a particular example of this.

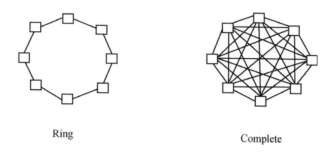

Ring   Complete

Figure 6.3.1: Ring and Complete Multiprocessors

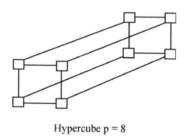

Hypercube p = 8

Figure 6.3.2: Hypercube Multiprocessor

Figure 6.3.1 contains the two extreme communication connections. The *ring* multiprocessor will have two communication links for each node, and the *complete* multiprocessor will have $p-1$ communications links per node where p is the number of nodes. If p is large, then the complete multiprocessor has a very complicated physical layout. Interconnection schemes are important because of certain types of applications. For example in a closed loop hydraulic system a ring interconnection might be the best. Or, if a problem requires a great deal of communication between processors, then the complete interconnection scheme might be appropriate. The *hypercube* depicted in Figure 6.3.2 is an attempt to work in between the extremes given by the ring and complete schemes. The hypercube has $p = 2^d$ nodes, and each node has $d = log_2(p)$ communication links.

Classification by data streams has two main categories: SIMD and MIMD. The first represents single instruction and multiple data, and an example is a vector pipeline. The second is multiple instruction and multiple data. The Cray Y-MP and the IBM/SP are examples of MIMD computers. One can send different data and different code to the various processors. However, MIMD computers are often programmed as SIMD computers, that is, the same code is executed, but different data is input to the various CPUs.

6.3.2 Applied Area

Multiprocessing computers have been introduced to obtain more rapid computations. Basically, there are two ways to do this: either use faster computers or use faster algorithms. There are natural limits on the speed of computers. Signals cannot travel any faster than the speed of light, where it takes about one nanosecond to travel one foot. In order to reduce communication times, the devices must be moved closer. Eventually, the devices will be so small that either uncertainty principles will become dominant or the fabrication of chips will become too expensive.

An alternative is to use more than one processor on those problems that have a number of independent calculations. One class of problems that have many matrix products, which are independent calculations, is to the area of visualization where the use of multiprocessors is very common. But, not all computations have a large number of independent calculations. Here it is important to understand the relationship between the number of processors and the number of independent parts in a calculation. Below we will present a timing model of this, as well as a model of 3D pollutant transfer in a deep lake.

6.3.3 Model

An important consideration is the number of processors to be used. In order to be able to effectively use p processors, one must have p independent tasks to be performed. Vary rarely is this exactly the case; parts of the code may have no independent parts, two independent parts and so forth. In order to model the effectiveness of a multiprocessor with p processors, *Amdahl's timing* model has been widely used. It makes the assumption that α is the fraction of the computations with p independent parts and the rest of the calculation $1 - \alpha$ has one independent part.

Amdahl's Timing Model.

Let p = the number of processors,
α = the fraction with p independent parts,
$1 - \alpha$ = the fraction with one independent part,
T_1 = serial execution time,
$(1 - \alpha)T_1$ = execution time for the 1 independent part and
$\alpha T_1/p$ = execution time for the p independent parts.

$$Speedup = Sp(\alpha) = \frac{T_1}{(1 - \alpha)T_1 + \alpha T_1/p} = \frac{1}{1 - \alpha + \alpha/p}. \tag{6.3.1}$$

Example. Consider a dot product of two vectors of dimension $n = 100$. There are 100 scalar products and 99 additions, and we may measure execution time in terms of operations so that $T_1 = 199$. If $p = 4$ and the dot product is broken into four smaller dot products of dimension 25, then the parallel part will have $4(49)$ operations and the serial part will require 3 operations to add the smaller

Table 6.3.1: Speedup and Efficiency

Processor	Speedup	Efficiency
2	1.8	.90
4	3.1	.78
8	4.7	.59
16	6.4	.40

dot products. Thus, $\alpha = 196/199$ and $S_4 = 199/52$. If the dimension increases to $n = 1000$, then α and S_4 will increase to $\alpha = 1996/1999$ and $S_4 = 1999/502$.

If $\alpha = 1$, then the speedup is p, the ideal case. If $\alpha = 0$, then the speedup is 1! Another parameter is the *efficiency*, and this is defined to be the speedup divided by the number of processors. Thus, for a fixed code α will be fixed, and the efficiency will decrease as the number of processors increases. Another way to view this is in Table 6.3.1 where $\alpha = .9$ and p varies from 2 to 16. If the problem size remains the same, then the decreasing efficiencies in this table are not optimistic. However, the trend is to have larger problem sizes, and so as in the dot product example one can expect the α to increase so that the efficiency may not decrease for larger problem sizes. Other important factors include communication and startup times, which are not part of Amdahl's timing model.

Finally, we are ready to present the model for the dispersion of a pollutant in a deep lake. Let $u(x, y, z, t)$ be the concentration of a pollutant. Suppose it is decaying at a rate equal to *dec* units per time, and it is being dispersed to other parts of the lake by a known fluid constant velocity vector equal to (v_1, v_2, v_3). Following the derivations in Section 1.4, but now consider all three directions, we obtain the continuous and discrete models. Assume the velocity components are nonnegative so that the concentration levels on the "upstream" sides (west, south and bottom) must be given. In the partial differential equation in the continuous 3D model the term $-v_3 u_z$ models the amount of the pollutant entering and leaving the top and bottom of the volume $\Delta x \Delta y \Delta z$. Also, assume the pollutant is also being transported by Fickian dispersion (diffusion) as modeled in Sections 5.1 and 5.2 where D is the dispersion constant. In order to keep the details a simple as possible, assume the lake is a 3D box.

Continuous 3D Pollutant Model for $u(x, y, z, t)$.

$$
\begin{aligned}
u_t \;=\;\; & D(u_{xx} + u_{yy} + u_{zz}) \\
& -v_1 u_x - v_2 u_y - v_3 u_z - dec\, u, & (6.3.2) \\
& u(x, y, z, 0) \text{ given and} & (6.3.3) \\
& u(x, y, z, t) \text{ given on the upwind boundary.} & (6.3.4)
\end{aligned}
$$

Explicit Finite Difference 3D Pollutant Model
  for $u_{i,j,l}^k \approx u(i\Delta x, j\Delta y, l\Delta z, k\Delta t)$.

$$
\begin{aligned}
u_{i,j,l}^{k+1} =\ & \Delta t D / \Delta x^2 (u_{i-1,j,l}^k + u_{i+1,j,l}^k) \\
& + \Delta t D / \Delta y^2 (u_{i,j-1,l}^k + u_{i,j+1,l}^k) \\
& + \Delta t D / \Delta z^2 (u_{i,j,l-1}^k + u_{i,j,l+1}^k) \\
& + v_1(\Delta t/\Delta x)u_{i-1,j,l}^k + v_2(\Delta t/\Delta y)u_{i,j-1,l}^k + v_3(\Delta t/\Delta z)u_{i,j,l-1}^k \\
& + (1 - \alpha - v_1(\Delta t/\Delta x) - v_2(\Delta t/\Delta y) - v_3(\Delta t/\Delta z) - \Delta t\ dec)u_{i,j,l}^k
\end{aligned}
$$
$$(6.3.5)$$

$$
\alpha \equiv \Delta t D \left(2/\Delta x^2 + 2/\Delta y^2 + 2/\Delta z^2 \right)
$$

$u_{i,j,l}^0$ given and $\hspace{6cm}$ (6.3.6)

$u_{0,j,l}^k,\ u_{i,0,l}^k,\ u_{i,j,0}^k$ given. $\hspace{4.5cm}$ (6.3.7)

Stability Condition.

$$
1 - \alpha - v_1(\Delta t/\Delta x) - v_2(\Delta t/\Delta y) - v_3(\Delta t/\Delta z) - \Delta t\ dec > 0.
$$

6.3.4   Method

In order to illustrate the use of multiprocessors, consider the 2D heat diffusion model as described in the previous section. The following makes use of High Performance Fortran (HPF), and for more details about HPF do a search on HPF at http://www.mcs.anl.gov. In the last three sections of this chapter and the next chapter we will more carefully describe the Message Passing Interface (MPI) as an alternative to HPF for parallel computations.

The following calculations were done on the Cray T3E at the North Carolina Supercomputing Center. The directives (!hpf$ ....) in lines 7-10 are for HPF. These directives disperse groups of columns in the arrays u, *unew* and *uold* to the various processors. The parallel computation is done in lines 24-28 using the *forall* "loop" where all the computations are independent with respect to the i and j indices. Also the array equalities in lines 19-21, 29 and 30 are intrinsic parallel operations

HPF Code heat2d.hpf

```
1.      program heat
2.      implicit none
3.      real, dimension(601,601,10):: u
4.      real, dimension(601,601):: unew,uold
5.      real :: f,cond,dt,dx,alpha,t0, timef,tend
6.      integer :: n,maxit,k,i,j
7.  !hpf$ processors num_proc(number_of_processors)
8.  !hpf$ distribute unew(*,block) onto num_proc
9.  !hpf$ distribute uold(*,block) onto num_proc
```

Table 6.3.2: HPF for 2D Diffusion

Processors	Times (sec.)
1	1.095
2	0.558
4	0.315

```
10.   !hpf$ distribute u(*,block,*) onto num_proc
11.      print*, 'n = ?'
12.      read*, n
13.      maxit = 09
14.      f = 1.0
15.      cond = .001
16.      dt = 2
17.      dx = .1
18.      alpha = cond*dt/(dx*dx)
19.      u =0.0
20.      uold = 0.0
21.      unew = 0.0
22.      t0 = timef
23.      do k =1,maxit
24.          forall (i=2:n,j=2:n)
25.              unew(i,j) = dt*f + alpha*(uold(i-1,j)+uold(i+1,j)
26.   $                  + uold(i,j-1) + uold(i,j+1))
27.   $                  + (1 - 4*alpha)*uold(i,j)
28.          end forall
29.          uold = unew
30.          u(:,:,k+1)=unew(:,:)
31.      end do
32.      tend =timef
33.      print*, 'time =', tend
34.      end
```

The computations given in Table 6.3.2 were for the 2D heat diffusion code. Reasonable speedups for 1, 2 and 4 processors were attained because most of the computation is independent. If the problem size is too small or if there are many users on the computer, then the timings can be uncertain or the speedups will decrease.

6.3.5   Implementation

The MATLAB code flow3d.m simulates a large spill of a pollutant, which has been buried in the bottom of a deep lake. The source of the spill is defined in lines 28-35. The MATLAB code flow3d generates the 3D array of the concentrations as a function of the x, y, z and time grid. The input data is given in lines

1-35, the finite difference method is executed in the four nested loops in lines 37-50, and the output is given in line 53 where the MATLAB command slice is used. The MATLAB commands slice and pause allow one to see the pollutant move through the lake, and this is much more interesting in color than in a single grayscale graph as in Figure 6.3.3. In experimenting with the parameters in flow3d one should be careful to choose the time step to be small enough so that the stability condition holds, that is, coeff in line 36 must be positive.

MATLAB Code flow3d.m

```
1.     % Flow with Fickian Dispersion in 3D
2.     % Uses the explicit method.
3.     % Given boundary conditions on all sides.
4.     clear;
5.     L = 4.0;
6.     W = 1.0;
7.     T = 1.0;
8.     Tend = 20.;
9.     maxk = 100;
10.    dt = Tend/maxk;
11.    nx = 10.;
12.    ny = 10;
13.    nz = 10;
14.    u(1:nx+1,1:ny+1,1:nz+1,1:maxk+1) = 0.;
15.    dx = L/nx;
16.    dy = W/ny;
17.    dz = T/nz;
18.    rdx2 = 1./(dx*dx);
19.    rdy2 = 1./(dy*dy);
20.    rdz2 = 1./(dz*dz);
21.    disp = .001;
22.    vel = [ .05 .1 .05];       % Velocity of fluid.
23.    dec = .001;       % Decay rate of pollutant.
24.    alpha = dt*disp*2*(rdx2+rdy2+rdz2);
25.    x = dx*(0:nx);
26.    y = dy*(0:ny);
27.    z = dz*(0:nz);
28.    for k=1:maxk+1       % Source of pollutant.
29.         time(k) = (k-1)*dt;
30.         for l=1:nz+1
31.              for i=1:nx+1
32.                   u(i,1,2,k) =10.*(time(k)<15);
33.              end
34.         end
35.    end
36.    coeff =1-alpha-vel(1)*dt/dx-vel(2)*dt/dy-vel(3)*dt/dz-dt*dec
```

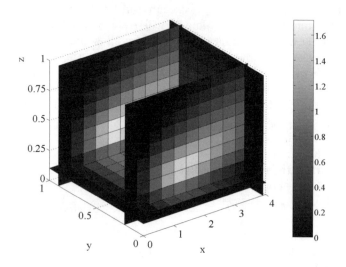

Figure 6.3.3: Concentration at t = 17

```
37.      for k=1:maxk % Explicit method.
38.          for l=2:nz
39.              for j = 2:ny
40.                  for i = 2:nx
41.                      u(i,j,l,k+1)=coeff*u(i,j,l,k) ...
42.                          +dt*disp*(rdx2*(u(i-1,j,l,k)+u(i+1,j,l,k))...
43.                          +rdy2*(u(i,j-1,l,k)+u(i,j+1,l,k))...
44.                          +rdz2*(u(i,j,l-1,k)+u(i,j,l+1,k)))...
45.                          +vel(1)*dt/dx*u(i-1,j,l,k)...
46.                          +vel(2)*dt/dy*u(i,j-1,l,k)...
47.                          +vel(3)*dt/dz*u(i,j,l-1,k);
48.                  end
49.              end
50.          end
51.          v=u(:,:,:,k);
52.          time(k)
53.          slice(x,y,z,v,3.9,[.2 .9],.1 )
54.          colorbar
55.          pause
56.      end
```

6.3.6 Assessment

The effective use of vector pipelines and multiprocessor computers will depend on the particular code being executed. There must exist independent calculations within the code. Some computer codes have a large number of independent parts and some have almost none. The use of timing models can give insight to possible performance of codes. Also, some codes can be restructured to have more independent parts.

In order for concurrent computation to occur in HPF, the arrays must be distributed and the code must be executed by either intrinsic array operations or by forall "loops" or by independent loops. There are number of provisions in HPF for distribution of the arrays among the processors, and this seems to be the more challenging step.

Even though explicit finite difference methods have many independent calculations, they do have a stability condition on the time step. Many computer simulations range over periods of years, and in such cases these restrictions on the time step may be too severe. The implicit time discretization is an alternative method, but as indicated in Section 4.5 an algebraic system must be solved at each time step.

6.3.7 Exercises

1.   Consider the dot product example of Amdahl's timing model. Repeat the calculations of the alphas, speedups and efficiencies for $n = 200$ and 400. Why does the efficiency increase?

2.   Duplicate the calculations in flow3d.m. Use the MATLAB commands mesh and contour to view the temperatures at different times.

3.   In flow3d.m experiment with different time mesh sizes, $maxk = 100, 200,$ and 400. Be sure to consider the stability constraint.

4.   In flow3d.m experiment with different space mesh sizes, nx or ny or $nz = 5, 10$ and 20. Be sure to consider the stability constraint.

5.   In flow3d.m experiment with different decay rates $dec = .01, .02$ and .04. Be sure to make any adjustments to the time step so that the stability condition holds.

6.   Experiment with the fluid velocity in the MATLAB code flow3d.m.

   (a).   Adjust the magnitudes of the velocity components and observe stability as a function of fluid velocity.

   (b).   Modify the MATLAB code flow3d.m to account for fluid velocity with negative components.

7.   Suppose pollutant is being generated at a rate of 3 units of heat per unit volume per unit time.

   (a).   How are the models for the 3D problem modified to account for this?

   (b).   Modify flow3d.m to implement this source of pollution.

   (c).   Experiment with different values for the heat source $f = 0, 1, 2$ and

3.

6.4   MPI and the IBM/SP

6.4.1   Introduction

In this section we give a very brief description of the IBM/SP multiprocessing computer that has been located at the North Carolina Supercomputing Center (http://www.ncsc.org/). One can program this computer by using MPI, and this will also be very briefly described. In this section we give an example of a Fortran code for numerical integration that uses MPI. In subsequent sections there will be MPI codes for matrix products and for heat and mass transfer.

6.4.2   IBM/SP Computer

The following material was taken, in part, from the NCSC USER GUIDE, see [18, Chapter 10]. The IBM/SP located at NCSC during early 2003 had 180 nodes. Each node contained four 375 MHz POWER3 processors, two gigabytes of memory, a high-speed switch network interface, a low-speed ethernet network interface, and local disk storage. Each node runs a standalone version of AIX, IBM's UNIX based operating system. The POWER3 processor can perform two floating-point multiply-add operations each clock cycle. For the 375 MHz processors this gives a peak floating-point performance of 1500 MFLOPS. The IBM/SP can be viewed as a distributed memory computer with respect to the nodes, and each node as a shared memory computer. Each node had four CPUs, and there are upgrades to 8 and 16 CPUs per node.

Various parallel programming models are supported on the SP system. Within a node either message passing or shared memory parallel programming models can be used. Between nodes only message passing programming models are supported. A hybrid model is to use message passing (MPI) between nodes and shared memory (OpenMP) within nodes. The latency and bandwidth performance of MPI is superior to that achieved using PVM. MPI has been optimized for the SP system with continuing development by IBM.

Shared memory parallelization is only available within a node. The C and FORTRAN compilers provide an option (-qsmp) to automatically parallelize a code using shared memory parallelization. Significant programmer intervention is generally required to produce efficient parallel programs. IBM, as well as most other computer vendors, have developed a set of compiler directives for shared memory parallelization. While the compilers continue to recognize these directives, they have largely been superseded by the OpenMP standard.

Jobs are scheduled for execution on the SP by submitting them to the Load Leveler system. Job limits are determined by user resource specifications and by the job class specification. Job limits affect the wall clock time the job can execute and the number of nodes available to the job. Additionally, the user can specify the number of tasks for the job to execute per node as well as other limits such as file size and memory limits. Load Leveler jobs are defined using a command file. Load Leveler is the recommended method for running message passing jobs. If the requested resources (wall clock time or nodes) exceed those available for the specified class, then Load Leveler will reject the job. The command file is submitted to Load Leveler with the llsubmit command. The status of the job in the queue can be monitored with the llq command.

6.4.3 Basic MPI

The MPI homepage is http://www-unix.mcs.anl.gov/mpi/index.html. There is a very nice tutorial called "MPI User Guide in Fortran" by Pacheco and Ming, which can be found at the above homepage as well as a number of other references including the text by P. S. Pacheco [21]. Here we will not present a tutorial, but we will give some very simple examples of MPI code that can be run on the IBM/SP. The essential subroutines of MPI are *include 'mpif.h'*, *mpi_init(), mpi_comm_rank(), mpi_comm_size(), mpi_send(), mpi_recv(), mpi_barrier()* and *mpi_finalize()*. Additional information about MPI's subroutines can be found in Chapter 7. The following MPI/Fortran code, trapmpi.f, is a slightly modified version of one given by Pacheco and Ming. This code is an implementation of the trapezoid rule for numerical approximation of an integral, which approximates the integral by a summation of areas of trapezoids.

The line 7 *include 'mpif.h'* makes the mpi subroutines available. The data defined in line 13 will be "hard wired" into any processors that will be used. The lines 16-18 *mpi_init(), mpi_comm_rank()* and *mpi_comm_size()* start mpi, get a processor rank (a number from 0 to p-1), and find out how many processors (p) there are available for this program. All processors will be able to execute the code in lines 22-40. The work (numerical integration) is done in lines 29-40 by grouping the trapezoids; loc_n, loc_a and loc_b depend on the processor whose identifier is my_rank. Each processor will have its own copy of loc_a, loc_b, and integral. In the i-loop in lines 31-34 the calculations are done by each processor but with different data. The partial integrations are communicated and summed by *mpi_reduce()* in lines 39-40. Line 41 uses *barrier()* to stop any further computation until all previous work is done. The call in line 55 to *mpi_finalize()* terminates the mpi segment of the Fortan code.

MPI/Fortran Code trapmpi.f

```
1.      program trapezoid
2.! This illustrates how the basic mpi commands
3.! can be used to do parallel numerical integration
4.! by partitioning the summation.
5.      implicit none
6.! Includes the mpi Fortran library.
7.      include 'mpif.h'
8.      real:: a,b,h,loc_a,loc_b,integral,total,t1,t2,x
9.      real:: timef
10.      integer:: my_rank,p,n,source,dest,tag,ierr,loc_n
11.      integer:: i,status(mpi_status_size)
```

```
12.! Every processor gets values for a,b and n.
13.       data a,b,n,dest,tag/0.0,100.0,1024000,0,50/
14.! Initializes mpi, gets the rank of the processor, my_rank,
15.! and number of processors, p.
16.      call mpi_init(ierr)
17.      call mpi_comm_rank(mpi_comm_world,my_rank,ierr)
18.      call mpi_comm_size(mpi_comm_world,p,ierr)
19.      if (my_rank.eq.0) then
20.            t1 = timef()
21.      end if
22.      h = (b-a)/n
23.! Each processor has unique value of loc_n, loc_a and loc_b.
24.      loc_n = n/p
25.      loc_a = a+my_rank*loc_n*h
26.      loc_b = loc_a + loc_n*h
27.! Each processor does part of the integration.
28.! The trapezoid rule is used.
29.      integral = (f(loc_a) + f(loc_b))*.5
30.      x = loc_a
31.      do i = 1,loc_n-1
32.            x=x+h
33.            integral = integral + f(x)
34.      end do
35.      integral = integral*h
36.! The mpi subroutine mpi_reduce() is used to communicate
37.! the partial integrations, integral, and then sum
38.! these to get the total numerical approximation, total.
39.      call mpi_reduce(integral,total,1,mpi_real,mpi_sum,0&
40.                            ,mpi_comm_world,ierr)
41.      call mpi_barrier(mpi_comm_world,ierr)
42.      if (my_rank.eq.0) then
43.            t2 = timef()
44.      end if
45.! Processor 0 prints the n,a,b,total
46.! and time for computation and communication.
47.      if (my_rank.eq.0) then
48.            print*,n
49.            print*,a
50.            print*,b
51.            print*,total
52.            print*,t2
53.      end if
54.! mpi is terminated.
55.      call mpi_finalize(ierr)
56.      contains
```

57.! This is the function to be integrated.
58.   real function f(x)
59.   implicit none
60.   real x
61.   f = x*x
62.   end function
63.   end program trapezoid

The communication command in lines 39-40 *mpi_reduce()* sends all the partial integrals to processor 0, processor 0 receives them, and sums them. This command is an efficient concatenation of following sequence of *mpi_send()* and *mpi_recv()* commands:

```
if (my_rank .eq. 0) then
     total = integral
     do source = 1, p-1
          call mpi_recv(integral, 1, mpi_real, source, tag,
                    mpi_comm_world, status, ierr)
          total = total + integral
     enddo
else
     call mpi_send(integral, 1, mpi_real, dest,
               tag, mpi_comm_world, ierr)
endif.
```

If there are a large number of processors, then the sequential source loop may take some significant time. In the *mpi_reduce()* subroutine a "tree" or "fan-in" scheme allows for the use of any available processors to do the communication. One "tree" scheme of communication is depicted in Figure 6.4.1. By going backward in time processor 0 can receive the partial integrals in $3 = \log_2(8)$ time steps. Also, by going forward in time processor 0 can send information to all the other processors in 3 times steps. In the following sections three additional collective communication subroutines (*mpi_bcast()*, *mpi_scatter()* and *mpi_gather()*) that utilize "fan-out" or "fan-in" schemes, see Figure 7.2.1, will be illustrated.

The code can be compiled and executed on the IBM/SP by the following commands:

        mpxlf90 –O3 trapmpi.f
        poe ./a.out –procs 2 –hfile cpus.

The mpxlf90 is a multiprocessing version of a Fortran 90 compiler. Here we have used a third level of optimization given by –O3. The execution of the a.out file, which was generated by the compiler, is done by the parallel operating environment command, poe. The –procs 2 indicates the number of processors to be used, and the –hfile cpus indicates that the processors in the file cpus are to be used.

A better alternative is to use Load Leveler given by the llsubmit command. In the following we simply used the command:

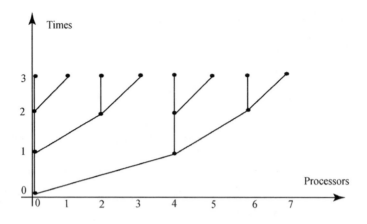

Figure 6.4.1: Fan-out Communication

Table 6.4.1: MPI Times for trapempi.f

p	Times(S_p) n = 102400	Times(S_p) n = 1024000
1	6.22(1.00)	61.59(1.00)
2	3.13(1.99)	30.87(1.99)
4	1.61(3.86)	15.48(3.98)
8	0.95(6.56)	07.89(7.81)
16	0.54(11.54)	04.24(14.54)

llsubmit envrmpi8.

This ran the compiled code with 2 nodes and 4 processors per node for a total of 8 processors. The output will be sent to the file mpijob2.out. One can check on the status of the job by using the command:

llq.

The code trapmpi.f generated Table 6.4.1 by using llsubmit with different numbers of CPUs or processors. The efficiencies (S_p/p) decrease as the number of processors increase. The amount of independent parallel computation increases as the number of trapezoids, n, increases, and so, one expects the better speedups in the third column than in the second column. If one decreases n to 10240, then the speedups for 8 and 16 processors will be very poor. This is because the communication times are very large relative to the computation times. The execution times will vary with the choice of optimization (see man xlf90) and with the number of other users (see who and llq). The reader will find it very interesting to experiment with these parameters as well as the number of trapezoids in trapmpi.f.

6.4.4 Exercises

1. Browse the www pages for the NCSC and MPI.
2. Experiment with different levels of optimization in the compiler mpxlf90.
3. Repeat the calculations in Table 6.4.1. Use additional p and n.
4. Experiment with the alternative to $mpi\_reduce()$, which uses a loop with $mpi\_send()$ and $mpi\_recv()$.
5. In trapmpi.f replace the trapezoid rule with Simpson's rule and repeat the calculations in Table 6.4.1.

6.5 MPI and Matrix Products

6.5.1 Introduction

In this section we will give examples of MPI/Fortran codes for matrix-vector and matrix-matrix products. Here we will take advantage of the column order of the arrays in Fortran. MPI communication subroutines $mpi\_reduce()$ and $mpi\_gather()$, and optimized BLAS (basic linear algebra subroutines) $sgemv()$ and $sgemm()$ will be illustrated.

6.5.2 Matrix-vector Products

The ij method uses products of rows in the $m \times n$ matrix times the column vector, and the ji method uses linear combinations of the columns in the $m \times n$ matrix. In Fortran $m \times n$ arrays are stored by columns, and so, the ji method is best because it retrieves components of the array in stride equal to one.

Matrix-Vector Product (ji version) $d^+ = d + Ax$.

$$\text{for } j = 1, n$$
$$\quad \text{for } i = 1, m$$
$$\quad\quad d_i = d_i + a_{i,j} x_j$$
$$\quad \text{endloop}$$
$$\text{endloop}.$$

A parallel version of this algorithm will group the columns of the matrix, that is, the j-loop will be partitioned so that the column sums are done by a particular processor. Let bn and en be the beginning and end of a subset of this partition, which is to be assigned to some processor. In parallel we will compute the following partial matrix-vector products

$$A(1:m, bn:en)x(bn:en).$$

Upon completion of all the partial products, they will be communicated to some processor, usually the root processor 0, and then summed.

In the MPI/Fortran code matvecmpi.f the arrays in lines 13-15 are initialized before MPI is initialized in lines 16-18, and therefore, each processor will have a copy of the array. Thus, there is no need to send data via *mpi_ bcast()* in lines 26-28; note the *mpi_ bcast()* subroutines are commented out, and they would only send the required data to the appropriate processors. The matrix-vector product is done by computing a linear combination of the columns of the matrix. The linear combination is partitioned to obtain the parallel computation. Here these calculations are done on each processor by either the BLAS2 subroutine *sgemv()* (see http://www.netlib.org /blas/sgemv.f) in line 29, or by the ji-loops in lines 30-34. Then *mpi_ reduce()* in line 36 is used to send n real numbers (a column vector) to processor 0, received by processor 0 and summed to the product vector. The mflops (million floating point operations per second) are computed in line 42 where the timings are in milliseconds and there are 1000 repetitions of the matrix-vector product.

MPI/Fortran Code matvecmpi.f

```
1.     program matvec
2.     implicit none
3.     include 'mpif.h'
4.     real,dimension(1:1024,1:4096):: a
5.     real,dimension(1:1024)::prod,prodt
6.     real,dimension(1:4096)::x
7.     real:: t1,t2,mflops
8.     real:: timef
9.     integer:: my_rank,p,n,source,dest,tag,ierr,loc_m
10.    integer:: i,status(mpi_status_size),bn,en,j,it,m
11.    data n,dest,tag/1024,0,50/
12.    m = 4*n
13.    a = 1.0
14.    prod = 0.0
15.    x = 3.0
16.    call mpi_init(ierr)
17.    call mpi_comm_rank(mpi_comm_world,my_rank,ierr)
18.    call mpi_comm_size(mpi_comm_world,p,ierr)
19.    loc_m = m/p
20.    bn = 1+(my_rank)*loc_m
21.    en = bn + loc_m - 1
22.    if (my_rank.eq.0) then
23.         t1 = timef()
24.    end if
25.    do it = 1,1000
26.    ! call mpi_bcast(a(1,bn),n*(en-bn+1),mpi_real,0,
                        mpi_comm_world,ierr)
27.    ! call mpi_bcast(prod(1),n,mpi_real,0,
                        mpi_comm_world,ierr)
28.    ! call mpi_bcast(x(bn),(en-bn+1),mpi_real,0,
                        mpi_comm_world,ierr)
```

Table 6.5.1: Matrix-vector Product mflops

p	sgemv, m = 2048	sgemv, m = 4096	ji-loops, m = 4096
1	430	395	328
2	890	843	683
4	1628	1668	1391
8	2421	2803	2522
16	3288	4508	3946

```
29.    ! call sgemv('N',n,loc_m,1.0,a(1,bn),n,x(bn),1,1.0,prod,1)
30.         do j = bn,en
31.           do i = 1,n
32.               prod(i) = prod(i) + a(i,j)*x(j)
33.           end do
34.         end do
35.       call mpi_barrier(mpi_comm_world,ierr)
36.       call mpi_reduce(prod(1),prodt(1),n,mpi_real,mpi_sum,0,
                         mpi_comm_world,ierr)
37.       end do
38.       if (my_rank.eq.0) then
39.            t2 = timef()
40.       end if
41.       if (my_rank.eq.0) then
42.            mflops =float(2*n*m)*1./t2
43.            print*,prodt(n/3)
44.            print*,prodt(n/2)
45.            print*,prodt(n/4)
46.            print*,t2,mflops
47.       end if
48.       call mpi_finalize(ierr)
49.       end program
```

Table 6.5.1 records the mflops for 1000 repetitions of a matrix-vector product where the matrix is $n \times m$ with $n = 1048$ and variable m. Columns two and three use the BLAS2 subroutine *sgemv()* with $m = 2048$ and 4096. The mflops are greater for larger m. The fourth column uses the ji-loops in place of the optimized *sgemv()*, and smaller mflops are recorded.

6.5.3 Matrix-matrix Products

Matrix-matrix products have three nested loops, and therefore, there are six possible ways to compute these products. Let A be the product B times C. The traditional order is the ijk method or dotproduct method, which computes row i times column j. The jki method computes column j of A by multiplying

B times column j of C, which is done by linear combinations of the columns of B. A is initialized to zero.

Matrix-matrix Product (jki version) $A^+= A + BC$.

```
for j = 1, n
    for k = 1, n
        for i = 1, m
            a_{i,j} = a_{i,j} + b_{i,k}c_{k,j}
        endloop
    endloop
endloop.
```

This is used in the following MPI/Fortran implementation of the matrix-matrix product. Here the outer j-loop can be partitioned and the smaller matrix-matrix products can be done concurrently. Let bn and en be the beginning and end of a subset of the partition. Then the following smaller matrix-matrix products can be done in parallel

$$B(1 : m, 1 : n)C(1 : n, bn : en).$$

Then the smaller products are gathered into the larger product matrix. The center k-loop can also be partitioned, and this could be done by any vector pipelines or by the CPUs within a node.

The arrays are initialized in lines 12-13 before MPI is initialized in lines 15-17, and therefore, each processor will have a copy of the array. The matrix-matrix products on the submatrices can be done by either a call to the optimized BLAS3 subroutine *sgemm()* (see http://www.netlib.org /blas/sgemm.f) in line 26, or by the jki-loops in lines 27-33. The *mpi_gather()* subroutine is used in line 34, and here nm real numbers are sent to processor 0, received by processor 0 and stored in the product matrix. The mflops (million floating point operations per second) are computed in line 41 where we have used the timings in milliseconds, and ten repetitions of the matrix-matrix product with nm dotproducts of vectors with n components.

MPI/Fortran Code mmmpi.f

```
1.      program mm
2.      implicit none
3.      include 'mpif.h'
4.      real,dimension(1:1024,1:512):: a,b,prodt
5.      real,dimension(1:512,1:512):: c
6.      real:: t1,t2
7.      real:: timef,mflops
8.      integer:: l, my_rank,p,n,source,dest,tag,ierr,loc_n
9.      integer:: i,status(mpi_status_size),bn,en,j,k,it,m
10.     data n,dest,tag/512,0,50/
11.     m = 2*n
```

```
12.      a = 0.0
13.      b = 2.0
14.      c = 3.0
15.      call mpi_init(ierr)
16.      call mpi_comm_rank(mpi_comm_world,my_rank,ierr)
17.      call mpi_comm_size(mpi_comm_world,p,ierr)
18.      loc_n = n/p
19.      bn = 1+(my_rank)*loc_n
20.      en = bn + loc_n - 1
21.      call mpi_barrier(mpi_comm_world,ierr)
22.      if (my_rank.eq.0) then
23.           t1 = timef()
24.      end if
25.      do it = 1,10
26.           call sgemm('N','N',m,loc_n,n,1.0,b(1,1),m,c(1,bn) &
                            ,n,1.0,a(1,bn),m)
27.      !      do j = bn,en
28.      !         do k = 1,n
29.      !            do i = 1,m
30.      !               a(i,j) = a(i,j) + b(i,k)*c(k,j)
31.      !            end do
32.      !         end do
33.      !      end do
34.           call mpi_barrier(mpi_comm_world,ierr)
35.           call mpi_gather(a(1,bn),m*loc_n,mpi_real,prodt, &
                            m*loc_n, mpi_real,0,mpi_comm_world,ierr)
36.      end do
37.      if (my_rank.eq.0) then
38.           t2= timef()
39.      end if
40.      if (my_rank.eq.0) then
41.           mflops = 2*n*n*m*0.01/t2
42.           print*,t2,mflops
43.      end if
44.      call mpi_finalize(ierr)
45.      end program
```

In Table 6.5.2 the calculations were for A with $m = 2n$ rows and n columns. The second and third columns use the jki-loops with $n = 256$ and 512, and the speedup is generally better for the larger n. Column four uses the sgemm to do the matrix-matrix products, and noticeable improvement in the mflops is recorded.

Table 6.5.2: Matrix-matrix Product mflops

p	jki-loops, n = 256	jki-loops, n = 512	sgemm, n = 512
1	384	381	1337
2	754	757	2521
4	1419	1474	4375
8	2403	2785	7572
16	4102	5038	10429

6.5.4   Exercise

1.   Browse the www for MPI sites.

2.   In matvecmpi.f experiment with different n and compare mflops.

3.   In matvecmpi.f experiment with the ij-loop method and compare mflops.

4.   In matvecmpi.f use *sgemv()* to compute the matrix-vector product. You may need to use a special compiler option for *sgemv()*, for example, on the IBM/SP use -lessl to gain access to the engineering and scientific subroutine library.

5.   In mmmpi.f experiment with different n and compare mflops.

6.   In mmmpi.f experiment with other variations of the jki-loop method and compare mflops.

7.   In mmmpi.f use *sgemm()* and to compute the matrix-matrix product. You may need to use a special compiler option for *sgemm()*, for example, on the IBM/SP use -lessl to gain access to the engineering and scientific subroutine library.

6.6   MPI and 2D Models

6.6.1   Introduction

In this section we will give examples of MPI/Fortran codes for heat diffusion and pollutant transfer in two directions. Both the discrete models generate 2D arrays for the temperature, or pollutant concentration, as a function of discrete space for each time step. These models could be viewed as a special case of the matrix-vector products where the matrix is sparse and the column vectors are represented as a 2D space grid array.

6.6.2   Heat Diffusion in Two Directions

The basic model for heat diffusion in two directions was formulated in Section 1.5.

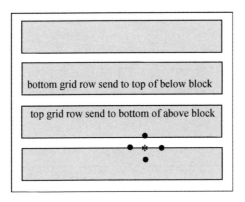

Figure 6.6.1: Space Grid with Four Subblocks

Explicit Finite Difference 2D Model: $\mathbf{u}_{i,j}^k \approx u(ih, jh, k\Delta t)$.

$$
\begin{aligned}
u_{i,j}^{k+1} &= (\Delta t/\rho c)f + \alpha(u_{i+1,j}^k + u_{i-1,j}^k + u_{i,j+1}^k + u_{i,j-1}^k) \\
&\quad +(1-4\alpha)u_{i,j}^k, \tag{6.6.1} \\
\alpha &= (K/\rho c)(\Delta t/h^2),\ i,j = 1,..,n-1 \text{ and } k = 0,..,maxk-1, \\
u_{i,j}^0 &= \text{given, } i,j = 1,..,n-1 \text{ and} \tag{6.6.2} \\
u_{i,j}^k &= \text{given, } k = 1,...,maxk, \text{ and } i,j \text{ on the boundary grid.} \tag{6.6.3}
\end{aligned}
$$

The execution of (6.6.1) requires at least a 2D array u(i,j) and three nested loops where the time loop (k-loop) must be on the outside. The two inner loops are over the space grid for x (i-loop) and y (j-loop). In order to distribute the work, we will partition the space grid into horizontal blocks by partitioning the j-loop. Then each processor will do the computations in (6.6.1) for some partition of the j-loop and all the i-loop, that is, over some horizontal block of the space grid. Because the calculations for each ij (depicted in Figure 6.6.1 by *) require inputs from the four adjacent space nodes (depicted in Figure 6.6.1 by •), some communication must be done so that the bottom and top rows of the partitioned space can be computed. See Figure 6.6.1 where there are four horizontal subblocks in the space grid, and three pairs of grid rows must be communicated.

The communication at each time step is done by a sequence of *mpi_send()* and *mpi_recv()* subroutines. Here one must be careful to avoid "deadlocked" communications, which can occur if two processors try to send data to each other at the same time. One needs *mpi_send()* and *mpi_recv()* to be coupled with respect to time. Figure 6.6.2 depicts one way of pairing the communications for eight processors associated with eight horizontal subblocks in the space grid. Each vector indicates a pair of *mpi_send()* and *mpi_recv()* where the processor at the beginning of the vector is sending data and the processor

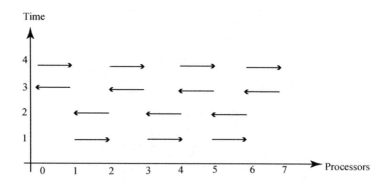

Figure 6.6.2: Send and Receive for Processors

at end with the arrow is receiving data. For example, at times 1, 2, 3 and 4 processor 1 will send to processor 2, receive from processor 2, send to processor 0, receive from processor 0, respectively.

In the heat2dmpi.f code lines 1-17 are the global initialization of the variables. Lines 18-20 start the multiprocessing, and lines 28-55 execute the explicit finite difference method where only the current temperatures are recorded. In lines 29-34 the computations for the processor my_rank are done for the horizontal subblock of the space grid associated with this processor. Note, the grid rows are associated with the columns in the array uold and unew. The communications between the processors, as outlined in Figure 6.6.2 for p = 8 processors, is executed in lines 35-54. In particular, processor 1 communications are done in lines 39–44 when my_rank = 1. After the last time step in line 55, lines 56-63 gather the computations from all the processors onto processor 0; this could have been done by the subroutine *mpi_gather().*

MPI/Fortran Code heat2dmpi.f

```
1.     program heat
2.     implicit none
3.     include 'mpif.h'
4.     real, dimension(2050,2050):: unew,uold
5.     real :: f,cond,dt,dx,alpha,t0, timef,tend
6.     integer :: my_rank,p,n,source,dest,tag,ierr,loc_n
7.     integer :: status(mpi_status_size),bn,en,j,k
8.     integer :: maxk,i,sbn
9.     n = 2049
10.    maxk = 1000
11.    f = 1000.0
12.    cond = .01
13.    dt = .01
14.    dx = 100.0/(n+1)
```

```
15.    alpha = cond*dt/(dx*dx)
16.    uold = 0.0
17.    unew = 0.0
18.    call mpi_init(ierr)
19.    call mpi_comm_rank(mpi_comm_world,my_rank,ierr)
20.    call mpi_comm_size(mpi_comm_world,p,ierr)
21.    loc_n = (n-1)/p
22.    bn = 2+(my_rank)*loc_n
23.    en = bn + loc_n -1
24.    call mpi_barrier(mpi_comm_world,ierr)
25.    if (my_rank.eq.0) then
26.        t0 = timef()
27.    end if
28.    do k =1,maxk
29.        do j = bn,en
30.            do i= 2,n
31.                unew(i,j) = dt*f + alpha*(uold(i-1,j)+uold(i+1,j)&
                        + uold(i,j-1) + uold(i,j+1))&
                        + (1- 4*alpha)*uold(i,j)
32.            end do
33.        end do
34.        uold(2:n,bn:en)= unew(2:n,bn:en)
35.        if (my_rank.eq.0) then
36.            call mpi_recv(uold(1,en+1),(n+1),mpi_real, &
                    my_rank+1,50, mpi_comm_world,status,ierr)
37.            call mpi_send(uold(1,en),(n+1),mpi_real, &
                    my_rank+1,50, mpi_comm_world,ierr)
38.        end if
39.        if ((my_rank.gt.0).and.(my_rank.lt.p-1) &
                    .and.(mod(my_rank,2).eq.1)) then
40.            call mpi_send(uold(1,en),(n+1),mpi_real, &
                    my_rank+1,50, mpi_comm_world,ierr)
41.            call mpi_recv(uold(1,en+1),(n+1),mpi_real, &
                    my_rank+1,50, mpi_comm_world,status,ierr)
42.            call mpi_send(uold(1,bn),(n+1),mpi_real, &
                    my_rank-1,50, mpi_comm_world,ierr)
43.            call mpi_recv(uold(1,bn-1),(n+1),mpi_real, &
                    my_rank-1,50,mpi_comm_world,status,ierr)
44.        end if
45.        if ((my_rank.gt.0).and.(my_rank.lt.p-1) &
                    .and.(mod(my_rank,2).eq.0)) then
46.            call mpi_recv(uold(1,bn-1),(n+1),mpi_real, &
                    my_rank-1,50, mpi_comm_world,status,ierr)
47.            call mpi_send(uold(1,bn),(n+1),mpi_real, &
                    my_rank-1,50, mpi_comm_world,ierr)
```

Table 6.6.1: Processor Times for Diffusion

p	Times	Speedups
2	87.2	1.0
4	41.2	2.1
8	21.5	4.1
16	11.1	7.9
32	06.3	13.8

```
48.            call mpi_recv(uold(1,en+1),(n+1),mpi_real, &
                   my_rank+1,50, mpi_comm_world,status,ierr)
49.            call mpi_send(uold(1,en),(n+1),mpi_real, &
                   my_rank+1,50, mpi_comm_world,ierr)
50.        end if
51.        if (my_rank.eq.p-1) then
52.            call mpi_send(uold(1,bn),(n+1),mpi_real, &
                   my_rank-1,50, mpi_comm_world,ierr)
53.            call mpi_recv(uold(1,bn-1),(n+1),mpi_real, &
                   my_rank-1,50, mpi_comm_world,status,ierr)
54.        end if
55.    end do
56.    if (my_rank.eq.0) then
57.        do source = 1,p-1
58.            sbn = 2+(source)*loc_n
59.            call mpi_recv(uold(1,sbn),(n+1)*loc_n,mpi_real, &
                   source,50, mpi_comm_world,status,ierr)
60.        end do
61.    else
62.        call mpi_send(uold(1,bn),(n+1)*loc_n,mpi_real, &
                   0,50, mpi_comm_world,ierr)
63.    end if
64.    if (my_rank.eq.0) then
65.        tend = timef()
66.        print*, 'time =', tend
67.        print*, uold(2,2),uold(3,3),uold(4,4),uold(500,500)
68.    end if
69.    call mpi_finalize(ierr)
70.    end
```

The code can be compiled and executed on the IBM/SP by the following:
mpxlf90 –O4 heat2dmpi.f where mpxlf90 is the compiler, and using the load
leveler llsubmit envrmpi2 where envmpi2 contains the job parameters such as
time and number of processors. Table 6.6.1 contains the times in seconds to
execute the above file with different numbers of processors $p = 2, 4, 8, 16$ and 32.
Good speedups relative to the execution time using two processors are recorded.

Table 6.6.2: Processor Times for Pollutant

p	Times	Speedups
2	62.9	1.0
4	28.9	2.2
8	15.0	4.2
16	07.9	8.0
32	04.6	13.7

6.6.3 Pollutant Transfer in Two Directions

A simple model for pollutant transfer in a shallow lake was formulated in Section 1.5.

Explicit Finite Difference 2D Pollutant Model: $u_{i,j}^k \approx u(i\Delta x, j\Delta y, k\Delta t)$.

$$u_{i,j}^{k+1} = v_1(\Delta t/\Delta x)u_{i-1,j}^k + v_2(\Delta t/\Delta y)u_{i,j-1}^k + \qquad (6.6.4)$$
$$(1 - v_1(\Delta t/\Delta x) - v_2(\Delta t/\Delta y) - \Delta t\, dec)u_{i,j}^k$$

$$u_{i,j}^0 = \text{given and} \qquad (6.6.5)$$

$$u_{0,j}^k \text{ and } u_{i,0}^k = \text{given.} \qquad (6.6.6)$$

The MPI/Fortran code poll2dmpi.f is only a slight modification of the above code for heat diffusion. We have kept the same communication scheme. This is not completely necessary because the wind is from the southwest, and therefore, the new concentration will depend only on two of the adjacent space nodes, the south and the west nodes. The initialization is similar, and the execution on the processors for (6.6.4) is

```
do j = bn,en
    do i= 2,n
        unew(i,j) = dt*f + dt*velx/dx*uold(i-1,j)&
                    + dt*vely/dy*uold(i,j-1) &
                    + (1- dt*velx/dx - dt*vely/dy &
                    - dt*dec)*uold(i,j)
    end do
end do.
```

The calculations that are recorded in Table 6.6.2 are for the number of processors $p = 2, 4, 8, 16$ and 32 and have good speedups relative to the two processors time.

6.6.4 Exercises

1.   In heat2dmpi.f carefully study the communication scheme, and verify the communications for the case of eight processors as depicted in Figure 6.6.2.

2.   In poll2dm.f study the communication scheme and delete any unused *mpi_send()* and *mpi_recv()* subroutines. Also, try to use any of the mpi collective subroutines such a *mpi_gather()*.

3.   In heat2dmpi.f and in poll2dmpi.f explain why the codes fail if only one processor is used.

4.   In poll2dm.f consider the case where the wind comes from the northwest. Modify the discrete model and the code.

5.   Duplicate the computations in Table 6.6.1. Experiment with different n and compare speedups.

6.   Duplicate the computations in Table 6.6.2. Experiment with different n and compare speedups.

Chapter 7

Message Passing Interface

In the last three sections in Chapter 6 several MPI codes were illustrated. In this chapter a more detailed discussion of MPI will be undertaken. The basic eight MPI commands and the four collective communication subroutines *mpi_bcast()*, *mpi_reduce()*, *mpi_gather()* and *mpi_scatter()* will be studied in the first three sections. These twelve commands/subroutines form a basis for all MPI programming, but there are many additional MPI subroutines. Section 7.4 describes three methods for grouping data so as to minimize the number of calls to communication subroutines, which can have significant startup times. Section 7.5 describes other possible communicators, which are just subsets of the processors that are allowed to have communications. In the last section these topics are applied to matrix-matrix products via Fox's algorithm. Each section has several short demonstration MPI codes, and these should be helpful to the first time user of MPI. This chapter is a brief introduction to MPI, and the reader should also consult other texts on MPI such as P. S. Pacheco [21] and W. Gropp, E. Lusk, A. Skjellum and R. Thahur [8].

7.1 Basic MPI Subroutines

7.1.1 Introduction

MPI programming can be done in either C or Fortran by using a library of MPI subroutines. In the text we have used Fortran 9x and the MPI library is called mpif.h. The text web site contains both the C and Fortran codes. The following is the basic structure for MPI codes:

```
include 'mpif.h'
⋮
call mpi_init(ierr)
call mpi_comm_rank(mpi_comm_world, my_rank, ierr)
call mpi_comm_size(mpi_comm_world, p, ierr)
```

275

$\vdots$

do parallel work

$\vdots$

call mpi_barrier(mpi_comm_world, ierr)

$\vdots$

do communications via mpi_send() and mpi_recv()

$\vdots$

call mpi_finalize(ierr).

The parameters my_rank, ierr, and p are integers where p is the number of processors, which are listed from 0 to p-1. Each processor is identified by my_rank ranging from 0 to p-1. Any error status is indicated by ierr. The parameter mpi_comm_world is a special MPI type, called a communicator, that is used to identify subsets of processors having communication patterns. Here the generic communicator, mpi_comm_world, is the set of all p processors and all processors are allowed to communicate with each other. Once the three calls to *mpi_init(), mpi_comm_rank()* and *mpi_comm_size()* have been made, each processor will execute the code before the call to *mpi_finalize()*, which terminates the parallel computations. Since each processor has a unique value for my_rank, the code or the input data may be different for each processor. The call to *mpi_barrier()* is used to insure that each processor has completed its computations. Any communications between processors may be done by calls to *mpi_send()* and *mpi_recv()*.

7.1.2 Syntax for mpi_send() and mpi_recv()

MPI has many different subroutines that can be used to do communications between processors. The communication subroutines *mpi_send()* and *mpi_recv()* are the most elementary, and they must be called in pairs. That is, if processor 0 wants to send data to processor 1, then a *mpi_send()* from processor 0 must be called as well as a *mpi_recv()* from processor 1 must be called.

```
mpi_send(senddata, count, mpi_datatype, dest, tag, mpi_comm, ierr)
      senddata           array(*)
      count              integer
      mpi_datatype       integer
      dest               integer
      tag                integer
      mpi_comm           integer
      ierr               integer
```

There a number of mpi_datatypes, and some of these are mpi_real, mpi_int, and mpi_char. The integer dest indicates the processor that data is to be sent. The parameter tag is used to clear up any confusion concerning multiple calls

to *mpi_send()*. The syntax for *mpi_recv()* is similar, but it does have one
additional parameter, mpi_status for the status of the communication.

mpi_rev(recvdata, count, mpi_datatype, source, tag
                , mpi_comm, status, ierr)

recvdata	array(*)
count	integer
mpi_datatype	integer
source	integer
tag	integer
mpi_comm	integer
status(mpi_status_size)	integer
ierr	integer

Suppose processor 0 needs to communicate the real number, a, and the
integer, n, to the other p-1 processors. Then there must be 2(p-1) pairs of
calls to *mpi_send()* and *mpi_recv()*, and one must be sure to use different tags
associated with a and n. The following if-else-endif will do this where the first
part of the if-else-endif is for only processor 0 and the second part has p-1 copies
with one for each of the other processors from 1 to p-1:

```
if (my_rank.eq.0) then
    do dest = 1,p-1
        taga = 0
        call mpi_send(a, 1, mpi_real, dest, taga
                    , mpi_comm_world, ierr)
        tagn = 1
        call mpi_send(n, 1, mpi_int, dest, tagn
                    , mpi_comm_world, ierr)
    end do
else
    taga = 0
    call mpi_recv(a, 1, mpi_real, 0, taga
                    , mpi_comm_world, status, ierr)
    tagn = 1
    call mpi_recv(n, 1, mpi_int, 0, tagn
                    , mpi_comm_world, status, ierr)
end if.
```

7.1.3 First MPI Code

This first MPI code simply partitions an interval from a to b into p equal
parts. The data in line 11 will be "hardwired" into all the processors because it
precedes the initialization of MPI in lines 14-15. Each processor will execute the
print commands in lines 17-19. Since my_rank will vary with each processor,
each processor will have unique values of loc_a and loc_b. The if_else_endif

in lines 31-40 communicates all the loc_a to processor 0 and stores them in the array a_list. The print commands in lines 26-28 and lines 43-47 verify this. The outputs for the print commands might not appear in sequential order that is indicated following the code listing. This output verifies the communications for p = 4 processors.

MPI/Fortran 9x Code basicmpi.f

```
1.      program basicmpi
2.! Illustrates the basic eight mpi commands.
3.      implicit none
4.! Includes the mpi Fortran library.
5.      include 'mpif.h'
6.      real:: a,b,h,loc_a,loc_b,total
7.      real, dimension(0:31):: a_list
8.      integer:: my_rank,p,n,source,dest,tag,ierr,loc_n
9.      integer:: i,status(mpi_status_size)
10.! Every processor gets values for a,b and n.
11.      data a,b,n,dest,tag/0.0,100.0,1024,0,50/
12.! Initializes mpi, gets the rank of the processor, my_rank,
13.! and number of processors, p.
14.      call mpi_init(ierr)
15.      call mpi_comm_rank(mpi_comm_world,my_rank,ierr)
16.      call mpi_comm_size(mpi_comm_world,p,ierr)
17.      print*,'my_rank =',my_rank, 'a = ',a
18.      print*,'my_rank =',my_rank, 'b = ',b
19.      print*,'my_rank =',my_rank, 'n = ',n
20.      h = (b-a)/n
21.! Each processor has unique value of loc_n, loc_a and loc_b.
22.      loc_n = n/p
23.      loc_a = a+my_rank*loc_n*h
24.      loc_b = loc_a + loc_n*h
25.! Each processor prints its loc_n, loc_a and loc_b.
26.      print*,'my_rank =',my_rank, 'loc_a = ',loc_a
27.      print*,'my_rank =',my_rank, 'loc_b = ',loc_b
28.      print*,'my_rank =',my_rank, 'loc_n = ',loc_n
29.! Processors p not equal 0 sends a_loc to an array, a_list,
30.! in processor 0, and processor 0 recieves these.
31.      if (my_rank.eq.0) then
32.          a_list(0) = loc_a
33.          do source = 1,p-1
34.              call mpi_recv(a_list(source),1,mpi_real,source &
35.                          ,50,mpi_comm_world,status,ierr)
36.          end do
37.      else
38.          call mpi_send(loc_a,1,mpi_real,0,50,&
```

```
39.                           mpi_comm_world,ierr)
40.     end if
41.     call mpi_barrier(mpi_comm_world,ierr)
42.! Processor 0 prints the list of all loc_a.
43.     if (my_rank.eq.0) then
44.          do i = 0,p-1
45.               print*, 'a_list(',i,') = ',a_list(i)
46.          end do
47.     end if
48.! mpi is terminated.
49.     call mpi_finalize(ierr)
50.     end program basicmpi
```

```
my_rank = 0 a = 0.0000000000E+00
my_rank = 0 b = 100.0000000
my_rank = 0 n = 1024.
my_rank = 1 a = 0.0000000000E+00
my_rank = 1 b = 100.0000000
my_rank = 1 n = 1024
my_rank = 2 a = 0.0000000000E+00
my_rank = 2 b = 100.0000000
my_rank = 2 n = 1024
my_rank = 3 a = 0.0000000000E+00
my_rank = 3 b = 100.0000000
my_rank = 3 n = 1024
!
my_rank = 0 loc_a = 0.0000000000E+00
my_rank = 0 loc_b = 25.00000000
my_rank = 0 loc_n = 256
my_rank = 1 loc_a = 25.00000000
my_rank = 1 loc_b = 50.00000000
my_rank = 1 loc_n = 256
my_rank = 2 loc_a = 50.00000000
my_rank = 2 loc_b = 75.00000000
my_rank = 2 loc_n = 256
my_rank = 3 loc_a = 75.00000000
my_rank = 3 loc_b = 100.0000000
my_rank = 3 loc_n = 256
!
a_list( 0 ) = 0.0000000000E+00
a_list( 1 ) = 25.00000000
a_list( 2 ) = 50.00000000
a_list( 3 ) = 75.00000000
```

7.1.4   Application to Dot Product

The dot product of two vectors is simply the sum of the products of the components of the two vectors. The summation can be partitioned and computed in parallel. Once the partial dot products have been computed, the results can be communicated to a root processor, usually processor 0, and the sum of the partial dot products can be computed. The data in lines 9-13 is "hardwired" to all the processors. In lines 18-20 each processor gets a unique beginning n, bn, and an ending n, en. This is verified by the print commands in lines 21-23. The local dot products are computed in lines 24-27. Lines 30-38 communicate these partial dot products to processor 0 and stores them in the array loc_dots. The local dot products are summed in lines 40-43. The output is for p = 4 processors.

MPI/Fortran 9x Code dot1mpi.f

```
1.      program dot1mpi
2.! Illustrates dot product via mpi_send and mpi_recv.
3.      implicit none
4.      include 'mpif.h'
5.      real:: loc_dot,dot
6.      real, dimension(0:31):: a,b, loc_dots
7.      integer:: my_rank,p,n,source,dest,tag,ierr,loc_n
8.      integer:: i,status(mpi_status_size),en,bn
9.      data n,dest,tag/8,0,50/
10.     do i = 1,n
11.          a(i) = i
12.          b(i) = i+1
13.     end do
14.     call mpi_init(ierr)
15.     call mpi_comm_rank(mpi_comm_world,my_rank,ierr)
16.     call mpi_comm_size(mpi_comm_world,p,ierr)
17.! Each processor computes a local dot product.
18.     loc_n = n/p
19.     bn = 1+(my_rank)*loc_n
20.     en = bn + loc_n-1
21.     print*,'my_rank =',my_rank, 'loc_n = ',loc_n
22.     print*,'my_rank =',my_rank, 'bn = ',bn
23.     print*,'my_rank =',my_rank, 'en = ',en
24.     loc_dot = 0.0
25.     do i = bn,en
26.          loc_dot = loc_dot + a(i)*b(i)
27.     end do
28.     print*,'my_rank =',my_rank, 'loc_dot = ',loc_dot
29.! The local dot products are sent and recieved to processor 0.
30.     if (my_rank.eq.0) then
```

```
31.           do source = 1,p-1
32.                   call mpi_recv(loc_dots(source),1,mpi_real,source,50,&
33.                                     50,mpi_comm_world,status,ierr)
34.               end do
35.       else
36.               call mpi_send(loc_dot,1,mpi_real,0,50,&
37.                                     mpi_comm_world,ierr)
38.       end if
39.! Processor 0 sums the local dot products.
40.       if (my_rank.eq.0) then
41.               dot = loc_dot + sum(loc_dots(1:p-1))
42.               print*, 'dot product = ',dot
43.       end if
44.       call mpi_finalize(ierr)
45.       end program dot1mpi
```

my_rank = 0 loc_n = 2
my_rank = 0 bn = 1
my_rank = 0 en = 2
my_rank = 1 loc_n = 2
my_rank = 1 bn = 3
my_rank = 1 en = 4
my_rank = 2 loc_n = 2
my_rank = 2 bn = 5
my_rank = 2 en = 6
my_rank = 3 loc_n = 2
my_rank = 3 bn = 7
my_rank = 3 en = 8
!
my_rank = 0 loc_dot = 8.000000000
my_rank = 1 loc_dot = 32.00000000
my_rank = 2 loc_dot = 72.00000000
my_rank = 3 loc_dot = 128.0000000
dot product = 240.0000000

Another application is numerical integration, and in this case a summation also can be partitioned and computed in parallel. See Section 6.4 where this is illustrated for the trapezoid rule, trapmpi.f. Also, the collective communication *mpi_reduce()* is introduced, and this will be discussed in more detail in the next section.

There are variations of *mpi_send()* and *mpi_recv()* such as *mpi_isend()*, *mpi_irecv()*, *mpi_sendrecv()* and *mpi_sendrecv_replace()*. The *mpi_isend()* and *mpi_irecv()* are nonblocking communications that attempt to use an intermediate buffer so as to avoid locking of the processors involved with the communications. The mpi_sendrecv() and *mpi_sendrecv_replace()* are compositions of mpi_send() and mpi_recv(), and more details on these can be found in the texts [21] and [8].

7.1.5 Exercises

1.   Duplicate the calculations for basicmpi.f and experiment with different numbers of processors.
2.   Duplicate the calculations for dot1mpi.f and experiment with different numbers of processors and different size vectors.
3.   Modify dot1mpi.f so that one can compute in parallel a linear combination of the two vectors, $\alpha x + \beta y$.
4.   Modify trapmpi.f to execute Simpson's rule in parallel.

7.2 Reduce and Broadcast

If there are a large number of processors, then the loop method for communicating information can be time consuming. An alternative is to use any available processors to execute some of the communications using either a fan-out (see Figure 6.4.1) or a fan-in (see Figure 7.2.1). As depicted in Figure 7.2.1, consider the dot product problem where there are p = 8 partial dot products that have been computed on processors 0 to 7. Processors 0, 2, 4, and 6 could receive the partial dot products from processors 1, 3, 5, and 7; in the next time step processors 0 and 4 receive two partial dot products from processors 2 and 6; in the third time step processor 0 receives the four additional partial dot products from processor 4. In general, if there are $p = 2^d$ processors, then fan-in and and fan-out communications can be executed in d time steps plus some startup time.

Four important collective communication subroutines that use these ideas are *mpi_reduce()*, *mpi_bcast()*, *mpi_gather()* and *mpi_scatter()*. These subroutines and their variations can significantly reduce communication and computation times, simplify MPI codes and reduce coding errors and times.

7.2.1 Syntax for mpi_reduce() and mpi_bcast()

The subroutine *mpi_reduce()* not only can send data to a root processor but it can also perform a number of additional operations with this data. It can add the data sent to the root processor or it can calculate the product of the sent data or the maximum of the sent data as well as other operations. The operations are indicated by the mpi_oper parameter. The data is collected from all the other processors in the communicator, and the call to *mpi_reduce()* must appear in all processors of the communicator.

mpi_reduce(loc_data, result, count, mpi_datatype, mpi_oper
, root, mpi_comm, ierr)

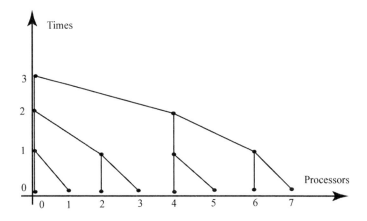

Figure 7.2.1: A Fan-in Communication

loc_data	array(*)
result	array(*)
count	integer
mpi_datatype	integer
mpi_oper	integer
root	integer
mpi_comm	integer
ierr	integer

The subroutine *mpi_bcast()* sends data from a root processor to all of the other processors in the communicator, and the call to *mpi_bcast()* must appear in all the processors of the communicator. The *mpi_bcast()* does not execute any computation, which is in contrast to *mpi_reduce()*.

mpi_bcast(data, count, mpi_datatype,
                                , root, mpi_comm, ierr)

data	array(*)
count	integer
mpi_datatype	integer
root	integer
mpi_comm	integer
ierr	integer

7.2.2 Illustrations of mpi_reduce()

The subroutine *mpi_reduce()* is used to collect results from other processors and then to perform additional computations on the root processor. The code reducmpi.f illustrates this for the additional operations of sum and product of the output from the processors. After MPI is initialized in lines 10-12, each processor computes local values of a and b in lines 16 and 17. The call to *mpi_reduce()* in line 23 sums all the loc_b to sum in processor 0 via the mpi_oper equal to mpi_sum. The call to *mpi_reduce()* in line 26 computes the product of all the loc_b to prod in processor 0 via the mpi_oper equal to mpi_prod. These results are verified by the print commands in lines 18-20 and 27-30.

MPI/Fortran 9x Code reducmpi.f

```
1.      program reducmpi
2.! Illustrates mpi_reduce.
3.      implicit none
4.      include 'mpif.h'
5.      real:: a,b,h,loc_a,loc_b,total,sum,prod
6.      real, dimension(0:31):: a_list
7.      integer:: my_rank,p,n,source,dest,tag,ierr,loc_n
8.      integer:: i,status(mpi_status_size)
9.      data a,b,n,dest,tag/0.0,100.0,1024,0,50/
10.     call mpi_init(ierr)
11.     call mpi_comm_rank(mpi_comm_world,my_rank,ierr)
12.     call mpi_comm_size(mpi_comm_world,p,ierr)
13.! Each processor has a unique loc_n, loc_a and loc_b.
14.     h = (b-a)/n
15.     loc_n = n/p
16.     loc_a = a+my_rank*loc_n*h
17.     loc_b = loc_a + loc_n*h
18.     print*,'my_rank =',my_rank, 'loc_a = ',loc_a
19.     print*,'my_rank =',my_rank, 'loc_b = ',loc_b
20.     print*,'my_rank =',my_rank, 'loc_n = ',loc_n
21.! mpi_reduce is used to compute the sum of all loc_b
22.! to sum on processor 0.
23.     call mpi_reduce(loc_b,sum,1,mpi_real,mpi_sum,0,&
                        mpi_comm_world,status,ierr)
24.! mpi_reduce is used to compute the product of all loc_b
25.! to prod on processor 0.
26.     call mpi_reduce(loc_b,prod,1,mpi_real,mpi_prod,0,&
                        mpi_comm_world,status,ierr)
27.     if (my_rank.eq.0) then
28.         print*, 'sum = ',sum
29.         print*, 'product = ',prod
30.     end if
31.     call mpi_finalize(ierr)
32.     end program reducmpi
```

```
my_rank = 0 loc_a = 0.0000000000E+00
my_rank = 0 loc_b = 25.00000000
my_rank = 0 loc_n = 256
```

```
       my_rank = 1 loc_a = 25.00000000
       my_rank = 1 loc_b = 50.00000000
       my_rank = 1 loc_n = 256
       my_rank = 2 loc_a = 50.00000000
       my_rank = 2 loc_b = 75.00000000
       my_rank = 2 loc_n = 256
       my_rank = 3 loc_a = 75.00000000
       my_rank = 3 loc_b = 100.0000000
       my_rank = 3 loc_n = 256
       !
       sum = 250.0000000
       product = 9375000.000
```

The next code is a second version of the dot product, and *mpi_reduce()* is now used to sum the partial dot products. As in dot1mpi.f the local dot products are computed in parallel in lines 24-27. The call to *mpi_reduce()* in line 31 sends the local dot products, loc_dot, to processor 0 and sums them to dot on processor 0. This is verified by the print commands in lines 28 and 32-34.

MPI/Fortran 9x Code dot2mpi.f

```
1.      program dot2mpi
2.! Illustrates dot product via mpi_reduce.
3.      implicit none
4.      include 'mpif.h'
5.      real:: loc_dot,dot
6.      real, dimension(0:31):: a,b, loc_dots
7.      integer:: my_rank,p,n,source,dest,tag,ierr,loc_n
8.      integer:: i,status(mpi_status_size),en,bn
9.      data n,dest,tag/8,0,50/
10.     do i = 1,n
11.        a(i) = i
12.        b(i) = i+1
13.     end do
14.     call mpi_init(ierr)
15.     call mpi_comm_rank(mpi_comm_world,my_rank,ierr)
16.     call mpi_comm_size(mpi_comm_world,p,ierr)
17.! Each processor computes a local dot product.
18.     loc_n = n/p
19.     bn = 1+(my_rank)*loc_n
20.     en = bn + loc_n-1
21.     print*,'my_rank =',my_rank, 'loc_n = ',loc_n
22.     print*,'my_rank =',my_rank, 'bn = ',bn
23.     print*,'my_rank =',my_rank, 'en = ',en
24.     loc_dot = 0.0
25.     do i = bn,en
```

```
26.              loc_dot = loc_dot + a(i)*b(i)
27.       end do
28.       print*,'my_rank =',my_rank, 'loc_dot = ',loc_dot
29.! mpi_reduce is used to sum all the local dot products
30.! to dot on processor 0.
31.       call mpi_reduce(loc_dot,dot,1,mpi_real,mpi_sum,0,&
                                mpi_comm_world,status,ierr)
32.       if (my_rank.eq.0) then
33.            print*, 'dot product = ',dot
34.       end if
35.       call mpi_finalize(ierr)
36.       end program dot2mpi
```

```
my_rank = 0 loc_dot = 8.000000000
my_rank = 1 loc_dot = 32.00000000
my_rank = 2 loc_dot = 72.00000000
my_rank = 3 loc_dot = 128.0000000
dot product = 240.0000000
```

Other illustrations of the subroutine *mpi_reduce()* are given in Sections 6.4 and 6.5. In trapmpi.f the partial integrals are sent to processor 0 and added to form the total integral. In matvecmpi.f the matrix-vector products are computed by forming linear combinations of the column vector of the matrix. Parallel computations are formed by computing partial linear combinations, and using *mpi_reduce()* with the count parameter equal to the number of components in the column vectors.

7.2.3   Illustrations of mpi_bcast()

The subroutine *mpi_bcast()* is a fan-out algorithm that sends data to the other processors in the communicator. The constants a = 0 and b = 100 are defined in lines 12-15 for processor 0. Lines 17 and 18 verify that only processor 0 has this information. Lines 21 and 22 use *mpi_bcast()* to send these values to all the other processors. This is verified by the print commands in lines 25 and 26. Like *mpi_send()* and *mpi_recv()*, *mpi_bcast()* must appear in the code for all the processors involved in the communication. Lines 29-33 also do this, and they enable the receiving processors to rename the sent data. This is verified by the print command in line 34.

MPI/Fortran 9x Code bcastmpi.f

```
1.       program bcastmpi
2.! Illustrates mpi_bcast.
3.       implicit none
4.       include 'mpif.h'
5.       real:: a,b,new_b
6.       integer:: my_rank,p,n,source,dest,tag,ierr,loc_n
```

```
 7.     integer:: i,status(mpi_status_size)
 8.     data n,dest,tag/1024,0,50/
 9.     call mpi_init(ierr)
10.     call mpi_comm_rank(mpi_comm_world,my_rank,ierr)
11.     call mpi_comm_size(mpi_comm_world,p,ierr)
12.     if (my_rank.eq.0) then
13.         a = 0
14.         b = 100.
15.     end if
16.! Each processor attempts to print a and b.
17.     print*,'my_rank =',my_rank, 'a = ',a
18.     print*,'my_rank =',my_rank, 'b = ',b
19.! Processor 0 broadcasts a and b to the other processors.
20.! The mpi_bcast is issued by all processors.
21.     call mpi_bcast(a,1,mpi_real,0,&
                           mpi_comm_world,ierr)
22.     call mpi_bcast(b,1,mpi_real,0,&
                           mpi_comm_world,ierr)
23.     call mpi_barrier(mpi_comm_world,ierr)
24.! Each processor prints a and b.
25.     print*,'my_rank =',my_rank, 'a = ',a
26.     print*,'my_rank =',my_rank, 'b = ',b
27.! Processor 0 broadcasts b to the other processors and
28.! stores it in new_b.
29.     if (my_rank.eq.0) then
30.         call mpi_bcast(b,1,mpi_real,0,&
                               mpi_comm_world,ierr)
31.     else
32.         call mpi_bcast(new_b,1,mpi_real,0,&
                               mpi_comm_world,ierr)
33.     end if
34.     print*,'my_rank =',my_rank, 'new_b = ',new_b
35.     call mpi_finalize(ierr)
36.     end program bcastmpi
```

```
my_rank = 0 a = 0.0000000000E+00
my_rank = 0 b = 100.0000000
my_rank = 1 a = -0.9424863232E+10
my_rank = 1 b = -0.1900769888E+30
my_rank = 2 a = -0.9424863232E+10
my_rank = 2 b = -0.1900769888E+30
my_rank = 3 a = -0.7895567565E+11
my_rank = 3 b = -0.4432889195E+30
!
my_rank = 0 a = 0.0000000000E+00
```

```
my_rank = 0 b = 100.0000000
my_rank = 1 a = 0.0000000000E+00
my_rank = 1 b = 100.0000000
my_rank = 2 a = 0.0000000000E+00
my_rank = 2 b = 100.0000000
my_rank = 3 a = 0.0000000000E+00
my_rank = 3 b = 100.0000000
!
my_rank = 0 new_b = 0.4428103147E-42
my_rank = 1 new_b = 100.0000000
my_rank = 2 new_b = 100.0000000
my_rank = 3 new_b = 100.0000000
```

The subroutines *mpi_reduce()* and *mpi_bcast()* are very effective when the count and mpi_oper parameters are used. Also, there are variations of these subroutines such as *mpi_allreduce()* and *mpi_alltoall()*, and for more details one should consult the texts [21] and [8].

7.2.4   Exercises

1.   Duplicate the calculations for reducmpi.f and experiment with different numbers of processors.
2.   Duplicate the calculations for dot2mpi.f and experiment with different numbers of processors and different size vectors.
3.   Modify dot2mpi.f so that one can compute in parallel a linear combination of the two vectors, $\alpha x + \beta y$.
4.   Use *mpi_reduce()* to modify trapmpi.f to execute Simpson's rule in parallel.
5.   Duplicate the calculations for bcastmpi.f and experiment with different numbers of processors.

7.3   Gather and Scatter

7.3.1   Introduction

When programs are initialized, often the root or host processor has most of the initial data, which must be distributed either to all the processors, or parts of the data must be distributed to various processors. The subroutine *mpi_scatter()* can send parts of the initial data to various processors. This differs from *mpi_bcast()*, because *mpi_bcast()* sends certain data to all of the processors in the communicator. Once the parallel computation has been executed, the parallel outputs must be sent to the host or root processor. This can be done by using *mpi_gather()*, which systematically stores the outputs from the nonroot processors. These collective subroutines use fan-in and fan-out schemes, and so they are effective for larger numbers of processors.

7.3.2 Syntax for mpi_scatter() and mpi_gather

The subroutine *mpi_scatter()* can send adjacent segments of data to local arrays in other processors. For example, an array a(1:16) defined on processor 0 may be distributed to loc_a(1:4) on each of four processors by a(1:4), a(5:8), a(9:12) and a(13:16). In this case, the count parameter is used where count = 4. The processors in the communicator are the destination processors.

    mpi_scatter(sourcedata, count, mpi_datatype,
                recvdata, count, mpi_datatype,
                source, mpi_comm, status, ierr)

sourcedata	array(*)
count	integer
mpi_datatype	integer
recvdata	array(*)
count	integer
mpi_datatype	integer
source	integer
mpi_comm	integer
status(mpi_status_size)	integer
ierr	integer

The subroutine *mpi_gather()* can act as an inverse of *mpi_scatter()*. For example, if loc_a(1:4) is on each of four processors, then
        processor 0 sends loc_a(1:4) to a(1:4) on processor 0,
        processor 1 sends loc_a(1:4) to a(5:8) on processor 0,
        processor 2 sends loc_a(1:4) to a(9:12) on processor 0 and
        processor 3 sends loc_a(1:4) to a(13:16) on processor 0.

    mpi_gather(locdata, count, mpi_datatype,
               destdata, count, mpi_datatype,
               source, mpi_comm, status, ierr)

locdata	array(*)
count	integer
mpi_datatype	integer
destdata	array(*)
count	integer
mpi_datatype	integer
dest	integer
mpi_comm	integer
status(mpi_status_size)	integer
ierr	integer

7.3.3 Illustrations of mpi_scatter()

In scatmpi.f the array a_list(0:7) is initialized for processor 0 in line 12-16. The scatmpi.f code scatters the arrary a_list(0:7) to four processors in groups of two components, which is dictated by the count parameter in *mpi_scatter()* in line 19. The two real numbers are stored in the first two components in the local arrays, a_loc. The components a_loc(2:7) are not defined, and the print commands in line 20 verify this.

MPI/Fortran 9x Code scatmpi.f

```
1.      program scatmpi
2.! Illustrates mpi_scatter.
3.      implicit none
4.      include 'mpif.h'
5.      real, dimension(0:7):: a_list,a_loc
6.      integer:: my_rank,p,n,source,dest,tag,ierr,loc_n
7.      integer:: i,status(mpi_status_size)
8.      data n,dest,tag/1024,0,50/
9.      call mpi_init(ierr)
10.     call mpi_comm_rank(mpi_comm_world,my_rank,ierr)
11.     call mpi_comm_size(mpi_comm_world,p,ierr)
12.     if (my_rank.eq.0) then
13.          do i = 0,7
14.              a_list(i) = i
15.          end do
16.     end if
17.! The array, a_list, is sent and received in groups of
18.! two to the other processors and stored in a_loc.
19.     call mpi_scatter(a_list,2,mpi_real,a_loc,2,mpi_real,0,&
                         mpi_comm_world,status,ierr)
20.     print*, 'my_rank =',my_rank,'a_loc = ', a_loc
21.     call mpi_finalize(ierr)
22.     end program scatmpi
```

```
   my_rank = 0 a_loc = 0.0000000000E+00 1.000000000
   !
   0.2347455187E-40  0.1010193260E-38 -0.8896380928E+10
   -0.2938472521E+30  0.3083417141E-40 0.1102030158E-38
   !
   my_rank = 1 a_loc = 2.000000000 3.000000000
   !
   .2347455187E-40  0.1010193260E-38  -0.8896380928E+10
   -0.2947757071E+30  0.3083417141E-40  0.1102030158E-38
   !
   my_rank = 2 a_loc = 4.000000000 5.000000000
   !
   0.2347455187E-40 0.1010193260E-38  -0.8896380928E+10
   -0.2949304496E+30 0.3083417141E-40  0.1102030158E-38
   !
   my_rank = 3 a_loc = 6.000000000 7.000000000
```

```
!
0.2347455187E-40 0.1010193260E-38  -0.8896380928E+10
-0.3097083589E+30 0.3083417141E-40 0.1102030158E-38
```

7.3.4 Illustrations of mpi_gather()

The second code gathmpi.f collects some of the data loc_n, loc_a, and loc_b, which is computed in lines 15-17 for each processor. In particular, all the values of loc_a are sent and stored in the array a_list on processor 0. This is done by *mpi_gather()* on line 23 where count is equal to one and the root processor is zero. This is verified by the print commands in lines 18-20 and 25-29.

MPI/Fortran 9x Code gathmpi.f

```
1.      program gathmpi
2.! Illustrates mpi_gather.
3.      implicit none
4.      include 'mpif.h'
5.      real:: a,b,h,loc_a,loc_b,total
6.      real, dimension(0:31):: a_list
7.      integer:: my_rank,p,n,source,dest,tag,ierr,loc_n
8.      integer:: i,status(mpi_status_size)
9.      data a,b,n,dest,tag/0.0,100.0,1024,0,50/
10.     call mpi_init(ierr)
11.     call mpi_comm_rank(mpi_comm_world,my_rank,ierr)
12.     call mpi_comm_size(mpi_comm_world,p,ierr)
13.     h = (b-a)/n
14.! Each processor has a unique loc_n, loc_a and loc_b
15.     loc_n = n/p
16.     loc_a = a+my_rank*loc_n*h
17.     loc_b = loc_a + loc_n*h
18.     print*,'my_rank =',my_rank, 'loc_a = ',loc_a
19.     print*,'my_rank =',my_rank, 'loc_b = ',loc_b
20.     print*,'my_rank =',my_rank, 'loc_n = ',loc_n
21.! The loc_a are sent and recieved to an array, a_list, on
22.! processor 0.
23.     call mpi_gather(loc_a,1,mpi_real,a_list,1,mpi_real,0,&
                            mpi_comm_world,status,ierr)
24.     call mpi_barrier(mpi_comm_world,ierr)
25.     if (my_rank.eq.0) then
26.         do i = 0,p-1
27.             print*, 'a_list(',i,') = ',a_list(i)
28.         nd do
29.     end if
30.     call mpi_finalize(ierr)
31.     end program gathmpi
```

```
my_rank = 0 loc_a = 0.0000000000E+00
my_rank = 0 loc_b = 25.00000000
my_rank = 0 loc_n = 256
my_rank = 1 loc_a = 25.00000000
my_rank = 1 loc_b = 50.00000000
my_rank = 1 loc_n = 256
my_rank = 2 loc_a = 50.00000000
my_rank = 2 loc_b = 75.00000000
my_rank = 2 loc_n = 256
my_rank = 3 loc_a = 75.00000000
my_rank = 3 loc_b = 100.0000000
my_rank = 3 loc_n = 256
!
a_list( 0 ) = 0.0000000000E+00
a_list( 1 ) = 25.00000000
a_list( 2 ) = 50.00000000
a_list( 3 ) = 75.00000000
```

The third version of a parallel dot product in dot3mpi.f uses *mpi_ gather()* to collect the local dot products that have been computed concurrently in lines 25-27. The local dot products, loc_dot, are sent and stored in the array loc_dots(0:31) on processor 0. This is done by the call to *mpi_ gather()* on line 31 where the count parameter is equal to one and the root processor is zero. Lines 33-36 sum the local dot products, and the print commands in lines 21-23 and 33-36 confirm this.

MPI/Fortran 9x Code dot3mpi.f

```
1.      program dot3mpi
2.! Illustrates dot product via mpi_gather.
3.      implicit none
4.      include 'mpif.h'
5.      real:: loc_dot,dot
6.      real, dimension(0:31):: a,b, loc_dots
7.      integer:: my_rank,p,n,source,dest,tag,ierr,loc_n
8.      integer:: i,status(mpi_status_size),en,bn
9.      data n,dest,tag/8,0,50/
10.     do i = 1,n
11.        a(i) = i
12.        b(i) = i+1
13.     end do
14.     call mpi_init(ierr)
15.     call mpi_comm_rank(mpi_comm_world,my_rank,ierr)
16.     call mpi_comm_size(mpi_comm_world,p,ierr)
17.! Each processor computes a local dot product
18.     loc_n = n/p
19.     bn = 1+(my_rank)*loc_n
```

```
20.     en = bn + loc_n-1
21.     print*,'my_rank =',my_rank, 'loc_n = ',loc_n
22.     print*,'my_rank =',my_rank, 'bn = ',bn
23.     print*,'my_rank =',my_rank, 'en = ',en
24.     loc_dot = 0.0
25.     do i = bn,en
26.         loc_dot = loc_dot + a(i)*b(i)
27.     end do
28.     print*,'my_rank =',my_rank, 'loc_dot = ',loc_dot
29.! mpi_gather sends and recieves all local dot products
30.! to the array loc_dots in processor 0.
31.     call mpi_gather(loc_dot,1,mpi_real,loc_dots,1,mpi_real,0,&
                                mpi_comm_world,status,ierr)
32.! Processor 0 sums the local dot products.
33.     if (my_rank.eq.0) then
34.         dot = loc_dot + sum(loc_dots(1:p-1))
35.         print*, 'dot product = ',dot
36.     end if
37.     call mpi_finalize(ierr)
38.     end program dot3mpi
```

```
my_rank = 0 loc_n = 2
my_rank = 0 bn = 1
my_rank = 0 en = 2
my_rank = 1 loc_n = 2
my_rank = 1 bn = 3
my_rank = 1 en = 4
my_rank = 2 loc_n = 2
my_rank = 2 bn = 5
my_rank = 2 en = 6
my_rank = 3 loc_n = 2
my_rank = 3 bn = 7
my_rank = 3 en = 8
!
my_rank = 0 loc_dot = 8.000000000
my_rank = 1 loc_dot = 32.00000000
my_rank = 2 loc_dot = 72.00000000
my_rank = 3 loc_dot = 128.0000000
dot product = 240.0000000
```

Another application of *mpi_gather()* is in the matrix-matrix product code mmmpi.f, which was presented in Section 6.5. Here the product BC was formed by computing in parallel $BC\,(bn:en)$, and these partial products were communicated via *mpi_gather()* to the root processor.

7.3.5   Exercises

1.   Duplicate the calculations for scatmpi.f and experiment with different numbers of processors.

2.   Duplicate the calculations for gathmpi.f and experiment with different numbers of processors.

3.   Duplicate the calculations for dot3mpi.f and experiment with different numbers of processors and different size vectors.

4.   Use *mpi_ gather()* to compute in parallel a linear combination of the two vectors, $\alpha x + \beta y$.

5.   Use *mpi_ gather()* to modify trapmpi.f to execute Simpson's rule in parallel.

7.4   Grouped Data Types

7.4.1   Introduction

There is some startup time associated with each MPI subroutine. So if a large number of calls to *mpi_ send()* and *mpi_ recv()* are made, then the communication portion of the code may be significant. By *collecting data in groups* a single communication subroutine may be used for large amounts of data. Here we will present three methods for the grouping of data: count, derived types and packed.

7.4.2   Count Type

The *count parameter* has already been used in some of the previous codes. The parameter count refers to the number of mpi_datatypes to be communicated. The most common data types are mpi_real or mpi_int, and these are usually stored in arrays whose components are addressed sequentially. In Fortran the two dimensional arrays components are listed by columns starting with the leftmost column. For example, if the array is b(1:2,1:3), then the list for b is b(1,1), b(2,1), b(1,2), b(2,2), b(1,3) and b(2,3). Starting at b(1,1) with count = 4 gives the first four components, and starting at b(1,2) with count = 4 gives the last four components.

The code countmpi.f illustrates the count parameter method when it is used in the subroutine *mpi_ bcast()*. Lines 14-24 initialize in processor 0 two arrays a(1:4) and b(1:2,1:3). All of the array a is broadcast, in line 29, to the other processors, and just the first four components of the two dimensional array b are broadcast, in line 30, to the other processors. This is confirmed by the print commands in lines 26, 32 and 33.

MPI/Fortran 9x Code countmpi.f

```
   1.     program countmpi
   2.! Illustrates count for arrays.
```

```
3.      implicit none
4.      include 'mpif.h'
5.      real, dimension(1:4):: a
6.      integer, dimension(1:2,1:3):: b
7.      integer:: my_rank,p,n,source,dest,tag,ierr,loc_n
8.      integer:: i,j,status(mpi_status_size)
9.      data n,dest,tag/4,0,50/
10.     call mpi_init(ierr)
11.     call mpi_comm_rank(mpi_comm_world,my_rank,ierr)
12.     call mpi_comm_size(mpi_comm_world,p,ierr)
13.! Define the arrays.
14.     if (my_rank.eq.0) then
15.         a(1) = 1.
16.         a(2) = exp(1.)
17.         a(3) = 4*atan(1.)
18.         a(4) = 186000.
19.         do j = 1,3
20.             do i = 1,2
21.                 b(i,j) = i+j
22.             end do
23.         end do
24.     end if
25.! Each processor attempts to print the array.
26.     print*,'my_rank =',my_rank, 'a = ',a
27.     call mpi_barrier(mpi_comm_world,ierr)
28.! The arrays are broadcast via count equal to four.
29.     call mpi_bcast(a,4,mpi_real,0,&
                       mpi_comm_world,ierr)
30.     call mpi_bcast(b,4,mpi_int,0,&
                       mpi_comm_world,ierr)
31.! Each processor prints the arrays.
32.     print*,'my_rank =',my_rank, 'a = ',a
33.     print*,'my_rank =',my_rank, 'b = ',b
34.     call mpi_finalize(ierr)
35.     end program countmpi
```

```
        my_rank = 0 a = 1.000000000 2.718281746
                       3.141592741 186000.0000
        my_rank = 1 a = -0.1527172301E+11 -0.1775718601E+30
                       0.8887595380E-40 0.7346867719E-39
        my_rank = 2 a = -0.1527172301E+11 -0.1775718601E+30
                       0.8887595380E-40 0.7346867719E-39
        my_rank = 3 a = -0.1527172301E+11 -0.1775718601E+30
                       0.8887595380E-40 0.7346867719E-39
    !
```

my_rank = 0 a = 1.000000000 2.718281746
                    3.141592741 186000.0000
my_rank = 0 b = 2 3 3 4 4 5
my_rank = 1 a = 1.000000000 2.718281746
                    3.141592741 186000.0000
my_rank = 1 b = 2 3 3 4 -803901184 -266622208
my_rank = 2 a = 1.000000000 2.718281746
                    3.141592741 186000.0000
my_rank = 2 b = 2 3 3 4 -804478720 -266622208
my_rank = 3 a = 1.000000000 2.718281746
                    3.141592741 186000.0000
my_rank = 3 b = 2 3 3 4 -803901184 -266622208

7.4.3   Derived Type

If the data to be communicated is either of mixed type or is not adjacent in the memory, then one can create a user defined mpi_type. For example, the data to be grouped may have some mpi_real, mpi_int and mpi_char entries and be in nonadjacent locations in memory. The *derived type* must have four items for each entry: blocks or count of each mpi_type, type list, address in memory and displacement. The address in memory can be gotten by a MPI subroutine called mpi_address(a,addresses(1),ierr) where a is one of the entries in the new data type.

The following code dertypempi.f creates a new data type, which is called data_mpi_type. It consists of four entries with one mpi_real, a, one mpi_real, b, one mpi_int, c and one mpi_int, d. These entries are initialized on processor 0 by lines 19-24. In order to communicate them as a single new data type via *mpi_bcast()*, the new data type is created in lines 26-43. The four arrays blocks, typelist, addresses and displacements are initialized. The call in line 42 to *mpi_type_struct*(4, blocks, displacements, typelist, data_mpi_type ,ierr) enters this structure and identifies it with the name data_mpi_type. Finally the call in line 43 to *mpi_type_commit*(data_mpi_type,ierr) finalizes this user defined data type. The call to *mpi_bcast()* in line 52 addresses the first entry of the data_mpi_type and uses count =1 so that the data a, b, c and d will be broadcast to the other processors. This is verified by the print commands in lines 46-49 and 54-57.

MPI/Fortran 9x Code dertypempi.f

```
1.     program dertypempi
2.! Illustrates a derived type.
3.     implicit none
4.     include 'mpif.h'
5.     real:: a,b
6.     integer::c,d
7.     integer::data_mpi_type
```

```
8.      integer::ierr
9.      integer, dimension(1:4)::blocks
10.     integer, dimension(1:4)::displacements
11.     integer, dimension(1:4)::addresses
12.     integer, dimension(1:4)::typelist
13.     integer:: my_rank,p,n,source,dest,tag,loc_n
14.     integer:: i,status(mpi_status_size)
15.     data n,dest,tag/4,0,50/
16.     call mpi_init(ierr)
17.     call mpi_comm_rank(mpi_comm_world,my_rank,ierr)
18.     call mpi_comm_size(mpi_comm_world,p,ierr)
19.     if (my_rank.eq.0) then
20.         a = exp(1.)
21.         b = 4*atan(1.)
22.         c = 1
23.         d = 186000
24.     end if
25.! Define the new derived type, data_mpi_type.
26.     typelist(1) = mpi_real
27.     typelist(2) = mpi_real
28.     typelist(3) = mpi_integer
29.     typelist(4) = mpi_integer
30.     blocks(1) = 1
31.     blocks(2) = 1
32.     blocks(3) = 1
33.     blocks(4) = 1
34.     call mpi_address(a,addresses(1),ierr)
35.     call mpi_address(b,addresses(2),ierr)
36.     call mpi_address(c,addresses(3),ierr)
37.     call mpi_address(d,addresses(4),ierr)
38.     displacements(1) = addresses(1) - addresses(1)
39.     displacements(2) = addresses(2) - addresses(1)
40.     displacements(3) = addresses(3) - addresses(1)
41.     displacements(4) = addresses(4) - addresses(1)
42.     call mpi_type_struct(4,blocks,displacements,&
                    .        typelist,data_mpi_type,ierr)
43.     call mpi_type_commit(data_mpi_type,ierr)
44.! Before the broadcast of the new type data_mpi_type
45.! try to print the data.
46.     print*,'my_rank =',my_rank, 'a = ',a
47.     print*,'my_rank =',my_rank, 'b = ',b
48.     print*,'my_rank =',my_rank, 'c = ',c
49.     print*,'my_rank =',my_rank, 'd = ',d
50.     call mpi_barrier(mpi_comm_world,ierr)
51.! Broadcast data_mpi_type.
```

```
52.     call mpi_bcast(a,1,data_mpi_type,0,&
                              mpi_comm_world,ierr)
53.! Each processor prints the data.
54.     print*,'my_rank =',my_rank, 'a = ',a
55.     print*,'my_rank =',my_rank, 'b = ',b
56.     print*,'my_rank =',my_rank, 'c = ',c
57.     print*,'my_rank =',my_rank, 'd = ',d
58.     call mpi_finalize(ierr)
59.     end program dertypempi
```

```
my_rank = 0 a = 2.718281746
my_rank = 0 b = 3.141592741
my_rank = 0 c = 1
my_rank = 0 d = 186000
my_rank = 1 a = 0.2524354897E-28
my_rank = 1 b = 0.1084320046E-18
my_rank = 1 c = 20108
my_rank = 1 d = 3
my_rank = 2 a = 0.2524354897E-28
my_rank = 2 b = 0.1084320046E-18
my_rank = 2 c = 20108
my_rank = 2 d = 3
my_rank = 3 a = 0.2524354897E-28
my_rank = 3 b = 0.1084320046E-18
my_rank = 3 c = 20108
my_rank = 3 d = 3
!
my_rank = 0 a = 2.718281746
my_rank = 0 b = 3.141592741
my_rank = 0 c = 1
my_rank = 0 d = 186000
my_rank = 1 a = 2.718281746
my_rank = 1 b = 3.141592741
my_rank = 1 c = 1
my_rank = 1 d = 186000
my_rank = 2 a = 2.718281746
my_rank = 2 b = 3.141592741
my_rank = 2 c = 1
my_rank = 2 d = 186000
my_rank = 3 a = 2.718281746
my_rank = 3 b = 3.141592741
my_rank = 3 c = 1
my_rank = 3 d = 186000
```

7.4.4 Packed Type

The subroutine *mpi_pack()* relocates data to a new array, which is addressed sequentially. Communication subroutines such as *mpi_bcast()* can be used with the count parameter to send the data to other processors. The data is then unpacked from the array created by *mpi_unpack()*

```
mpi_pack(locdata, count, mpi_datatype,
            packarray, position, mpi_comm, ierr)
     locdata          array(*)
     count            integer
     mpi_datatype     integer
     packarray        array(*)
     packcount        integer
     position         integer
     mpi_comm         integer
     ierr             integer

mpi_unpack(destarray, count, mpi_datatype,
            locdata, position, mpi_comm, ierr)
     packarray        array(*)
     packcount        integer
     mpi_datatype     integer
     locdata          array(*)
     count            integer
     position         integer
     mpi_comm         integer
     ierr             integer
```

In packmpi.f four variables on processor 0 are initialized in lines 17-18 and packed into the array numbers in lines 21-25. Then in lines 26 and 28 the array number is broadcast to the other processors. In lines 30-34 this data is unpacked to the original local variables, which are duplicated on each of the other processors. The print commands in lines 37-40 verify this.

MPI/Fortran 9x Code packmpi.f

```
1.     program packmpi
2.! Illustrates mpi_pack and mpi_unpack.
3.     implicit none
4.     include 'mpif.h'
5.     real:: a,b
6.     integer::c,d,location
7.     integer::ierr
8.     character, dimension(1:100)::numbers
9.     integer:: my_rank,p,n,source,dest,tag,loc_n
10.     integer:: i,status(mpi_status_size)
11.     data n,dest,tag/4,0,50/
```

```
12.     call mpi_init(ierr)
13.     call mpi_comm_rank(mpi_comm_world,my_rank,ierr)
14.     call mpi_comm_size(mpi_comm_world,p,ierr)
15.! Processor 0 packs and broadcasts the four number.
16.     if (my_rank.eq.0) then
17.         a = exp(1.)
18.         b = 4*atan(1.)
19.         c = 1
20.         d = 186000
21.         location = 0
22.         call mpi_pack(a,1,mpi_real,numbers,100,location,&
                                    mpi_comm_world, ierr)
23.         call mpi_pack(b,1,mpi_real,numbers,100,location,&
                                    mpi_comm_world, ierr)
24.         call mpi_pack(c,1,mpi_integer,numbers,100,location,&
                                    mpi_comm_world, ierr)
25.         call mpi_pack(d,1,mpi_integer,numbers,100,location,&
                                    mpi_comm_world, ierr)
26.         call mpi_bcast(numbers,100,mpi_packed,0,&
                                    mpi_comm_world,ierr)
27.     else
28.         call mpi_bcast(numbers,100,mpi_packed,0,&
                                    mpi_comm_world,ierr)
29.! Each processor unpacks the numbers.
30.         location = 0
31.         call mpi_unpack(numbers,100,location,a,1,mpi_real,&
                                    mpi_comm_world, ierr)
32.         call mpi_unpack(numbers,100,location,b,1,mpi_real,&
                                    mpi_comm_world, ierr)
33.         call mpi_unpack(numbers,100,location,c,1,mpi_integer,&
                                    mpi_comm_world, ierr)
34.         call mpi_unpack(numbers,100,location,d,1,mpi_integer,&
                                    mpi_comm_world, ierr)
35.     end if
36.! Each processor prints the numbers.
37.     print*,'my_rank =',my_rank, 'a = ',a
38.     print*,'my_rank =',my_rank, 'b = ',b
39.     print*,'my_rank =',my_rank, 'c = ',c
40.     print*,'my_rank =',my_rank, 'd = ',d
41.     call mpi_finalize(ierr)
42.     end program packmpi

            !
            my_rank = 0 a = 2.718281746
            my_rank = 0 b = 3.141592741
```

```
my_rank = 0 c = 1
my_rank = 0 d = 186000
my_rank = 1 a = 2.718281746
my_rank = 1 b = 3.141592741
my_rank = 1 c = 1
my_rank = 1 d = 186000
my_rank = 2 a = 2.718281746
my_rank = 2 b = 3.141592741
my_rank = 2 c = 1
my_rank = 2 d = 186000
my_rank = 3 a = 2.718281746
my_rank = 3 b = 3.141592741
my_rank = 3 c = 1
my_rank = 3 d = 186000
```

7.4.5 Exercises

1.   Duplicate the calculations for countmpi.f and experiment with different size arrays and numbers of processors.
2.   Duplicate the calculations for dertypempi.f and experiment with different data types.
3.   Duplicate the calculations for packmpi.f and experiment with different numbers of processors and different size vectors.
4.   Consider a one dimensional array that has many nonzero numbers. Use *mpi_pack()* and *mpi_unpack()* to communicate the nonzero entries in the array.
5.   Repeat exercise 4 for a two dimensional array.

7.5 Communicators

7.5.1 Introduction

The generic communicator that has been used in all of the previous codes is called *mpi_comm_world*. It is the set of all p processors, and all processors can communicate with each other. When collective subroutines are called and the communicator is mpi_comm_world, then the data is to be communicated among the other p-1 processors. In many applications it may not be necessary to communicate with all other processors. Two similar examples were given in the Section 6.6, where the two space dimension heat and pollutant models are considered. In these cases each processor is associated with a horizontal portion of space, and each processor is required to exchange information with the adjacent processors.

7.5.2   A Grid Communicator

In this and the next section *grid communicators* will be used to do matrix-matrix products. In order to motivate this discussion, consider a block 3×3 matrix times a block 3×1 matrix where the blocks are $n \times n$

$$\begin{bmatrix} A_{11} & A_{12} & A_{13} \\ A_{21} & A_{22} & A_{23} \\ A_{31} & A_{32} & A_{33} \end{bmatrix} \begin{bmatrix} X_1 \\ X_2 \\ X_3 \end{bmatrix} = \begin{bmatrix} A_{11}X_1 + A_{12}X_2 + A_{13}X_3 \\ A_{21}X_1 + A_{22}X_2 + A_{23}X_3 \\ A_{31}X_1 + A_{32}X_2 + A_{33}X_3 \end{bmatrix}.$$

Consider $p = 9$ processors and associate then with a 3×3 grid. Assume the matrices A_{ij} are stored on grid processor ij. Then the 9 matrix products $A_{ij}X_j$ could be done concurrently. The overall process is as follows:

> broadcast X_j to column j of processors,
> in parallel compute the matrix products $A_{ij}X_j$ and
> sum the products in row i of processors.

We wish to use collective subroutines restricted to either columns or rows of this grid of processors. In particular, start with the generic communicator, and then create a two dimension grid communicator as well as three row and three column subgrid communicators. The MPI subroutines *mpi_ cart_ create()*, *mpi_ cart_ coords* and *mpi_ cart_ sub()* will help us do this. The subroutine *Setup_ grid(grid)* uses these three MPI subroutines, and it is used in gridcommpi.f, and in foxmpi.f of the next section.

The subroutine *mpi_ cart_ create()* will generate a $d = dim$ dimensional grid communicator from $p = q^d$ processors in the original communicator called mpi_comm. The *dimsize $= q$* with periodic parameter in each dimension is set equal to TRUE, and the numbering of the processor in each row or column of processor will begin at 0 and end at $q - 1$. The new communicator is called grid_comm.

```
call mpi_cart_create(mpi_comm_world, dim,&
        dimsize, periods, .TRUE. , grid_comm, ierr)
    mpi_comm      integer
    dim           integer
    dimsize       integer(*)
    periods       logical(*)
    reorder       logical
    grid_comm     integer
    ierr          logical
```

The subroutine *mpi_ cart_ coords()* associates with each processor, given by grid_my_rank, in grid_comm, a grid_row = coordinates(0) or grid_col = coordinates(1) for *dim $= 2$*.

```
call mpi_cart_coords(grid_comm, grid_my_rank, 2,&
        coordinates, ierr )
```

```
grid_comm        integer
grid_my_rank     integer
dim              integer
coordinates      integer(*)
ierr             logical
```

Subcommunicators can easily be formed by a call to *mpi_cart_sub()*. The subcommunicators are associated with the grid row or grid columns of processors for dim = 2.

```
call mpi_cart_sub(grid_comm, vary_coords, &
       sub_comm, ierr)
    grid_comm     integer
    vary_coords   logical(*)
    sub_comm      integer
    ierr          logical
```

7.5.3   Illustration gridcommpi.f

First, we examine the subroutine *Setup_grid(grid)* in lines 56-91. This subroutine and subsequent subroutines on communicators are Fortran 9x variations of those in Pacheco [21]. This is a two dimensional grid and we have assumed $p = q^2$. The parameter grid is of type GRID_INFO_TYPE as defined in lines 5-14. The integer array dimension and logical array periods are defined in lines 70-73. The call to *mpi_cart_create()* is done in line 74 where a grid%comm is defined. In line 76 grid%my_rank is identified for the communicator grid%comm. Lines 79-80 identify the grid%my_row and grid%my_col. In lines 84 and 88 *mpi_cart_sub()* define the communicators grid%row_comm and grid%col_comm.

Second, the main part of gridcommpi.f simply defines a 6×6 array and uses $p = 3^2$ processors so that the array can be defined by nine processors as given in lines 26-32. The local arrays A are 2×2, and there is a version on each of the nine processors. After line 32 the 6×6 array, which is distributed over the grid communicator, is

$$
\begin{bmatrix}
1 & 1 & 2 & 2 & 3 & 3 \\
1 & 1 & 2 & 2 & 3 & 3 \\
2 & 2 & 3 & 3 & 4 & 4 \\
2 & 2 & 3 & 3 & 4 & 4 \\
3 & 3 & 4 & 4 & 5 & 5 \\
3 & 3 & 4 & 4 & 5 & 5
\end{bmatrix}.
$$

In line 48 *mpi_bcast()* from column processors 1 (corresponds to the second block column in the above matrix) to the other processors in row_comm. This means the new distribution of the matrix will be

$$\begin{bmatrix} 2 & 2 & 2 & 2 & 2 & 2 \\ 2 & 2 & 2 & 2 & 2 & 2 \\ 3 & 3 & 3 & 3 & 3 & 3 \\ 3 & 3 & 3 & 3 & 3 & 3 \\ 4 & 4 & 4 & 4 & 4 & 4 \\ 4 & 4 & 4 & 4 & 4 & 4 \end{bmatrix}.$$

The output from the print command in lines 50-53 verifies this.

MPI/Fortran 9x gridcommpi.f

```
1.      program gridcommpi
2.! Illustrates grid communicators.
3.      include 'mpif.h'
4.      IMPLICIT NONE
5.      type GRID_INFO_TYPE
6.          integer p       ! total number of processes.
7.          integer comm       ! communicator for the entire grid.
8.          integer row_comm       ! communicator for my row.
9.          integer col_comm       ! communicator for my col.
10.         integer q       ! order of grid.
11.         integer my_row       ! my row number.
12.         integer my_col       ! my column number.
13.         integer my_rank       ! my rank in the grid communicator.
14.     end type GRID_INFO_TYPE
15.     TYPE (GRID_INFO_TYPE) :: grid_info
16.     integer :: my_rank, ierr
17.     integer, allocatable, dimension(:,:) :: A,B,C
18.     integer :: i,j,k,n, n_bar
19.     call mpi_init(ierr)
20.     call Setup_grid(grid_info)
21.     call mpi_comm_rank(mpi_comm_world, my_rank, ierr)
22.     if (my_rank == 0) then
23.          n=6
24.     endif
25.     call mpi_bcast(n,1,mpi_integer, 0, mpi_comm_world, ierr)
26.     n_bar = n/(grid_info%q)
27.! Allocate local storage for local matrix.
28.     allocate( A(n_bar,n_bar) )
29.     allocate( B(n_bar,n_bar) )
30.     allocate( C(n_bar,n_bar) )
31.     A = 1 + grid_info%my_row + grid_info%my_col
32.     B = 1 - grid_info%my_row - grid_info%my_col
33.     if (my_rank == 0) then
34.          print*,'n = ',n,'n_bar = ',n_bar,&
```

```
35.                      'grid%p = ',grid_info%p, 'grid%q = ',grid_info%q
36.      end if
37.      print*, 'my_rank = ',my_rank,&
38.                          'grid_info%my_row = ',grid_info%my_row,&
39.                          'grid_info%my_col = ',grid_info%my_col
40.      call mpi_barrier(mpi_comm_world, ierr)
41.      print*, 'grid_info%my_row =',grid_info%my_row,&
42.              'grid_info%my_col =',grid_info%my_col,&
43.              'A = ',A(1,:),&
44.              ' ; ',A(2,:)
45.! Uses mpi_bcast to send and receive parts of the array, A,
46.! to the processors in grid_info%row_com, which was defined
47.! in the call to the subroutine Setup_grid(grid_info).
48.      call mpi_bcast(A,n_bar*n_bar,mpi_integer,&
49.                       1, grid_info%row_comm, ierr)
50.      print*, 'grid_info%my_row =',grid_info%my_row,&
51.              'grid_info%my_col =',grid_info%my_col,&
52.              ' new_A = ',A(1,:),&
53.              ' ; ',A(2,:)
54.      call mpi_finalize(ierr)
55.      contains
!
56.      subroutine Setup_grid(grid)
57.! This subroutine defines a 2D grid communicator.
58.! And for each grid row and grid column additional
59.! communicators are defined.
60.      TYPE (GRID_INFO_TYPE), intent(inout) :: grid
61.          integer old_rank
62.          integer dimensions(0:1)
63.          logical periods(0:1)
64.          integer coordinates(0:1)
65.          logical varying_coords(0:1)
66.          integer ierr
67.      call mpi_comm_size(mpi_comm_world, grid%p, ierr)
68.      call mpi_comm_rank(mpi_comm_world, old_rank, ierr )
69.      grid%q = int(sqrt(dble(grid%p)))
70.      dimensions(0) = grid%q
71.      dimensions(1) = grid%q
72.      periods(0) = .TRUE.
73.      periods(1) = .TRUE.
74.      call mpi_cart_create(mpi_comm_world, 2,&
75.                      dimensions, periods, .TRUE. , grid%comm, ierr)
76.      call mpi_comm_rank (grid%comm, grid%my_rank, ierr )
77.      call mpi_cart_coords(grid%comm, grid%my_rank, 2,&
78.                          coordinates, ierr )
```

```
79.      grid%my_row = coordinates(0)
80.      grid%my_col = coordinates(1)
81.! Set up row and column communicators.
82.      varying_coords(0) = .FALSE.
83.      varying_coords(1) = .TRUE.
84.      call mpi_cart_sub(grid%comm,varying_coords,&
85.                          grid%row_comm,ierr)
86.      varying_coords(0) = .TRUE.
87.      varying_coords(1) = .FALSE.
88.      call mpi_cart_sub(grid%comm,varying_coords,&
89.                          grid%col_comm,ierr)
90.      end subroutine Setup_grid
91.      end program gridcommpi
!
     n = 6 n_bar = 2 grid%p = 9 grid%q = 3
     !
     my_rank = 0 grid_info%my_row = 2 grid_info%my_col = 2
     my_rank = 1 grid_info%my_row = 2 grid_info%my_col = 1
     my_rank = 2 grid_info%my_row = 2 grid_info%my_col = 0
     my_rank = 3 grid_info%my_row = 1 grid_info%my_col = 0
     my_rank = 4 grid_info%my_row = 1 grid_info%my_col = 2
     my_rank = 5 grid_info%my_row = 0 grid_info%my_col = 2
     my_rank = 6 grid_info%my_row = 0 grid_info%my_col = 1
     my_rank = 7 grid_info%my_row = 0 grid_info%my_col = 0
     my_rank = 8 grid_info%my_row = 1 grid_info%my_col = 1
     !
     grid_info%my_row = 0 grid_info%my_col = 0 A = 1 1 ; 1 1
     grid_info%my_row = 1 grid_info%my_col = 0 A = 2 2 ; 2 2
     grid_info%my_row = 2 grid_info%my_col = 0 A = 3 3 ; 3 3
     grid_info%my_row = 0 grid_info%my_col = 1 A = 2 2 ; 2 2
     grid_info%my_row = 1 grid_info%my_col = 1 A = 3 3 ; 3 3
     grid_info%my_row = 2 grid_info%my_col = 1 A = 4 4 ; 4 4
     grid_info%my_row = 0 grid_info%my_col = 2 A = 3 3 ; 3 3
     grid_info%my_row = 1 grid_info%my_col = 2 A = 4 4 ; 4 4
     grid_info%my_row = 2 grid_info%my_col = 2 A = 5 5 ; 5 5
     !
     grid_info%my_row = 0 grid_info%my_col = 0 new_A = 2 2 ; 2 2
     grid_info%my_row = 1 grid_info%my_col = 0 new_A = 3 3 ; 3 3
     grid_info%my_row = 2 grid_info%my_col = 0 new_A = 4 4 ; 4 4
     grid_info%my_row = 0 grid_info%my_col = 1 new_A = 2 2 ; 2 2
     grid_info%my_row = 1 grid_info%my_col = 1 new_A = 3 3 ; 3 3
     grid_info%my_row = 2 grid_info%my_col = 1 new_A = 4 4 ; 4 4
     grid_info%my_row = 0 grid_info%my_col = 2 new_A = 2 2 ; 2 2
     grid_info%my_row = 1 grid_info%my_col = 2 new_A = 3 3 ; 3 3
     grid_info%my_row = 2 grid_info%my_col = 2 new_A = 4 4 ; 4 4
```

7.5.4 Exercises

1.   Duplicate the computations for gridcommmpi.f. Change $mpi\_bcast()$ to $mpi\_bcast(A,n\_bar*n\_bar,mpi\_real,x,grid\_info\%row\_comm,ierr)$ where x is 0 and 2. Explain the outputs.

2.   In gridcommpi.f change the communicator from row_comm to col_comm.by using $mpi\_bcast(A,n\_bar*n\_bar,mpi\_real,x,grid\_info\%col\_comm,ierr)$ where x is 0, 1 and 2. Explain the outputs.

7.6   Fox Algorithm for AB

7.6.1   Introduction

In this section the block matrix-matrix product AB will be done where A and B are both $q \times q$ block matrices. The number of processors used to do this will be $p = q^2$, and the grid communicator that was defined in the subroutine Setup_grid() will be used. Fox's algorithm follows a similar pattern as in AX in the previous section where A is 3×3 and X is 3×1. The numbering of the block rows and columns of the matrices will start at 0 and end at $q - 1$.

7.6.2   Matrix-Matrix Product

The classical definition of matrix product $C = AB$ is block ij of C equals block row i of A times block column j of B

$$C_{ij} = \sum_{k=0}^{q-1} A_{ik} B_{kj}.$$

The summation can be done in any order, and the matrix products for a fixed ij can be done by the grid processor ij. If the matrices A_{ij}, B_{ij}, and C_{ij} are stored on the grid processor ij, then the challenge is to communicate the required matrices to the grid processor ij.

7.6.3   Parallel Fox Algorithm

In order to motivate the Fox algorithm, consider the block 3×3 case

$$\begin{bmatrix} C_{00} & C_{01} & C_{02} \\ C_{10} & C_{11} & C_{12} \\ C_{20} & C_{21} & C_{22} \end{bmatrix} = \begin{bmatrix} A_{00} & A_{01} & A_{02} \\ A_{10} & A_{11} & A_{12} \\ A_{20} & A_{21} & A_{22} \end{bmatrix} \begin{bmatrix} B_{00} & B_{01} & B_{02} \\ B_{10} & B_{11} & B_{12} \\ B_{20} & B_{21} & B_{22} \end{bmatrix}.$$

Assume that processor ij in the *grid communicator* has stored the matrices A_{ij} and B_{ij}. Consider the computation of the second block row of C, which can be reordered as follows

$$
\begin{aligned}
C_{10} &= A_{11}B_{10} + A_{12}B_{20} + A_{10}B_{00} \\
C_{11} &= A_{11}B_{11} + A_{12}B_{21} + A_{10}B_{01} \\
C_{12} &= A_{11}B_{12} + A_{12}B_{22} + A_{10}B_{02}.
\end{aligned}
$$

Grid processor $1j$ can compute the first term on the right side if the matrix A_{11} has been broadcast to grid processor $1j$. In order for grid processor $1j$ to compute the second matrix product on the right side, the matrix B_{1j} must be replaced by B_{2j}, and matrix A_{12} must be broadcast to the grid processor $1j$. The last step is for $q - 1 = 2$ where the matrix B_{2j} must be replaced by B_{0j}, and the matrix A_{10} must be broadcast to the grid processor $1j$. For the $q \times q$ block matrices there are q matrix products for each grid processor, and this can be done in a loop whose index is step as in the following algorithm.

Fox Algorithm for the Matrix Product $C = C + AB$

> $q = p^{1/2}$, $source = (mod(i+1, q), j)$, $dest = (mod(i-1, q), j)$
> concurrently with $0 \le i, j \le q - 1$
> for $step = 0, q - 1$
> $k\_bar = mod(i + step, q)$
>   broadcast A_{i,k_bar} to grid row i of processors
> $C_{ij} = C_{ij} + A_{i,k\_bar} B_{k\_bar,j}$
>   send $B_{k\_bar,j}$ to processor dest
>   receive $B_{kk,j}$ from source where $kk = mod(k\_bar + 1, q)$
> endloop.

7.6.4   Illustration foxmpi.f

This implementation of the Fox algorithm is a variation on that given by Pacheco [21], but here Fortran 9x is used, and the matrix products for the submatrices are done either by a call to the BLAS3 subroutine *sgemm()* or by the jki loops. The input to this is given in lines 1-33, the call to the fox subroutine is in line 34, and the output in given in lines 36-42. The subroutine *Setup_ grid(grid)* is the same as listed in the previous section. The Fox subroutine is listed in lines 48-96. The step loop of the Fox algorithm is executed in lines 60-95. The matrix products may be done by either *sgemm()* or the jki loops, which are listed here as commented out of execution. The broadcast of the matrix A_{i,k_bar} over the grid_row communicator is done in lines 63-64 and 79-80; note how one stores local_A from bcast_root in temp_A so as not to overwrite local_A in destination. Then the matrix products $A_{i,k\_bar} B_{k\_bar,j}$ are done in lines 66-67 and 81-82. The subroutine *mpi_ sendrecv_replace()* in lines 92-94 is used to communicate $B_{k\_bar,j}$ within the grid_col communicator.

MPI/Fortran 9x Code foxmpi.f

    1.   program foxmpi

```
2.      include 'mpif.h'
3.      IMPLICIT NONE
4.      type GRID_INFO_TYPE
5.      integer p ! total number of processes.
6.      integer comm ! communicator for the entire grid.
7.      integer row_comm ! communicator for my row.
8.      integer col_comm ! communicator for my col.
9.      integer q ! order of grid.
10.     integer my_row ! my row number.
11.     integer my_col ! my column number.
12.     integer my_rank ! my rank in the grid communicator.
13.     end type GRID_INFO_TYPE
14.     TYPE (GRID_INFO_TYPE) :: grid_info
15.     integer :: my_rank, ierr
16.     real, allocatable, dimension(:,:) :: A,B,C
17.     integer :: i,j,k,n, n_bar
18.     real:: mflops,t1,t2,timef
19.     call mpi_init(ierr)
20.     call Setup_grid(grid_info)
21.     call mpi_comm_rank(mpi_comm_world, my_rank, ierr)
22.     if (my_rank == 0) then
23.         n = 800 !n = 6
24.         t1 = timef()
25.     endif
26.     call mpi_bcast(n,1,mpi_integer, 0, mpi_comm_world,ierr)
27.     n_bar = n/(grid_info%q)
28.     ! Allocate storage for local matrix.
29.     allocate( A(n_bar,n_bar) )
30.     allocate( B(n_bar,n_bar) )
31.     allocate( C(n_bar,n_bar) )
32.     A = 1.0 + grid_info%my_row + grid_info%my_col
33.     B = 1.0 - grid_info%my_row - grid_info%my_col
34.     call Fox(n,grid_info,A,B,C,n_bar)
35.     ! print*,grid_info%my_row, grid_info%my_col, 'C = ',C
36.     if (my_rank == 0) then
37.         t2 = timef()
38.         print*,t2
39.         print*,n,n_bar,grid_info%q
40.         mflops = (2*n*n*n)*.001/t2
41.         print*, mflops
42.     endif
43.     call mpi_finalize(ierr)
44.     contains
45.     !
46.     !subroutine Setup_grid ....see chapter 7.5 and gridcommpi.f
```

```
47.     !
48.     subroutine Fox(n,grid,local_A,local_B,local_C,n_bar)
49.     integer, intent(in) :: n, n_bar
50.     TYPE(GRID_INFO_TYPE), intent(in) :: grid
51.     real, intent(in) , dimension(:,:) :: local_A, local_B
52.     real, intent(out), dimension (:,:):: local_C
53.     real, dimension(1:n_bar,1:n_bar) :: temp_A
54.     integer:: step, source, dest, request,i,j
55.     integer:: status(MPI_STATUS_SIZE), bcast_root
56.     temp_A = 0.0
57.     local_C = 0.0
58.     source = mod( (grid%my_row + 1), grid%q )
59.     dest = mod( (grid%my_row - 1 + grid%q), grid%q )
60.     do step = 0, grid%q -1
61.         bcast_root = mod( (grid%my_row + step), grid%q )
62.         if (bcast_root == grid%my_col) then
63.             call mpi_bcast(local_A, n_bar*n_bar, mpi_real,&
64.                            bcast_root, grid%row_comm, ierr)
65.         ! print*, grid%my_row, grid%my_col, 'local_A = ',local_A
66.             call sgemm('N','N',n_bar,n_bar,n_bar,1.0,&
67.                 local_A,n_bar,local_B,n_bar,1.0,local_C,n_bar)
68.         ! do j = 1,n_bar
69.         !     do k = 1,n_bar
70.         !         do i = 1,n_bar
71.         !             local_C(i,j)=local_C(i,j) + local_A(i,k)*&
72.         !                          local_B(k,j)
73.         !         end do
74.         !     end do
75.         ! end do
76.         else
77.             ! Store local_A from bcast_root in temp_A so as
78.             ! not to overwrite local_A in destination.
79.             call mpi_bcast(temp_A, n_bar*n_bar, mpi_real,&
80.                            bcast_root, grid%row_comm, ierr)
81.             call sgemm('N','N',n_bar,n_bar,n_bar,1.0,&
82.                 temp_A,n_bar,local_B,n_bar,1.0,local_C,n_bar)
83.         ! do j = 1,n_bar
84.         !     do k = 1,n_bar
85.         !         do i = 1,n_bar
86.         !             local_C(i,j)=local_C(i,j) + temp_A(i,k)*&
87.         !                          local_B(k,j)
88.         !             enddo
89.         !         enddo
90.         ! enddo
91.         endif
```

Table 7.6.1: Fox Times

Sub Product	Dimension	Processors	Time	mflops
sgemm()	800	2 × 2	295	4,173
sgemm()	800	4 × 4	121	8,193
sgemm()	1600	2 × 2	1,635	5,010
sgemm()	1600	4 × 4	578	14,173
jki loops	800	2 × 2	980	1,043
jki loops	800	4 × 4	306	3,344
jki loops	1600	2 × 2	7,755	1,056
jki loops	1600	4 × 4	2,103	3,895

```
92.          call mpi_sendrecv_replace(local_B,n_bar*n_bar,mpi_real,&
93.                               dest, 0,source, 0, &
94.                               grid%col_comm,status, ierr)
95.     end do
96.     end subroutine Fox
97.     !
98.     end program foxmpi
```

Eight executions were done with foxmpi.f, and the outputs are recorded in Table 7.6.1. The first four used the *sgemm()* and the second four used jki loops. The optimized *sgemm()* was about four times faster than the jki loops. The time units were in milliseconds, and the mflops for the larger dimensions were always the largest.

7.6.5 Exercises

1. Verify for n = 6 that the matrix product is correct. See lines 23 and 35 in foxmpi.f.

2. Duplicate the computations for foxmpi.f, and also use a 8 × 8 grid of processors.

3. Compare the matrix product scheme used in Section 6.5 mmmpi.f with the Fox algorithm in foxmpi.f.

Chapter 8

Classical Methods for Ax = d

The first three sections contain a description of direct methods based on the Schur complement and domain decomposition. After the first section the coefficient matrix will be assumed to be symmetric positive definite (SPD). In Section 8.3 an MPI code will be studied that illustrates domain decomposition. Iterative methods based on P-regular splittings and domain decompositions will be described in the last three sections. Here convergence analysis will be given via the minimization of the equivalent quadratic functional. An MPI version of SOR using domain decomposition will be presented in Section 8.5. This chapter is more analysis-oriented and less application-driven.

8.1   Gauss Elimination

Gauss elimination method, which was introduced in Section 2.2, requires A to be factored into a product of a lower and upper triangular matrices that have inverses. This is not always possible, for example, consider the 2×2 matrix where the a_{11} is zero. Then one can interchange the second and first rows

$$\begin{bmatrix} 0 & 1 \\ 1 & 0 \end{bmatrix} \begin{bmatrix} 0 & a_{12} \\ a_{21} & a_{22} \end{bmatrix} = \begin{bmatrix} a_{21} & a_{22} \\ 0 & a_{12} \end{bmatrix}.$$

If a_{11} is not zero, then one can use an elementary row operation

$$\begin{bmatrix} 1 & 0 \\ -a_{21}/a_{11} & 1 \end{bmatrix} \begin{bmatrix} a_{11} & a_{12} \\ a_{21} & a_{22} \end{bmatrix} = \begin{bmatrix} a_{11} & a_{12} \\ 0 & a_{22} - (a_{21}/a_{11})a_{12} \end{bmatrix}.$$

Definition. *If there is a permutation matrix P such that $PA = LU$ where L and U are invertible lower and upper triangular matrices, respectively, then the matrix PA is said to have an LU factorization.*

Gaussian elimination method for the solution of $Ax = d$ uses permutations of the rows and elementary row operations to find the LU factorization so that

$$
\begin{aligned}
PAx &= L(Ux) = Pd \\
\text{solve } Ly &= Pd \text{ and} \\
\text{solve } Ux &= y.
\end{aligned}
$$

Example. Consider the 3×3 matrix

$$
\begin{bmatrix} 1 & 2 & 0 \\ 1 & 2 & 1 \\ 0 & 1 & 3 \end{bmatrix}.
$$

Since the component in the first row and column is not zero, no row interchange is necessary for the first column. The elementary row operation on the first column is

$$
\begin{bmatrix} 1 & 0 & 0 \\ -1 & 1 & 0 \\ 0 & 0 & 1 \end{bmatrix}
\begin{bmatrix} 1 & 2 & 0 \\ 1 & 2 & 1 \\ 0 & 1 & 3 \end{bmatrix}
=
\begin{bmatrix} 1 & 2 & 0 \\ 0 & 0 & 1 \\ 0 & 1 & 3 \end{bmatrix}.
$$

For column two we must interchange rows two and three

$$
\begin{bmatrix} 1 & 0 & 0 \\ 0 & 0 & 1 \\ 0 & 1 & 0 \end{bmatrix}
\begin{bmatrix} 1 & 0 & 0 \\ -1 & 1 & 0 \\ 0 & 0 & 1 \end{bmatrix}
\begin{bmatrix} 1 & 2 & 0 \\ 1 & 2 & 1 \\ 0 & 1 & 3 \end{bmatrix}
=
\begin{bmatrix} 1 & 2 & 0 \\ 0 & 1 & 3 \\ 0 & 0 & 1 \end{bmatrix}.
$$

Note the first two factors on the left side can be rewritten as

$$
\begin{bmatrix} 1 & 0 & 0 \\ 0 & 0 & 1 \\ 0 & 1 & 0 \end{bmatrix}
\begin{bmatrix} 1 & 0 & 0 \\ -1 & 1 & 0 \\ 0 & 0 & 1 \end{bmatrix}
=
\begin{bmatrix} 1 & 0 & 0 \\ 0 & 1 & 0 \\ -1 & 0 & 1 \end{bmatrix}
\begin{bmatrix} 1 & 0 & 0 \\ 0 & 0 & 1 \\ 0 & 1 & 0 \end{bmatrix}.
$$

This gives the desired factorization of the matrix

$$
\begin{bmatrix} 1 & 0 & 0 \\ 0 & 1 & 0 \\ -1 & 0 & 1 \end{bmatrix}
\begin{bmatrix} 1 & 0 & 0 \\ 0 & 0 & 1 \\ 0 & 1 & 0 \end{bmatrix}
\begin{bmatrix} 1 & 2 & 0 \\ 1 & 2 & 1 \\ 0 & 1 & 3 \end{bmatrix}
=
\begin{bmatrix} 1 & 2 & 0 \\ 0 & 1 & 3 \\ 0 & 0 & 1 \end{bmatrix}
$$

$$
\begin{bmatrix} 1 & 0 & 0 \\ 0 & 0 & 1 \\ 0 & 1 & 0 \end{bmatrix}
\begin{bmatrix} 1 & 2 & 0 \\ 1 & 2 & 1 \\ 0 & 1 & 3 \end{bmatrix}
=
\begin{bmatrix} 1 & 0 & 0 \\ 0 & 1 & 0 \\ 1 & 0 & 1 \end{bmatrix}
\begin{bmatrix} 1 & 2 & 0 \\ 0 & 1 & 3 \\ 0 & 0 & 1 \end{bmatrix}.
$$

In order to extend this to $n \times n$ matrices, as in Section 2.4 consider just a 2×2 block matrix where the diagonal blocks are square but may not have the same dimension

$$
A = \begin{bmatrix} B & E \\ F & C \end{bmatrix}. \tag{8.1.1}
$$

In general A is $n \times n$ with $n = k + m$, B is $k \times k$, C is $m \times m$, E is $k \times m$ and F is $m \times k$. If B has an inverse, then we can multiply block row one by FB^{-1} and subtract it from block row two. This is equivalent to multiplication of A by a *block elementary matrix* of the form

$$\begin{bmatrix} I_k & 0 \\ -FB^{-1} & I_m \end{bmatrix}.$$

If $Ax = d$ is viewed in block form, then

$$\begin{bmatrix} B & E \\ F & C \end{bmatrix} \begin{bmatrix} X_1 \\ X_2 \end{bmatrix} = \begin{bmatrix} D_1 \\ D_2 \end{bmatrix}. \tag{8.1.2}$$

The above block elementary matrix multiplication gives

$$\begin{bmatrix} B & E \\ 0 & C - FB^{-1}E \end{bmatrix} \begin{bmatrix} X_1 \\ X_2 \end{bmatrix} = \begin{bmatrix} D_1 \\ D_2 - FB^{-1}D_1 \end{bmatrix}. \tag{8.1.3}$$

So, if the block upper triangular matrix has an inverse, then this last block equation can be solved.

The following basic properties of square matrices play an important role in the solution of (8.1.1). These properties follow directly from the definition of an inverse matrix.

Theorem 8.1.1 *(Basic Matrix Properties) Let B and C be square matrices that have inverses. Then the following equalities hold:*

1. $\begin{bmatrix} B & 0 \\ 0 & C \end{bmatrix}^{-1} = \begin{bmatrix} B^{-1} & 0 \\ 0 & C^{-1} \end{bmatrix}$,

2. $\begin{bmatrix} I_k & 0 \\ F & I_m \end{bmatrix}^{-1} = \begin{bmatrix} I_k & 0 \\ -F & I_m \end{bmatrix}$,

3. $\begin{bmatrix} B & 0 \\ F & C \end{bmatrix} = \begin{bmatrix} B & 0 \\ 0 & C \end{bmatrix} \begin{bmatrix} I_k & 0 \\ C^{-1}F & I_m \end{bmatrix}$ and

4. $\begin{bmatrix} B & 0 \\ F & C \end{bmatrix}^{-1} = \begin{bmatrix} B^{-1} & 0 \\ -C^{-1}FB^{-1} & C^{-1} \end{bmatrix}$.

Definition. *Let A have the form in (8.1.1) and B be nonsingular. The Schur complement of B in A is $C - FB^{-1}E$.*

Theorem 8.1.2 *(Schur Complement Existence) Consider A as in (8.1.1) and let B have an inverse. A has an inverse if and only if the Schur complement of B in A has an inverse.*

Proof.   The proof of the Schur complement theorem is a direct consequence of using a block elementary row operation to get a zero matrix in the block row 2 and column 1 position

$$
\begin{bmatrix} I_k & 0 \\ -FB^{-1} & I_m \end{bmatrix} \begin{bmatrix} B & E \\ F & C \end{bmatrix} = \begin{bmatrix} B & E \\ 0 & C - FB^{-1}E \end{bmatrix}.
$$

Assume that A has an inverse and show the Schur complement must have an inverse. Since the two matrices on the left side have inverses, the matrix on the right side has an inverse. Because B has an inverse, the Schur complement must have an inverse. Conversely, A may be factored as

$$
\begin{bmatrix} B & E \\ F & C \end{bmatrix} = \begin{bmatrix} I_k & 0 \\ FB^{-1} & I_m \end{bmatrix} \begin{bmatrix} B & E \\ 0 & C - FB^{-1}E \end{bmatrix}.
$$

If both B and the Schur complement have inverses, then both matrices on the right side have inverses so that A also has an inverse. ■

   The choice of the blocks B and C can play a very important role. Often the choice of the physical object, which is being modeled, does this. For example consider the airflow over an aircraft. Here we might partition the aircraft into wing, rudder, fuselage and "connecting" components. Such partitions of the physical object or the matrix are called *domain decompositions*.

   Physical problems often should have a unique solution, and so we shall assume that for $Ax = d$, if it has a solution, then the solution is unique. This means if $Ax = d$ and $A\widehat{x} = d$, then $x = \widehat{x}$. This will be true if the only solution of $Az = 0$ is the zero vector $z = 0$ because for $z = x - \widehat{x}$

$$
Ax - A\widehat{x} = A(x - \widehat{x}) = d - d = 0.
$$

Another important consequence of $Az = 0$ implies $z = 0$ is that no column of A can be the zero column vector. This follows from contradiction, if the column j of A is a zero vector, then let z be the j unit column vector so that $Az = 0$ and z is not a zero vector.

   The condition, $Az = 0$ implies $z = 0$, is very helpful in finding the factorization $PA = LU$. For example, if A is 2×2 and $Az = 0$ implies $z = 0$, then column one must have at least one nonzero component so that either $PA = U$ or $A = LU$, see the first paragraph of this section. If A has an inverse, then $Az = 0$ implies $z = 0$ and either $PA = U$ or $L^{-1}A = U$ so that U must have an inverse. This generalizes via mathematical induction when A is $n \times n$.

Theorem 8.1.3 *(LU Factorization) If A has an inverse so that $Az = 0$ implies $z = 0$, then there exist permutation matrix P, and invertible lower and upper triangular matrices L and U, respectively, such that $PA = LU$.*

Proof.   We have already proved the $n = 2$ case. Assume it is true for any $(n - 1) \times (n - 1)$ matrix. If A is invertible, then column one must have some nonzero component, say in row i. So, if the first component is not zero, interchange the first row with the row i. Let P be the associated permutation matrix so that

$$PA = \begin{bmatrix} b & e \\ f & C \end{bmatrix}$$

where $b \neq 0$ is $1 \times 1, e$ is $1 \times (n-1), f$ is $(n-1) \times 1$ and C is $(n-1) \times (n-1)$. Apply the Schur complement analysis and the block elementary matrix operation

$$\begin{bmatrix} 1 & 0 \\ -fb^{-1} & I_{n-1} \end{bmatrix} \begin{bmatrix} b & e \\ f & C \end{bmatrix} = \begin{bmatrix} b & e \\ 0 & \widehat{C} \end{bmatrix}$$

where $\widehat{C} = C - fb^{-1}e$ is the $(n-1) \times (n-1)$ Schur complement. By the Schur complement theorem $\widehat{C}$ must have an inverse. Use the inductive assumption to write $\widehat{C} = \widehat{P}\widehat{L}\widehat{U}$.

$$\begin{bmatrix} b & e \\ 0 & \widehat{C} \end{bmatrix} = \begin{bmatrix} b & e \\ 0 & \widehat{P}\widehat{L}\widehat{U} \end{bmatrix}$$

$$= \begin{bmatrix} 1 & 0 \\ 0 & \widehat{P} \end{bmatrix} \begin{bmatrix} 1 & 0 \\ 0 & \widehat{L} \end{bmatrix} \begin{bmatrix} b & e \\ 0 & \widehat{U} \end{bmatrix}.$$

Since $\widehat{P}$ is a permutation matrix,

$$\begin{bmatrix} 1 & 0 \\ 0 & \widehat{P} \end{bmatrix} \begin{bmatrix} 1 & 0 \\ -fb^{-1} & I_{n-1} \end{bmatrix} PA = \begin{bmatrix} 1 & 0 \\ 0 & \widehat{L} \end{bmatrix} \begin{bmatrix} b & e \\ 0 & \widehat{U} \end{bmatrix}.$$

Note

$$\begin{bmatrix} 1 & 0 \\ 0 & \widehat{P} \end{bmatrix} \begin{bmatrix} 1 & 0 \\ -fb^{-1} & I_{n-1} \end{bmatrix} = \begin{bmatrix} 1 & 0 \\ \widehat{P}(-fb^{-1}) & \widehat{P} \end{bmatrix}$$

$$= \begin{bmatrix} 1 & 0 \\ \widehat{P}(-fb^{-1}) & I_{n-1} \end{bmatrix} \begin{bmatrix} 1 & 0 \\ 0 & \widehat{P} \end{bmatrix}.$$

Then

$$\begin{bmatrix} 1 & 0 \\ \widehat{P}(-fb^{-1}) & I_{n-1} \end{bmatrix} \begin{bmatrix} 1 & 0 \\ 0 & \widehat{P} \end{bmatrix} PA = \begin{bmatrix} 1 & 0 \\ 0 & \widehat{L} \end{bmatrix} \begin{bmatrix} b & e \\ 0 & \widehat{U} \end{bmatrix}.$$

Finally, multiply by the inverse of the left factor on the left side to get the desired factorization

$$\begin{bmatrix} 1 & 0 \\ 0 & \widehat{P} \end{bmatrix} PA = \begin{bmatrix} 1 & 0 \\ \widehat{P}fb^{-1} & I_{n-1} \end{bmatrix} \begin{bmatrix} 1 & 0 \\ 0 & \widehat{L} \end{bmatrix} \begin{bmatrix} b & e \\ 0 & \widehat{U} \end{bmatrix}$$

$$= \begin{bmatrix} 1 & 0 \\ \widehat{P}fb^{-1} & \widehat{L} \end{bmatrix} \begin{bmatrix} b & e \\ 0 & \widehat{U} \end{bmatrix}.$$

■

In order to avoid significant roundoff errors due to small diagonal components, the row interchanges can be done by choosing the row with the largest

possible nonzero component. If the matrix is symmetric positive definite, then
the diagonals will be positive and the row interchanges may not be necessary.
In either case one should give careful consideration to using the subroutines in
LAPACK [1].

8.1.1   Exercises

1.   Consider the 3×3 example. Use the Schur complement as in (8.1.3) to
solve $Ax = d = [1 \ 2 \ 3]^T$ where $B = [1]$.
2.   Consider the 3×3 example. Identify the steps in the existence theorem
for the factorization of the matrix.
3.   In the proof of the existence theorem for the factorization of the matrix,
write the factorization in component form using the Schur complement

$$\widehat{C} = C - (f/b)e = [c_{ij} - (f_i/b)e_j]$$

where e_j is jth component of the $1 \times (n-1)$ array e. Write this as either the ij
or ji version and explain why these could be described as the row and column
versions, respectively.
4.   Assume B is an invertible $k \times k$ matrix.
    (a).   Verify the following for $\widehat{C} = C - FB^{-1}F$

$$\begin{bmatrix} I_k & 0 \\ -FB^{-1} & I_m \end{bmatrix} \begin{bmatrix} B & E \\ F & C \end{bmatrix} \begin{bmatrix} I_k & -B^{-1}E \\ 0 & I_m \end{bmatrix} = \begin{bmatrix} B & 0 \\ 0 & \widehat{C} \end{bmatrix}.$$

    (b).   Use this to show A has an inverse if and only if $\widehat{C}$ has an inverse.
5.   Assume A is an $n \times n$ matrix and prove the following are equivalent:
    (i). A has an inverse,
    (ii).   there exist permutation matrix P, and
             invertible lower and upper triangular matrices L and U,
             respectively, such that $PA = LU$ and
    (iii). $Az = 0$ implies $z = 0$.

8.2   Symmetric Positive Definite Matrices

In this section we will restrict the matrices to symmetric positive definite matrices. Although this restriction may seem a little severe, there are a number of important applications, which include some classes of partial differential
equations and some classes of least squares problems. The advantage of this
restriction is that the number of operations to do Gaussian elimination can be
cut in half.

Definition. *Let A be an $n \times n$ real matrix. A is a real symmetric positive
definite matrix (SPD) if and only if $A = A^T$ and for all $x \neq 0$, $x^T Ax > 0$.*

Examples.

1. Consider the 2×2 matrix $\begin{bmatrix} 2 & -1 \\ -1 & 2 \end{bmatrix}$ and note

$$x^T A x = x_1^2 + (x_1 - x_2)^2 + x_2^2 > 0.$$

A similar $n \times n$ matrix is positive definite

$$\begin{bmatrix} 2 & -1 & & \\ -1 & 2 & \ddots & \\ & \ddots & \ddots & -1 \\ & & -1 & 2 \end{bmatrix}.$$

2. The matrix A for which $A = A^T$ with $a_{ii} > \sum_{j \neq i} |a_{ij}|$ is positive definite. The symmetry implies that the matrix is also column *strictly diagonally dominant*, that is, $a_{jj} > \sum_{i \neq j} |a_{ij}|$. Now use this and the inequality $|ab| \leq \frac{1}{2}(a^2 + b^2)$ to show for all $x \neq 0$, $x^T A x > 0$.

3. Consider the normal equation from the least squares problem where A is $m \times n$ where $m > n$. Assume A has *full column rank* $(Ax = 0$ implies $x = 0)$, then the normal equation $A^T A x = A^T d$ is equivalent to finding the least squares solution of $Ax = d$. Here $A^T A$ is SPD because if $x \neq 0$, then $Ax \neq 0$, and $x^T(A^T A)x = (Ax)^T(Ax) > 0$.

Theorem 8.2.1 *(Basic Properties of SPD Matrices) If A is an $n \times n$ SPD matrix, then*

1. *The diagonal components of A are positive, $a_{ii} > 0$,*

2. *If $A = \begin{bmatrix} B & F^T \\ F & C \end{bmatrix}$, then B and C are SPD,*

3. *$Ax = 0$ implies $x = 0$ so that A has an inverse and*

4. *If S is $m \times n$ with $m \geq n$ and has full column rank, then $S^T A S$ is positive definite.*

Proof. 1. Choose $x = e_i$, the unit vector with 1 in component i so that $x^T A x = a_{ii} > 0$.

2. Choose $x = \begin{bmatrix} X_1 \\ 0 \end{bmatrix}$ so that $x^T A x = X_1^T B X_1 > 0$.

3. Let $x \neq 0$ so that by the positive definite assumption $x^T A x > 0$ and $Ax \neq 0$. This is equivalent to item 3 and the existence of an inverse matrix.

4. Let $x \neq 0$ so that by the full rank assumption on S $Sx \neq 0$. By the positive definite assumption on A

$$(Sx)^T A(Sx) = x^T(S^T A S)x > 0.$$

The next theorem uses the Schur complement to give a characterization of the block 2×2 SPD matrix

$$A = \begin{bmatrix} B & F^T \\ F & C \end{bmatrix}. \tag{8.2.1}$$

Theorem 8.2.2 *(Schur Complement Characterization) Let A as in (8.2.1) be symmetric. A is SPD if and only if B and the Schur complement of B in A, $\hat{C} = C - FB^{-1}F^T$, are SPD.*

Proof.   Assume A is SPD so that B is also SPD and has an inverse. Then one can use block row and column elementary operations to show

$$\begin{bmatrix} I_k & 0 \\ -FB^{-1} & I_m \end{bmatrix} \begin{bmatrix} B & F^T \\ F & C \end{bmatrix} \begin{bmatrix} I_k & -B^{-1}F^T \\ 0 & I_m \end{bmatrix} = \begin{bmatrix} B & 0 \\ 0 & \hat{C} \end{bmatrix}.$$

Since B is SPD, $\left(B^{-1}\right)^T = \left(B^T\right)^{-1} = B^{-1}$ and thus

$$S = \begin{bmatrix} I_k & -B^{-1}F^T \\ 0 & I_m \end{bmatrix} \text{ and } S^T = \begin{bmatrix} I_k & -B^{-1}F^T \\ 0 & I_m \end{bmatrix}^T = \begin{bmatrix} I_k & 0 \\ -FB^{-1} & I_m \end{bmatrix}.$$

Then

$$S^T A S = \begin{bmatrix} B & 0 \\ 0 & \hat{C} \end{bmatrix}.$$

Since S has an inverse, it has full rank and B and $\hat{C}$ must be SPD. The converse is also true by reversing the above argument. ■

Example. Consider the 3×3 matrix

$$A = \begin{bmatrix} 2 & -1 & 0 \\ -1 & 2 & -1 \\ 0 & -1 & 2 \end{bmatrix}.$$

The first elementary row operation is

$$\begin{bmatrix} 1 & 0 & 0 \\ 1/2 & 1 & 0 \\ 0 & 0 & 1 \end{bmatrix} \begin{bmatrix} 2 & -1 & 0 \\ -1 & 2 & -1 \\ 0 & -1 & 2 \end{bmatrix} = \begin{bmatrix} 2 & -1 & 0 \\ 0 & 3/2 & -1 \\ 0 & -1 & 2 \end{bmatrix}.$$

The Schur complement of the 1×1 matrix $B = [2]$ is

$$\hat{C} = C - FB^{-1}F^T = \begin{bmatrix} 3/2 & -1 \\ -1 & 2 \end{bmatrix}.$$

It clearly is SPD and so one can do another elementary row operation, but now on the second column

$$\begin{bmatrix} 1 & 0 & 0 \\ 0 & 1 & 0 \\ 0 & 2/3 & 1 \end{bmatrix} \begin{bmatrix} 2 & -1 & 0 \\ 0 & 3/2 & -1 \\ 0 & -1 & 2 \end{bmatrix} = \begin{bmatrix} 2 & -1 & 0 \\ 0 & 3/2 & -1 \\ 0 & 0 & 4/3 \end{bmatrix}.$$

Thus the matrix A can be factored as

$$
A = \begin{bmatrix} 1 & 0 & 0 \\ -1/2 & 1 & 0 \\ 0 & -2/3 & 1 \end{bmatrix} \begin{bmatrix} 2 & -1 & 0 \\ 0 & 3/2 & -1 \\ 0 & 0 & 4/3 \end{bmatrix}
$$

$$
= \begin{bmatrix} 1 & 0 & 0 \\ -1/2 & 1 & 0 \\ 0 & -2/3 & 1 \end{bmatrix} \begin{bmatrix} 2 & 0 & 0 \\ 0 & 3/2 & 0 \\ 0 & 0 & 4/3 \end{bmatrix} \begin{bmatrix} 1 & -1/2 & 0 \\ 0 & 1 & -2/3 \\ 0 & 0 & 1 \end{bmatrix}
$$

$$
= \begin{bmatrix} \sqrt{2} & 0 & 0 \\ -1/\sqrt{2} & \sqrt{3}/\sqrt{2} & 0 \\ 0 & -\sqrt{2}/\sqrt{3} & 2/\sqrt{3} \end{bmatrix} \begin{bmatrix} \sqrt{2} & 0 & 0 \\ -1/\sqrt{2} & \sqrt{3}/\sqrt{2} & 0 \\ 0 & -\sqrt{2}/\sqrt{3} & 2/\sqrt{3} \end{bmatrix}^T .
$$

Definition. *The Cholesky factorization of A is $A = GG^T$ where G is a lower triangular matrix with positive diagonal components.*

Any SPD has a Cholesky factorization. The proof is again by mathematical induction on the dimension of the matrix.

Theorem 8.2.3 *(Cholesky Factorization) If A is SPD, then it has a Cholesky factorization.*

Proof. The $n = 2$ case is clearly true. Let $b = a_{11} > 0$ and apply a row and column elementary operation to $A = \begin{bmatrix} b & f^T \\ f & C \end{bmatrix}$

$$
\begin{bmatrix} 1 & 0 \\ -fb^{-1} & I \end{bmatrix} \begin{bmatrix} b & f^T \\ f & C \end{bmatrix} \begin{bmatrix} 1 & -b^{-1}f^T \\ 0 & I \end{bmatrix} = \begin{bmatrix} b & 0 \\ 0 & C - fb^{-1}f^T \end{bmatrix} .
$$

The Schur complement $\widehat{C} = C - fb^{-1}f^T$ must be SPD and has dimension $n-1$. Therefore, by the mathematical induction assumption it must have a Cholesky factorization $\widehat{C} = \widehat{G}\widehat{G}^T$. Then

$$
\begin{bmatrix} b & 0 \\ 0 & C - fb^{-1}f^T \end{bmatrix} = \begin{bmatrix} b & 0 \\ 0 & \widehat{G}\widehat{G}^T \end{bmatrix}
$$

$$
= \begin{bmatrix} \sqrt{b} & 0 \\ 0 & \widehat{G} \end{bmatrix} \begin{bmatrix} \sqrt{b} & 0 \\ 0 & \widehat{G}^T \end{bmatrix} .
$$

$$
A = \begin{bmatrix} 1 & 0 \\ fb^{-1} & I \end{bmatrix} \begin{bmatrix} \sqrt{b} & 0 \\ 0 & \widehat{G} \end{bmatrix} \begin{bmatrix} \sqrt{b} & 0 \\ 0 & \widehat{G}^T \end{bmatrix} \begin{bmatrix} 1 & b^{-1}f^T \\ 0 & I \end{bmatrix}
$$

$$
= \begin{bmatrix} \sqrt{b} & 0 \\ f/\sqrt{b} & \widehat{G} \end{bmatrix} \begin{bmatrix} \sqrt{b} & 0 \\ f/\sqrt{b} & \widehat{G} \end{bmatrix}^T .
$$

■

The mathematical induction proofs are not fully constructive, but they do imply that the Schur complement is either invertible or is SPD. This allows one to continue with possible permutations and elementary column operations. This process can be done until the upper triangular matrix is obtained. In the case of the SPD matrix, the Schur complement is also SPD and the first pivot must be positive and so no row interchanges are required.

The following Fortran 90 subroutine solves $AX = D$ where A is SPD, X and D may have more than one column and with no row interchanges. The lower triangular part of A is overwritten by the lower triangular factor. The matrix is factored only one time in lines 19-26 where the column version of the loops is used. The subsequent lower and upper triangular solves are done for each column of D. The column versions of the lower triangular solves are done in lines 32-40, and the column versions of the upper triangular solves are done in lines 45-57. This subroutine will be used in the next section as part of a direct solver based on domain decomposition.

Fortran 90 Code for subroutine gespd()

```
1.      Subroutine gespd(a,rhs,sol,n,m)
2.!
3.! Solves Ax = d with A a nxn SPD and d a nxm.
4.!
5.      implicit none
6.      real, dimension(n,n), intent(inout):: a
7.      real, dimension(n,m), intent(inout):: rhs
8.      real, dimension(n,n+m):: aa
9.      real, dimension(n,m) :: y
10.     real, dimension(n,m),intent(out)::sol
11.     integer ::k,i,j,l
12.     integer,intent(in)::n,m
13.     aa(1:n,1:n)= a
14.     aa(1:n,(n+1):n+m) = rhs
15.!
16.! Factor A via column version and
17.! write over the matrix.
18.!
19.     do k=1,n-1
20.         aa(k+1:n,k) = aa(k+1:,k)/aa(k,k)
21.         do j=k+1,n
22.             do i=k+1,n
23.                 aa(i,j) = aa(i,j) - aa(i,k)*aa(k,j)
24.             end do
25.         end do
26.     end do
27.!
28.! Solve Ly = d via column version and
```

```
29.! multiple right sides.
30.!
31.     do j=1,n-1
32.          do l =1,m
33.               y(j,l)=aa(j,n+l)
34.          end do
35.          do i = j+1,n
36.               do l=1,m
37.                    aa(i,n+l) = aa(i,n+l) - aa(i,j)*y(j,l)
38.               end do
39.          end do
40.     end do
41.!
42.! Solve Ux = y via column version and
43.! multiple right sides.
44.!
45.     do j=n,2,-1
46.          do l = 1,m
47.               sol(j,l) = aa(j,n+l)/aa(j,j)
48.          end do
49.          do i = 1,j-1
50.               do l=1,m
51.                    aa(i,n+l)=aa(i,n+l)-aa(i,j)*sol(j,l)
52.               end do
53.          end do
54.     end do
55.     do l=1,m
56.          sol(1,l) = aa(1,n+l)/a(1,1)
57.     end do
58.     end subroutine
```

8.2.1 Exercises

1. Complete the details showing Example 2 is an SPD matrix.
2. By hand find the Cholesky factorization for

$$A = \begin{bmatrix} 3 & -1 & 0 \\ -1 & 3 & -1 \\ 0 & -1 & 3 \end{bmatrix}.$$

3. In Theorem 8.2.1, part 2, prove C is SPD.
4. For the matrix in problem 2 use Theorem 8.2.2 to show it is SPD.
5. Prove the converse part of Theorem 8.2.2.
6. In Theorem 8.2.3 prove the $n = 2$ case.

7.   For the matrix in exercise 2 trace through the steps in the subroutine gespd() to solve

$$AX = \begin{bmatrix} 1 & 4 \\ 2 & 5 \\ 3 & 6 \end{bmatrix}.$$

8.3   Domain Decomposition and MPI

Domain decomposition order can be used to directly solve certain algebraic systems. This was initially described in Sections 2.4 and 4.6. Consider the Poisson problem where the spatial domain is partitioned into three blocks with the first two big blocks separated by a smaller interface block. If the interface block for the Poisson problem is listed last, then the algebraic system may have the form

$$\begin{bmatrix} A_{11} & 0 & A_{13} \\ 0 & A_{22} & A_{23} \\ A_{31} & A_{32} & A_{33} \end{bmatrix} \begin{bmatrix} U_1 \\ U_2 \\ U_3 \end{bmatrix} = \begin{bmatrix} F_1 \\ F_2 \\ F_3 \end{bmatrix}.$$

In the Schur complement B is the 2×2 block given by the block diagonal from A_{11} and A_{22}, and C is A_{33}. Therefore, all the solves with B can be done concurrently, in this case with two processors. By partitioning the domain into more blocks one can take advantage of additional processors. In the 3D space model the big block solves will be smaller 3D subproblems, and here one may need to use iterative methods such as SOR or conjugate gradient. Note the conjugate gradient algorithm has a number of vector updates, dot products and matrix-vector products, and all these steps have independent parts.

In order to be more precise about the above, consider the $(p + 1) \times (p + 1)$ block matrix equation in block component form with $1 \le k \le p$

$$A_{k,k}U_k + A_{k,p+1}U_{p+1} \quad = \quad F_k \tag{8.3.1}$$

$$\sum_{k=1}^{p} A_{p+1,k}U_k + A_{p+1,p+1}U_{p+1} \quad = \quad F_{p+1}. \tag{8.3.2}$$

Now solve (8.3.1) for U_k, and note the computations for $A_{k,k}^{-1}A_{k,p+1}$ and $A_{k,k}^{-1}F_k$, can be done concurrently. Put U_k into (8.3.2) and solve for U_{p+1}

$$\widehat{A}_{p+1,p+1}U_{p+1} \quad = \quad \widehat{F}_{p+1} \text{ where}$$

$$\widehat{A}_{p+1,p+1} \quad = \quad A_{p+1,p+1} - \sum_{k=1}^{p} A_{p+1,k}A_{k,k}^{-1}A_{k,p+1}$$

$$\widehat{F}_{p+1} \quad = \quad F_{p+1} - \sum_{k=1}^{p} A_{p+1,k}A_{k,k}^{-1}F_k.$$

Then concurrently solve for $U_k = A_{k,k}^{-1}F_k - A_{k,k}^{-1}A_{k,p+1}U_{p+1}.$

In order to do the above calculations, the matrices $A_{k,k}$ for $1 \leq k \leq p$, and $\widehat{A}_{p+1,p+1}$ must be invertible. Consider the 2×2 block version of the $(p+1) \times (p+1)$ matrix

$$A = \begin{bmatrix} B & E \\ F & C \end{bmatrix}$$

where B is the block diagonal of $A_{k,k}$ for $1 \leq k \leq p$, and C is $A_{k+1,k+1}$. In this case the Schur complement of B is $\widehat{A}_{p+1,p+1}$. According to Theorem 8.1.2, if the matrices A and $A_{k,k}$ for $1 \leq k \leq p$ have inverses, then $\widehat{A}_{p+1,p+1}$ will have an inverse. Or, according to Theorem 8.2.2, if the matrix A is SPD, then the matrices $A_{k,k}$ for $1 \leq k \leq p$, and $\widehat{A}_{p+1,p+1}$ must be SPD and have inverses.

Consider the 2D steady state heat diffusion problem as studied in Section 4.6. The MATLAB code gedd.m uses block Gaussian elimination where the B matrix, in the 2×2 block matrix of the Schur complement formulation, is a block diagonal matrix with four $(p = 4)$ blocks on its diagonal. The $C = A_{55}$ matrix is for the coefficients of the three interface grid rows between the four big blocks

$$\begin{bmatrix} A_{11} & 0 & 0 & 0 & A_{15} \\ 0 & A_{22} & 0 & 0 & A_{25} \\ 0 & 0 & A_{33} & 0 & A_{35} \\ 0 & 0 & 0 & A_{44} & A_{45} \\ A_{51} & A_{52} & A_{53} & A_{54} & A_{55} \end{bmatrix} \begin{bmatrix} U_1 \\ U_2 \\ U_3 \\ U_4 \\ U_5 \end{bmatrix} = \begin{bmatrix} F_1 \\ F_2 \\ F_3 \\ F_4 \\ F_5 \end{bmatrix}.$$

The following MPI code is a parallel implementation of the MATLAB code gedd.m. It uses three subroutines, which are not listed. The subroutine matrix_def() initializes the above matrix for the Poisson problem, and the dimension of the matrix is $4n^2 + 3n$ where $n = 30$ is the number of unknowns in the x direction and $4n + 3$ is the number of unknowns in the y direction. The subroutine gespd() is the same as in the previous section, and it assumes the matrix is SPD and does Gaussian elimination for multiple right hand sides. The subroutine cgssor3() is a sparse implementation of the preconditioned conjugate gradient method with SSOR preconditioner. It is a variation of cgssor() used in Section 4.3, but now it is for multiple right hand sides. For the larger solves with $A_{k,k}$ for $1 \leq k \leq p$, this has a much shorter computation time than when using gespd(). Because the Schur complement, $\widehat{A}_{p+1,p+1}$, is not sparse, gespd() is used to do this solve for U_{p+1}.

The arrays for $n = 30$ are declared and initialized in lines 8-27. MPI is started in lines 28-37 where up to four processors can be used. Lines 38-47 concurrently compute the arrays that are used in the Schur complement. These are gathered onto processor 0 in lines 49-51, and the Schur complement array is formed in lines 52-58. The Schur complement equation is solved by a call to gespd() in line 60. In line 62 the solution is broadcast from processor 0 to the other processors. Lines 64-70 concurrently solve for the big blocks of unknowns. The results are gathered in lines 72-82 onto processor 0 and partial results are printed.

MPI/Fortran Code geddmpi.f

```
1.      program schurdd
2.! Solves algebraic system via domain decomposition.
3.! This is for the Poisson equation with 2D space grid nx(4n+3).
4.! The solves may be done either by GE or PCG. Use either PCG
5.! or GE for big solves, and GE for the Schur complement solve.
6.      implicit none
7.      include 'mpif.h'
8.      real, dimension(30,30):: A,Id
9.! AA is only used for the GE big solve.
10.     real, dimension(900,900)::AA
11.     real, dimension(900,91,4)::AI,ZI
12.     real, dimension(900,91):: AII
13.     real, dimension(90,91) :: Ahat
14.     real, dimension(90,91,4) :: WI
15.     real, dimension(900) :: Ones
16.     real, dimension(90) :: dhat,xO
17.     real, dimension(900,4) :: xI, dI
18.     real:: h
19.     real:: t0,t1,timef
20.     integer:: n,i,j,loc_n,bn,en,bn1,en1
21.     integer:: my_rank,p,source,dest,tag,ierr,status(mpi_status_size)
22.     integer :: info
23.! Define the nonzero parts of the coefficient matrix with
24.! domain decomposition ordering.
25.     n = 30
26.     h = 1./(n+1)
27.     call matrix_def(n,A,AA,Ahat,AI,AII,WI,ZI,dhat)
28.! Start MPI
29.     call mpi_init(ierr)
30.     call mpi_comm_rank(mpi_comm_world,my_rank,ierr)
31.     call mpi_comm_size(mpi_comm_world,p,ierr)
32.     if (my_rank.eq.0) then
33.         t0 = timef()
34.     end if
35.     loc_n = 4/p
36.     bn = 1+my_rank*loc_n
37.     en = bn + loc_n -1
38.! Concurrently form the Schur complement matrices.
39.     do i = bn,en
40.        ! call gespd(AA,AI(1:n*n,1:3*n+1,i),&
41.        !                     ZI(1:n*n,1:3*n+1,i),n*n,3*n+1)
42.        call cgssor3(AI(1:n*n,1:3*n+1,i),&
43.                            ZI(1:n*n,1:3*n+1,i),n*n,3*n+1,n)
```

```
44.            AII(1:n*n,1:3*n) = AI(1:n*n,1:3*n,i)
45.            WI(1:3*n,1:3*n+1,i)=matmul(transpose(AII(1:n*n,1:3*n))&
46.                              ,ZI(1:n*n,1:3*n+1,i))
47.       end do
48.       call mpi_barrier(mpi_comm_world,ierr)
49.       call mpi_gather(WI(1,1,bn),3*n*(3*n+1)*(en-bn+1),mpi_real,&
50.                         WI,3*n*(3*n+1)*(en-bn+1),mpi_real,0,&
51.                         mpi_comm_world,status ,ierr)
52.       if (my_rank.eq.0) then
53.            Ahat(1:3*n,1:3*n) = Ahat(1:3*n,1:3*n)-&
54.                         WI(1:3*n,1:3*n,1)-WI(1:3*n,1:3*n,2)-&
55.                         WI(1:3*n,1:3*n,3)-WI(1:3*n,1:3*n,4)
56.            dhat(1:3*n) = dhat(1:3*n) -&
57.                         WI(1:3*n,1+3*n,1)-WI(1:3*n,1+3*n,2)-&
58.                         WI(1:3*n,1+3*n,3) -WI(1:3*n,1+3*n,4)
59.! Solve the Schur complement system via GE
60.            call gespd(Ahat(1:3*n,1:3*n),dhat(1:3*n),xO(1:3*n),3*n,1)
61.       end if
62.       call mpi_bcast(xO,3*n,mpi_real,0,mpi_comm_world,ierr)
63.! Concurrently solve for the big blocks.
64.       do i = bn,en
65.            dI(1:n*n,i) = AI(1:n*n,3*n+1,i)-&
66.                           matmul(AI(1:n*n,1:3*n,i),xO(1:3*n))
67.            ! call gespd(AA,dI(1:n*n,i),XI(1:n*n,i),n*n,1)
68.            call cgssor3(dI(1:n*n,i),&
69.                           xI(1:n*n,i),n*n,1,n)
70.       end do
71.       call mpi_barrier(mpi_comm_world,ierr)
72.       call mpi_gather(xI(1,bn),n*n*(en-bn+1),mpi_real,&
73.                         xI,n*n*(en-bn+1),mpi_real,0,&
74.                         mpi_comm_world,status ,ierr)
75.       call mpi_barrier(mpi_comm_world,ierr)
76.       if (my_rank.eq.0) then
77.            t1 = timef()
78.            print*, t1
79.            print*, xO(n/2),xO(n+n/2),xO(2*n+n/2)
80.            print*, xI(n*n/2,1),xI(n*n/2,2),&
81.                           xI(n*n/2,3),xI(n*n/2,4)
82.       end if
83.       call mpi_finalize(ierr)
84.       end program
```

The code was run for 1, 2 and 4 processors with both the gespd() and cgssor3() subroutines for the four large solves with $A_{k,k}$ for $1 \le k \le p = 4$. The computation times using gespd() were about 14 to 20 times longer than

Table 8.3.1: MPI Times for geddmpi.f

p	gespd()	cgssor3()
1	18.871	.924
2	09.547	.572
4	04.868	.349

the time with cgssor3(). The computation times (sec.) are given in Table 8.3.1, and they indicate good speedups close to the number of processors. The speedups with gespd() are better than those with cgssor3() because the large solves are a larger proportion of the computations, which include the same time for communication.

8.3.1   Exercises

1.   Verify the computations in Table 8.3.1.
2.   Experiment with the convergence criteria in the subroutine cgssor3().
3.   Experiment with n, the number of unknowns in the x direction.
4.   Experiment with k, the number of large spatial blocks of unknowns. Vary the number of processors that divide k.

8.4   SOR and P-regular Splittings

SPD matrices were initially introduced in Section 3.5 where the steady state membrane model was studied. Two equivalent models were introduced: the deformation must satisfy a particular partial differential operation, or it must minimize the potential energy of the membrane. The discrete forms of these are for x the approximation of the deformation and $J(y)$ the approximation of the potential energy

$$Ax \;\; = \;\; d \tag{8.4.1}$$

$$J(x) \;\; = \;\; \min_{y} J(y) \text{ where } J(y) \equiv \frac{1}{2}y^T Ay - y^T d. \tag{8.4.2}$$

When A is an SPD matrix, (8.4.1) and (8.4.2) are equivalent. Three additional properties are stated in the following theorem.

Theorem 8.4.1 *(SPD Equivalence Properties) If A is an SPD matrix, then*

1. *the algebraic problem (8.4.1) and the minimum problem (8.4.2) are equivalent,*

2. *there is a constant $c_0 > 0$ such that $x^T Ax \geq c_0 x^T x$,*

3. *$\left| x^T Ay \right| \leq \left(x^T Ax \right)^{\frac{1}{2}} \left(y^T Ay \right)^{\frac{1}{2}}$ (Cauchy inequality) and*

4. $\|x\|_A \equiv (x^T A x)^{\frac{1}{2}}$ is a norm.

Proof. 1. First, we will show if A is SPD and $Ax = d$, then $J(x) \le J(y)$ for all y. Let $y = x + (y - x)$ and use $A = A^T$ to derive

$$
\begin{aligned}
J(y) &= \frac{1}{2}(x + (y - x))^T A(x + (y - x)) - (x + (y - x))^T d \\
&= \frac{1}{2} x^T A x + (y - x)^T A x \\
&\quad + \frac{1}{2}(y - x)^T A(y - x) - x^T d - (y - x)^T d \\
&= J(x) - r(x)^T (y - x) + \frac{1}{2}(y - x)^T A(y - x). \qquad (8.4.3)
\end{aligned}
$$

Since $r(x) = d - Ax = 0$, (8.4.3) implies

$$
J(y) = J(x) + \frac{1}{2}(y - x)^T A(y - x).
$$

Because A is positive definite, $(y - x)^T A(y - x)$ is greater than or equal to zero. Thus, $J(y)$ is greater than or equal to $J(x)$.

Second, prove the converse by assuming $J(x) \le J(y)$ for all $y = x + tr(x)$ where t is any real number. From (8.4.3)

$$
\begin{aligned}
J(y) &= J(x) - r(x)^T (y - x) + \frac{1}{2}(y - x)^T A(y - x) \\
&= J(x) - r(x)^T (tr(x)) + \frac{1}{2}(tr(x))^T A(tr(x)) \\
&= J(x) - tr(x)^T r(x) + \frac{1}{2} t^2 r(x)^T A r(x).
\end{aligned}
$$

Since $0 \le J(y) - J(x) = -tr(x)^T r(x) + \frac{1}{2} t^2 r(x)^T A r(x)$. If $r(x)$ is not zero, then $r(x)^T r(x)$ and $r(x)^T A r(x)$ are positive. Choose

$$
t = \epsilon \frac{r(x)^T r(x)}{r(x)^T A r(x)} > 0.
$$

This gives the following inequality

$$
\begin{aligned}
0 &\le -r(x)^T r(x) + \frac{1}{2} tr(x)^T A r(x) \\
&\le -r(x)^T r(x) + \frac{1}{2} \epsilon \frac{r(x)^T r(x)}{r(x)^T A r(x)} r(x)^T A r(x) \\
&\le -r(x)^T r(x) + \frac{1}{2} \epsilon r(x)^T r(x) \\
&\le r(x)^T r(x)(-1 + \frac{1}{2} \epsilon).
\end{aligned}
$$

For $0 < \epsilon < 2$ this is a contradiction so that $r(x)$ must be zero.

2. The function $f(x) = x^T A x$ is a continuous real valued function. Since the set of y such that $y^T y = 1$ is closed and bounded, f restricted to this set will attain its minimum, that is, there exists $\widehat{y}$ with $\widehat{y}^T \widehat{y} = 1$ such that

$$\min_{y^T y = 1} f(y) = f(\widehat{y}) = \widehat{y}^T A \widehat{y} > 0.$$

Now let $y = x/(x^T x)^{\frac{1}{2}}$ and $c_0 = f(\widehat{y}) = \widehat{y}^T A \widehat{y}$ so that

$$
\begin{aligned}
f(x/(x^T x)^{\frac{1}{2}}) &\geq f(\widehat{y}) \\
(x/(x^T x)^{\frac{1}{2}})^T A (x/(x^T x)^{\frac{1}{2}}) &\geq c_0 \\
x^T A x &\geq c_0 \, x^T x.
\end{aligned}
$$

3. Consider the real valued function of the real number α and use the SPD property of A

$$
\begin{aligned}
f(\alpha) &\equiv (x + \alpha y)^T A (x + \alpha y) \\
&= x^T A x + 2\alpha x^T A y + \alpha^2 y^T A y.
\end{aligned}
$$

This quadratic function of α attains its nonnegative minimum at

$$
\begin{aligned}
\widehat{\alpha} &\equiv -x^T A y / y^T A y \text{ and} \\
0 &\leq f(\widehat{\alpha}) = x^T A x - (x^T A y)^2 / y^T A y.
\end{aligned}
$$

This implies the desired inequality.

4. Since A is SPD, $\|x\|_A \equiv (x^T A x)^{\frac{1}{2}} \geq 0$, and $\|x\|_A = 0$ if and only if $x = 0$. Let α be a real number.

$$\|\alpha x\|_A \equiv ((\alpha x)^T A (\alpha x))^{\frac{1}{2}} = (\alpha^2 x^T A x)^{\frac{1}{2}} = |\alpha| \, (x^T A x)^{\frac{1}{2}} = |\alpha| \, \|x\|_A.$$

The triangle inequality is given by the symmetry of A and the Cauchy inequality

$$
\begin{aligned}
\|x + y\|_A^2 &= (x + y)^T A (x + y) \\
&= x^T A x + 2 x^T A y + y^T A y \\
&\leq \|x\|_A^2 + 2 \left| x^T A y \right| + \|y\|_A^2 \\
&\leq \|x\|_A^2 + 2 \|x\|_A \|y\|_A + \|y\|_A^2 \\
&\leq (\|x\|_A + \|y\|_A)^2.
\end{aligned}
$$

■

We seek to solve the system $Ax = d$ by an iterative method which utilizes *splitting the coefficient matrix* A into a difference of two matrices

$$A = M - N$$

where M is assumed to have an inverse. Substituting this into the equation $Ax = d$, we have

$$Mx - Nx = d.$$

Solve for x to obtain a fixed point problem

$$x = M^{-1}Nx + M^{-1}d.$$

The iterative method based on the splitting is

$$
\begin{aligned}
x^{m+1} &= M^{-1}Nx^m + M^{-1}d \\
&= x^m + M^{-1}r(x^m).
\end{aligned}
\tag{8.4.4}
$$

Here x^0 is some initial guess for the solution, and the solve step with the matrix M is assumed to be relatively easy. The analysis of convergence can be either in terms of some norm of $M^{-1}N$ being less than one (see Section 2.5), or for A an SPD matrix one can place conditions on the splitting so that the quadratic function continues to decrease as the iteration advances. In particular, we will show

$$J(x^{m+1}) = J(x^m) - \frac{1}{2}(x^{m+1} - x^m)^T(M^T + N)(x^{m+1} - x^m).
\tag{8.4.5}$$

So, if $M^T + N$ is positive definite, then the sequence of real numbers $J(x^m)$ is decreasing. For more details on the following splitting consult J. Ortega [20].

Definition. $A = M - N$ is called a P-regular splitting if and only if M has an inverse and $M^T + N$ is positive definite.

Note, if A is SPD, then $A = A^T = M^T - N^T$ and $(M^T + N)^T = M + N^T = M + M^T - A = M^T + N$. Thus, if $A = M - N$ is P-regular and A is SPD, then $M^T + N$ is SPD.

Examples.

1. Jacobi splitting for $A = D - (L+U)$ where $M = D$ is the diagonal of A. $M^T + N = D + (L+U)$ should be positive definite.

2. Gauss-Seidel splitting for $A = (D - L) + U$ where $M = D - L$ is lower triangular part of A. $M^T + N = (D - L)^T + U = D - L^T + U$ should be positive definite. If A is SPD, then $L^T = U$ and the D will have positive diagonal components. In this case $M^T + N = D$ is SPD.

3. SOR splitting for $0 < \omega < 2$ and A SPD with

$$A = \frac{1}{\omega}(D - \omega L) - \frac{1}{\omega}((1-\omega)D + \omega U).$$

Here $M = \frac{1}{\omega}(D - \omega L)$ has an inverse because it is lower triangular with positive diagonal components. This also gives a P-regular splitting because

$$
\begin{aligned}
M^T + N &= (\frac{1}{\omega}(D - \omega L))^T + \frac{1}{\omega}((1-\omega)D + \omega U) \\
&= (\frac{2}{\omega} - 1)D.
\end{aligned}
$$

Theorem 8.4.2 *(P-regular Splitting Convergence) If A is SPD and $A = M - N$ is a P-regular splitting, then the iteration in (8.4.4) will satisfy (8.4.5) and will converge to the solution of $Ax = d$.*

Proof. First, the equality in (8.4.5) will be established by using (8.4.4) and (8.4.3) with $y = x^{m+1} = x^m + M^{-1}r(x^m)$ and $x = x^m$

$$
\begin{aligned}
J(x^{m+1}) &= J(x^m) - r(x^m)^T M^{-1} r(x^m) + \frac{1}{2}(M^{-1}r(x^m))^T A(M^{-1}r(x^m)) \\
&= J(x^m) - r(x^m)^T [M^{-1} - \frac{1}{2}M^{-T}AM^{-1}]r(x^m) \\
&= J(x^m) - r(x^m)^T M^{-T}[M^T - \frac{1}{2}A]M^{-1}r(x^m) \\
&= J(x^m) - (x^{m+1} - x^m)^T [M^T - \frac{1}{2}(M - N)](x^{m+1} - x^m) \\
&= J(x^m) - \frac{1}{2}(x^{m+1} - x^m)^T [2M^T - M + N](x^{m+1} - x^m).
\end{aligned}
$$

Since $z^T M^T z = z^T M z$, (8.4.5) holds.

Second, we establish that $J(x)$ is bounded from below. Use the inequality in part 2 of Theorem 8.4.1, $x^T A x \geq c_0 x^T x$, to write

$$
\begin{aligned}
J(x) &= \frac{1}{2}x^T A x - x^T d \\
&\geq \frac{1}{2}c_0 x^T x - x^T d.
\end{aligned}
$$

Next use the Cauchy inequality with $A = I$ so that $\left| x^T d \right| \leq \left(x^T x \right)^{\frac{1}{2}} \left(d^T d \right)^{\frac{1}{2}}$, and thus

$$
\begin{aligned}
J(x) &\geq \frac{1}{2}c_0 x^T x - \left(x^T x \right)^{\frac{1}{2}} \left(d^T d \right)^{\frac{1}{2}} \\
&\geq \frac{1}{2}c_0[((x^T x)^{\frac{1}{2}} - \frac{(d^T d)^{\frac{1}{2}}}{c_0})^2 - (\frac{(d^T d)^{\frac{1}{2}}}{c_0})^2] \\
&\geq \frac{1}{2}c_0[0 - (\frac{(d^T d)^{\frac{1}{2}}}{c_0})^2].
\end{aligned}
$$

Third, note that $J(x^m)$ is a decreasing sequence of real numbers that is bounded from below. Since the real numbers are complete, $J(x^m)$ must converge to some real number and $J(x^m) - J(x^{m+1}) = \frac{1}{2}(x^{m+1} - x^m)^T(M^T + N)(x^{m+1} - x^m)$ must converge to zero. Consider the norm associated with the SPD matrix $M^T + N$

$$
\left\| x^{m+1} - x^m \right\|_{M^T+N}^2 = 2\left(J(x^m) - J(x^{m+1}) \right) \longrightarrow 0.
$$

Thus, $x^{m+1} - x^m$ converges to the zero vector.

Fourth, $x^{m+1} = x^m + M^{-1}r(x^m)$ so that $M^{-1}r(x^m)$ converges to the zero vector. Since M is continuous, $r(x^m)$ also converges to the zero vector. Since A is SPD, there exists a solution of $Ax = d$, that is, $r(x) = 0$. Thus

$$r(x^m) - r(x) = (d - Ax^m) - (d - Ax) = A(x^m - x) \longrightarrow 0.$$

Since A^{-1} is continuous, $x^m - x$ converges to the zero vector. ∎

The next two sections will give additional examples of P-regular splittings where the solve step $Mz = r(x^m)$ can be done in parallel. Such schemes can be used as stand-alone algorithms or as preconditioners in the conjugate gradient method.

8.4.1   Exercises

1.   In the proof of part 3 in Theorem 8.4.1 prove the assertion about $\widehat{\alpha} \equiv -x^T Ay/y^T Ay$.

2.   Consider the following 2×2 SPD matrix and the indicated splitting

$$A = \begin{bmatrix} B & F^T \\ F & C \end{bmatrix} = \begin{bmatrix} M_1 & 0 \\ 0 & M_2 \end{bmatrix} - \begin{bmatrix} N_1 & -F^T \\ -F & N_2 \end{bmatrix}.$$

(a).   Find conditions on this splitting so that it will be P-regular.

(b).   Apply this to SOR on the first and second blocks.

8.5   SOR and MPI

Consider a block 3×3 matrix with the following form

$$A = \begin{bmatrix} A_{11} & A_{12} & A_{13} \\ A_{21} & A_{22} & A_{23} \\ A_{31} & A_{32} & A_{33} \end{bmatrix}.$$

One can think of this as a generalization of the Poisson problem with the first two blocks of unknowns separated by a smaller interface block. When finite differences or finite element methods are used, often the A_{12} and A_{21} are either zero or sparse with small nonzero components. Consider splittings of the three diagonal blocks

$$A_{ii} = M_i - N_i.$$

Then one can associate a number of splittings with the large A matrix.

Examples.

1.   In this example the M in the splitting of A is block diagonal. The inversion of M has three independent computations

$$A = \begin{bmatrix} M_1 & 0 & 0 \\ 0 & M_2 & 0 \\ 0 & 0 & M_3 \end{bmatrix} - \begin{bmatrix} N_1 & -A_{12} & -A_{13} \\ -A_{21} & N_2 & -A_{23} \\ -A_{31} & -A_{32} & N_3 \end{bmatrix}. \qquad (8.5.1)$$

2. The second example is slightly more complicated, and the M is block lower triangular. The inversion of M has two independent computations and some computations with just one processor.

$$A = \begin{bmatrix} M_1 & 0 & 0 \\ 0 & M_2 & 0 \\ A_{31} & A_{32} & M_3 \end{bmatrix} - \begin{bmatrix} N_1 & -A_{12} & -A_{13} \\ -A_{21} & N_2 & -A_{23} \\ 0 & 0 & N_3 \end{bmatrix}. \tag{8.5.2}$$

3. This M has the form from domain decomposition. The inversion of M can be done by the Schur complement where B is the block diagonal of M_1 and M_2, and $C = M_3$.

$$A = \begin{bmatrix} M_1 & 0 & A_{13} \\ 0 & M_2 & A_{23} \\ A_{31} & A_{32} & M_3 \end{bmatrix} - \begin{bmatrix} N_1 & -A_{12} & 0 \\ -A_{21} & N_2 & 0 \\ 0 & 0 & N_3 \end{bmatrix}. \tag{8.5.3}$$

Theorem 8.5.1 *(Block P-regular Splitting) Consider the iteration given by* (8.5.2). *If A is SPD, and*

$$M^T + N = \begin{bmatrix} M_1^T + N_1 & -A_{12} & 0 \\ -A_{21} & M_2^T + N_2 & 0 \\ 0 & 0 & M_3^T + N_3 \end{bmatrix}$$

is SPD, then the iteration will converge to the solution of $Ax = d$.

Proof. The proof is an easy application of Theorem 8.4.2. Since A is symmetric $A_{ij}^T = A_{ji}$,

$$\begin{aligned}
M^T + N &= \begin{bmatrix} M_1 & 0 & 0 \\ 0 & M_2 & 0 \\ A_{31} & A_{32} & M_3 \end{bmatrix}^T + \begin{bmatrix} N_1 & -A_{12} & -A_{13} \\ -A_{21} & N_2 & -A_{23} \\ 0 & 0 & N_3 \end{bmatrix} \\
&= \begin{bmatrix} M_1^T & 0 & A_{31}^T \\ 0 & M_2^T & A_{32}^T \\ 0 & 0 & M_3^T \end{bmatrix} + \begin{bmatrix} N_1 & -A_{12} & -A_{13} \\ -A_{21} & N_2 & -A_{23} \\ 0 & 0 & N_3 \end{bmatrix} \\
&= \begin{bmatrix} M_1^T + N_1 & -A_{12} & 0 \\ -A_{21} & M_2^T + N_2 & 0 \\ 0 & 0 & M_3^T + N_3 \end{bmatrix}.
\end{aligned}$$

∎

If the $A_{12} = A_{21} = 0$, then $M^T + N$ will be P-regular if and only if each splitting $A_{ii} = M_i - N_i$ is P-regular. A special case is the SOR algorithm applied to the large blocks of unknowns, updating the unknowns, and doing SOR of the interface block, which is listed last. In this case $M_i = \frac{1}{\omega}(D_i - \omega L_i)$ so that $M_i^T + N_i = (\frac{2}{\omega} - 1)D_i$.

The following MPI code solves the Poisson problem where the spatial blocks of nodes include horizontal large blocks with $n = 447$ unknowns in the x direction and $(n - p + 1)/p$ unknowns in the y direction. There are $p = 2, 4, 8, 16$

or 32 processors with $(n - p + 1)/p = 223, 111, 55, 27$ and 13 in unknowns in the y direction, respectively. There are $p - 1$ smaller interface blocks with one row each of n unknowns so that the $p + 1$ block of the matrix A has $n \times (p - 1)$ unknowns.

The code is initialized in lines 6-30, and the SOR while loop is in lines 32-120. SOR is done concurrently in lines 38-48 for the larger blocks, the results are communicated to the adjacent processors in lines 50-83, then SOR is done concurrently in lines 84-96 for the smaller interface blocks, and finally in lines 98-105 the results are communicated to adjacent blocks. The communication scheme used in lines 50-83 is similar to that used in Section 6.6 and illustrated in Figures 6.6.1 and 6.6.2. Here we have stored the kth big block and the top interface block on processor $k - 1$ where $0 < k < p$; the last big block is stored on processor $p - 1$. In lines 107-119 the global error is computed and broadcast to all the processors. If it satisfies the convergence criteria, then the while loop on each processor is exited. One can use the MPI subroutine *mpi_allreduce()* to combine the gather and broadcast operations, see Section 9.3 and the MPI code cgssormpi.f. The results in lines 123-133 are gathered onto processor 0 and partial results are printed.

MPI/Fortran Code sorddmpi.f

```
1.      program sor
2.!
3.! Solve Poisson equation via SOR.
4.! Uses domain decomposition to attain parallel computation.
5.!
6.      implicit none
7.      include 'mpif.h'
8.      real ,dimension (449,449)::u,uold
9.      real ,dimension (1:32)::errora
10.     real :: w, h, eps,pi,error,utemp,to,t1,timef
11.     integer :: n,maxk,maxit,it,k,i,j,jm,jp
12.     integer :: my_rank,p,source,dest,tag,ierr,loc_n
13.     integer :: status(mpi_status_size),bn,en,sbn
14.     n = 447
15.     w = 1.99
16.     h = 1.0/(n+1)
17.     u = 0.
18.     errora(:) = 0.0
19.     error = 1.
20.     uold = 0.
21.     call mpi_init(ierr)
22.     call mpi_comm_rank(mpi_comm_world,my_rank,ierr)
23.     call mpi_comm_size(mpi_comm_world,p,ierr)
24.     if (my_rank.eq.0) then
25.         to = timef()
```

```
26.    end if
27.    pi = 3.141592654
28.    maxit = 2000
29.    eps = .001
30.    it = 0
31.! Begin the while loop for the parallel SOR iterations.
32.        do while ((it.lt.maxit).and.(error.gt.eps))
33.            it = it + 1
34.            loc_n = (n-p+1)/p
35.            bn = 2+(my_rank)*(loc_n+1)
36.            en = bn + loc_n -1
37.! Do SOR for big blocks.
38.            do j=bn,en
39.                do i =2,n+1
40.                    utemp = (1000.*sin((i-1)*h*pi)*sin((j-1)*h*pi)*h*h&
41.                             + u(i-1,j) + u(i,j-1) &
42.                             + u(i+1,j) + u(i,j+1))*.25
43.                    u(i,j) = (1. -w)*u(i,j) + w*utemp
44.                end do
45.            end do
46.            errora(my_rank+1) = maxval(abs(u(2:n+1,bn:en)-&
47.                                uold(2:n+1,bn:en)))
48.            uold(2:n+1,bn:en) = u(2:n+1,bn:en)
49.! Communicate computations to adjacent blocks.
50.            if (my_rank.eq.0) then
51.                call mpi_recv(u(1,en+2),(n+2),mpi_real,my_rank+1,50,&
52.                             mpi_comm_world,status,ierr)
53.                call mpi_send(u(1,en+1),(n+2),mpi_real,my_rank+1,50,&
54.                             mpi_comm_world,ierr)
55.            end if
56.            if ((my_rank.gt.0).and.(my_rank.lt.p-1)&
57.                             .and.(mod(my_rank,2).eq.1)) then
58.                call mpi_send(u(1,en+1),(n+2),mpi_real,my_rank+1,50,&
59.                             mpi_comm_world,ierr)
60.                call mpi_recv(u(1,en+2),(n+2),mpi_real,my_rank+1,50,&
61.                             mpi_comm_world,status,ierr)
62.                call mpi_send(u(1,bn),(n+2),mpi_real,my_rank-1,50,&
63.                             mpi_comm_world,ierr)
64.                call mpi_recv(u(1,bn-1),(n+2),mpi_real,my_rank-1,50,&
65.                             mpi_comm_world,status,ierr)
66.            end if
67.            if ((my_rank.gt.0).and.(my_rank.lt.p-1)&
68.                             .and.(mod(my_rank,2).eq.0)) then
69.                call mpi_recv(u(1,bn-1),(n+2),mpi_real,my_rank-1,50,&
70.                             mpi_comm_world,status,ierr)
```

```
71.                call mpi_send(u(1,bn),(n+2),mpi_real,my_rank-1,50,&
72.                          mpi_comm_world,ierr)
73.                call mpi_recv(u(1,en+2),(n+2),mpi_real,my_rank+1,50,&
74.                          mpi_comm_world,status,ierr)
75.                call mpi_send(u(1,en+1),(n+2),mpi_real,my_rank+1,50,&
76.                          mpi_comm_world,ierr)
77.           end if
78.           if (my_rank.eq.p-1) then
79.                call mpi_send(u(1,bn),(n+2),mpi_real,my_rank-1,50,&
80.                          mpi_comm_world,ierr)
81.                call mpi_recv(u(1,bn-1),(n+2),mpi_real,my_rank-1,50,&
82.                          mpi_comm_world,status,ierr)
83.           end if
84.           if (my_rank.lt.p-1) then
85.                j = en +1
86.! Do SOR for smaller interface blocks.
87.                do i=2,n+1
88.                     utemp = (1000.*sin((i-1)*h*pi)*sin((j-1)*h*pi)*h*h&
89.                          + u(i-1,j) + u(i,j-1)&
90.                          + u(i+1,j) + u(i,j+1))*.25
91.                     u(i,j) = (1. -w)*u(i,j) + w*utemp
92.                end do
93.                errora(my_rank+1) = max1(errora(my_rank+1),&
94.                          maxval(abs(u(2:n+1,j)-uold(2:n+1,j))))
95.                uold(2:n+1,j) = u(2:n+1,j)
96.           endif
97.! Communicate computations to adjacent blocks.
98.           if (my_rank.lt.p-1) then
99.                call mpi_send(u(1,en+1),(n+2),mpi_real,my_rank+1,50,&
100.                          mpi_comm_world,ierr)
101.           end if
102.          if (my_rank.gt.0) then
103.                call mpi_recv(u(1,bn-1),(n+2),mpi_real,my_rank-1,50,&
104.                          mpi_comm_world,status,ierr)
105.          end if
106.! Gather local errors to processor 0.
107.          call mpi_gather(errora(my_rank+1),1,mpi_real,&
108.                          errora,1,mpi_real,0,&
109.                          mpi_comm_world,ierr)
110.          call mpi_barrier(mpi_comm_world,ierr)
111.! On processor 0 compute the maximum of the local errors.
112.          if (my_rank.eq.0) then
113.                error = maxval(errora(1:p))
114.          end if
115.! Send this global error to all processors so that
```

Table 8.5.1: MPI Times for sorddmpi.f

p	time	iteration
2	17.84	673
4	07.93	572
8	03.94	512
16	02.18	483
32	01.47	483

```
116.! they will exit the while loop when the global error
117.! test is satisfied.
118.        call mpi_bcast(error,1,mpi_real,0,&
119.                        mpi_comm_world,ierr)
120.     end do
121.! End of the while loop.
122.! Gather the computations to processor 0
123.     call mpi_gather(u(1,2+my_rank*(loc_n+1)),&
124.                     (n+2)*(loc_n+1),mpi_real,&
125.                     u,(n+2)*(loc_n+1),mpi_real,0,&
126.                     mpi_comm_world,ierr)
127.     if (my_rank.eq.0) then
128.         t1 = timef()
129.         print*, 'sor iterations = ',it
130.         print*, 'time = ', t1
131.         print*, 'error = ', error
132.         print*, 'center value of solution = ', u(225,225)
133.     end if
134.     call mpi_finalize(ierr)
135.     end
```

The computations in Table 8.5.1 use 2, 4, 8, 16 and 32 processors. Since the orderings of the unknowns are different, the SOR iterations required for convergence vary; in this case they decrease as the number of processors increase. Also, as the number of processors increase, the number of interface blocks increase as does the amount of communication. The execution times (sec.) reflect reasonable speedups given the decreasing iterations for convergence and increasing communication.

8.5.1   Exercises

1.   Establish analogues of Theorem 8.5.1 for the splittings in lines (8.5.1) and (8.5.3).
2.   In Theorem 8.5.1 use the Schur complement to give conditions so that $M^T + N$ will be positive definite.

3. Verify the computations in Table 8.5.1. Experiment with the convergence criteria.

4. Experiment with sorddmpi.f by varying the number of unknowns and SOR parameter. Note the effects of using different numbers of processors.

8.6 Parallel ADI Schemes

Alternating direction implicit (ADI) iterative methods can be used to approximate the solution to a two variable Poisson problem by a sequence of ordinary differential equations solved in the x or y directions

$$
\begin{aligned}
-u_{xx} - u_{yy} &= f \\
-u_{xx} &= f + u_{yy} \\
-u_{yy} &= f + u_{xx}.
\end{aligned}
$$

One may attempt to approximate this by the following scheme

for m = 0, maxm
  for each y solve
    $-u_{xx}^{m+\frac{1}{2}} = f + u_{yy}^m$
  for each x solve
    $-u_{yy}^{m+1} = f + u_{xx}^{m+\frac{1}{2}}$
  test for convergence
endloop.

The discrete version of this uses the finite difference method to discretize the ordinary differential equations

$$
-\frac{u_{i+1,j}^{m+\frac{1}{2}} - 2u_{i,j}^{m+\frac{1}{2}} + u_{i-1,j}^{m+\frac{1}{2}}}{\Delta x^2} = f_{i,j} + \frac{u_{i,j+1}^m - 2u_{i,j}^m + u_{i,j-1}^m}{\Delta y^2} \quad (8.6.1)
$$

$$
-\frac{u_{i,j+1}^{m+1} - 2u_{i,j}^{m+1} + u_{i,j-1}^{m+1}}{\Delta y^2} = f_{i,j} + \frac{u_{i+1,j}^{m+\frac{1}{2}} - 2u_{i,j}^{m+\frac{1}{2}} + u_{i-1,j}^{m+\frac{1}{2}}}{\Delta x^2}. \quad (8.6.2)
$$

The discrete version of the above scheme is

for m = 0, maxm
  for each j solve
    tridiagonal problem (8.6.1) for $u_{i,j}^{m+\frac{1}{2}}$
  for each i solve
    tridiagonal problem (8.6.2) for $u_{i,j}^{m+1}$
  test for convergence
endloop.

The solve steps may be done by the tridiagonal algorithm. The i or j solves are independent, that is, the solves for $u_{i,j}^{m+\frac{1}{2}}$ and $u_{i,j}^{m+\frac{1}{2}}$ can be done concurrently for $j \neq \hat{j}$, and also the solves for $u_{i,j}^{m+1}$ and $u_{\hat{i},j}^{m+1}$ can be done concurrently for $i \neq \hat{i}$.

The matrix form of this scheme requires the coefficient matrix to be written as $A = H + V$ where H and V reflect the discretized ordinary differential equations in the x and y directions, respectively. For example, suppose $\Delta x = \Delta y = h$ and there are n unknowns in each direction of a square domain. The A will be a block $n \times n$ matrix and using the classical order of bottom grid row first and moving from left to right in the grid rows

$$A = H + V$$

$$\begin{bmatrix} B & -I & Z \\ -I & B & -I \\ Z & -I & B \end{bmatrix} = \begin{bmatrix} C & Z & Z \\ Z & C & Z \\ Z & Z & C \end{bmatrix} + \begin{bmatrix} 2I & -I & Z \\ -I & 2I & -I \\ Z & -I & 2I \end{bmatrix}$$

where I is a $n \times n$ identity matrix, Z is a $n \times n$ zero matrix and C is a $n \times n$ tridiagonal matrix, for example for $n = 3$

$$C = \begin{bmatrix} 2 & -1 & 0 \\ -1 & 2 & -1 \\ 0 & -1 & 2 \end{bmatrix}.$$

The ADI algorithm for the solution of $Ax = d$ is based on two splittings of A

$$A = (\alpha I + H) - (\alpha I - V)$$
$$A = (\alpha I + V) - (\alpha I - H).$$

The positive constant α is chosen so that $\alpha I + H$ and $\alpha I + V$ have inverses and to accelerate convergence. More generally, αI is replaced by a diagonal matrix with positive diagonal components chosen to obtain an optimal convergence rate.

ADI Algorithm for $Ax = d$ with $A = H + V$.

> for m = 0, maxm
> > solve $(\alpha I + H)x^{m+\frac{1}{2}} = d + (\alpha I - V)x^m$
> > solve $(\alpha I + V)x^{m+1} = d + (\alpha I - H)x^{m+\frac{1}{2}}$
> > test for convergence
> endloop.

This algorithm may be written in terms of a single splitting.

$$\begin{aligned} x^{m+1} &= (\alpha I + V)^{-1}[d + (\alpha I - H)x^{m+\frac{1}{2}}] \\ &= (\alpha I + V)^{-1}[d + (\alpha I - H)(\alpha I + H)^{-1}(d + (\alpha I - V)x^m)] \\ &= (\alpha I + V)^{-1}[d + (\alpha I - H)(\alpha I + H)^{-1}d] + \\ &\quad (\alpha I + V)^{-1}(\alpha I - H)(\alpha I + H)^{-1}(\alpha I - V)x^m \\ &= M^{-1}d + M^{-1}Nx^m \end{aligned}$$

where

$$M^{-1} \equiv (\alpha I + V)^{-1}[I + (\alpha I - H)(\alpha I + H)^{-1}] \text{ and}$$
$$M^{-1}N = (\alpha I + V)^{-1}(\alpha I - H)(\alpha I + H)^{-1}(\alpha I - V).$$

Thus, the convergence can be analyzed by either requiring some norm of $M^{-1}N$ to be less than one, or by requiring A to be SPD and $A = M - N$ to be a P-regular splitting. The following theorem uses the P-regular splitting approach.

Theorem 8.6.1 *(ADI Splitting Convergence)* Consider the ADI algorithm where α is some positive constant. If A, H and V are SPD and α is such that $\alpha I + \frac{1}{2\alpha}(VH + HV)$ is positive definite, then the ADI splitting is P-regular and must converge to the solution.

Proof. Use the above splitting associated with the ADI algorithm

$$\begin{aligned} M^{-1} &= (\alpha I + V)^{-1}[I + (\alpha I - H)(\alpha I + H)^{-1}] \\ &= (\alpha I + V)^{-1}[(\alpha I + H) + (\alpha I - H)](\alpha I + H)^{-1} \\ &= (\alpha I + V)^{-1}2\alpha(\alpha I + H)^{-1}. \end{aligned}$$

Thus, $M = \frac{1}{2\alpha}(\alpha I + H)(\alpha I + V)$ and $N = -A + M = -V - H + M$ so that by the symmetry of H and V

$$\begin{aligned} M^T + N &= (\frac{1}{2\alpha}(\alpha I + H)(\alpha I + V))^T - V - H + \frac{1}{2\alpha}(\alpha I + H)(\alpha I + V) \\ &= \frac{1}{2\alpha}(\alpha I + V^T)(\alpha I + H^T) - V - H + \frac{1}{2\alpha}(\alpha I + H)(\alpha I + V) \\ &= \alpha I + \frac{1}{2\alpha}(VH + HV). \end{aligned}$$

∎

Example 2 in Section 8.2 implies that, for suitably large α, $M^T + N = \alpha I + \frac{1}{2\alpha}(VH + HV)$ will be positive definite. The parallel aspects of this method are that $\alpha I + H$ and $\alpha I + V$ are essentially block diagonal with tridiagonal matrices, and therefore, the solve steps have many independent substeps. The ADI method may also be applied to three space dimension problems, and also there are several domain decomposition variations of the tridiagonal algorithm.

ADI in Three Space Variables.
Consider the partial differential equation in three variables $-u_{xx} - u_{yy} - u_{zz} = f$. Discretize this so that the algebraic problem $Ax = d$ can be broken into parts associated with three directions and $A = H + V + W$ where W is associated with the z direction. Three splittings are

$$\begin{aligned} A &= (\alpha I + H) - (\alpha I - V - W) \\ A &= (\alpha I + V) - (\alpha I - H - W) \\ A &= (\alpha I + W) - (\alpha I - H - V). \end{aligned}$$

The ADI scheme has the form

    for m = 0, maxm
        solve $(\alpha I + H)x^{m+1/3} = d + (\alpha I - V - W)x^m$
        solve $(\alpha I + V)x^{m+2/3} = d + (\alpha I - H - W)x^{m+1/3}$
        solve $(\alpha I + W)x^{m+1} = d + (\alpha I - V - H)x^{m+2/3}$
        test for convergence
    endloop.

For this three variable case one can easily prove an analogue to Theorem 8.6.1.

Tridiagonal with Domain Decomposition and Interface Blocks.
    Consider the tridiagonal matrix with dimension $n = 3m + 2$. Reorder the unknowns so that unknowns for $i = m + 1$ and $i = 2m + 2$ are listed last. The reordered coefficient matrix for $m = 3$ will have the following nonzero pattern

$$
\begin{bmatrix}
x & x & & & & & & & & & \\
x & x & x & & & & & & & & \\
 & x & x & & & & & & & x & \\
 & & & x & x & & & & & x & \\
 & & & x & x & x & & & & & \\
 & & & & x & x & & & & & x \\
 & & & & & & x & x & & & x \\
 & & & & & & x & x & x & & \\
 & & & & & & & x & x & & \\
 & & x & x & & & & & & x & \\
 & & & & & x & x & & & & x
\end{bmatrix}.
$$

This the form associated with the Schur complement where B is the block diagonal with three tridiagonal matrices. There will then be three independent tridiagonal solves as substeps for solving this domain decomposition or "arrow" matrix.

Tridiagonal with Domain Decomposition and No Interface Blocks.
    Consider the tridiagonal matrix with dimension $n = 3m$. Multiply this matrix by the inverse of the block diagonal of the matrices $A(1 : m, 1 : m)$, $A(m + 1 : 2m, m + 1 : 2m)$ and $A(2m + 1 : 3m, 2m + 1 : 3m)$. The new matrix has the following nonzero pattern for $m = 3$

$$
\begin{bmatrix}
1 & & & x & & & & & \\
 & 1 & & x & & & & & \\
 & & 1 & x & & & & & \\
 & & x & 1 & & & x & & \\
 & & x & & 1 & & x & & \\
 & & x & & & 1 & x & & \\
 & & & & & x & 1 & & \\
 & & & & & x & & 1 & \\
 & & & & & x & & & 1
\end{bmatrix}.
$$

Now reorder the unknowns so that unknowns for $i = m, m+1, 2m$ and $2m+1$ are listed last. The new coefficient matrix has the form for $m = 3$

$$\begin{bmatrix} 1 & & & & & & x & & \\ & 1 & & & & & x & & \\ & & 1 & & & x & & & x \\ & & & 1 & & & & x & \\ & & & & 1 & & & x & \\ & & & & & 1 & x & & \\ & & & & & x & 1 & & x \\ & & & & & x & & 1 & x \\ & & & & & & & x & 1 \end{bmatrix}.$$

This matrix has a bottom diagonal block, which can be permuted to a 4×4 tridiagonal matrix. The top diagonal block is the identity matrix with dimension $3m - 4$ so that the new matrix is a 2×2 block upper triangular matrix. More details on this approach can be found in the paper by N. Mattor, T. J. Williams and D. W. Hewett [15].

8.6.1 Exercises

1.   Generalize Theorem 8.6.1 to the case where α is a diagonal matrix with positive diagonal components.

2.   Generalize Theorem 8.6.1 to the three space variable case.

3.   Generalize the tridiagonal algorithm with $p - 1$ interface nodes and $n = pm + p - 1$ where there are p blocks with m unknowns per block. Implement this in an MPI code using p processors.

4.   Generalize the tridiagonal algorithm with no interface nodes and $n = pm$ where there are p blocks with m unknowns per block. Implement this in an MPI code using p processors.

Chapter 9

Krylov Methods for Ax = d

The conjugate gradient method was introduced in Chapter 3. In this chapter we show each iterate of the conjugate gradient method can be expressed as the initial guess plus a linear combination of the $A^i r(x^0)$ where $r^0 = r(x^0) = d - Ax^0$ is the initial residual and $A^i r(x^0)$ are called the Krylov vectors. Here $x^m = x^0 + c_0 r^0 + c_1 A r^0 + \cdots + c_{m-1} A^{m-1} r^0$ and the coefficients are the unknowns. They can be determined by either requiring $J(x^m) = \frac{1}{2}(x^m)^T A x^m - (x^m)^T d$, where A is a SPD matrix, to be a minimum, or to require $R(x^m) = r(x^m)^T r(x^m)$ to be a minimum. The first case gives rise to the conjugate gradient method (CG), and the second case generates the generalized minimum residual method (GMRES). Both methods have many variations, and they can be accelerated by using suitable preconditioners, M, where one applies these methods to $M^{-1} Ax = M^{-1}d$. A number of preconditioners will be described, and the parallel aspects of these methods will be illustrated by MPI codes for preconditioned CG and a restarted version of GMRES.

9.1 Conjugate Gradient Method

The conjugate gradient method as described in Sections 3.5 and 3.6 is an enhanced version of the method of steepest descent. First, the multiple residuals are used so as to increase the dimensions of the underlying search set of vectors. Second, conjugate directions are used so that the resulting algebraic systems for the coefficients will be diagonal. As an additional benefit to using the conjugate directions, all the coefficients are zero except for the last one. This means the next iterate in the conjugate gradient method is the previous iterate plus a constant times the last search direction. Thus, not all the search directions need to be stored. This is in contrast to the GMRES method, which is applicable to problems where the coefficient matrix is not SPD.

The implementation of the conjugate gradient method has the form given below. The steepest descent in the direction p^m is given by using the parameter α, and the new conjugate direction p^{m+1} is given by using the parameter β. For each iteration there are two dot products, three vector updates and a matrix-vector product. These substages can be done in parallel, and often one tries to avoid storage of the full coefficient matrix.

Conjugate Gradient Method.

Let x^0 be an initial guess
$r^0 = d - Ax^0$
$p^0 = r^0$
for m = 0, maxm
$\quad \alpha = (r^m)^T r^m / (p^m)^T A p^m$
$\quad x^{m+1} = x^m + \alpha p^m$
$\quad r^{m+1} = r^m - \alpha A p^m$
$\quad$ test for convergence
$\quad \beta = (r^{m+1})^T r^{m+1} / (r^m)^T r^m$
$\quad p^{m+1} = r^{m+1} + \beta p^m$
endloop.

The connection with the *Krylov vectors* $A^i r(x^0)$ evolves from expanding the conjugate gradient loop.

$$
\begin{aligned}
x^1 &= x^0 + \alpha_0 p^0 = x^0 + \alpha_0 r^0 \\
r^1 &= r^0 - \alpha_0 A p^0 = r^0 - \alpha_0 A r^0 \\
p^1 &= r^1 + \beta_0 p^0 = r^1 + \beta_0 r^0 \\
x^2 &= x^1 + \alpha_1 p^1 \\
&= x^1 + \alpha_1 \left(r^1 + \beta_0 r^0 \right) \\
&= x^0 + \alpha_0 r^0 + \alpha_1 \left(r^0 - \alpha_0 A r^0 + \beta_0 r^0 \right) \\
&= x^0 + c_0 r^0 + c_1 A r^0 \\
&\vdots \\
x^m &= x^0 + c_0 r^0 + c_1 A r^0 + \cdots + c_{m-1} A^{m-1} r^0.
\end{aligned}
$$

An alternative definition of the conjugate gradient method is to choose the coefficients of the Krylov vectors so as to minimize $J(x)$

$$
J(x^{m+1}) = \min_c J(x^0 + c_0 r^0 + c_1 A r^0 + \cdots + c_m A^m r^0).
$$

Another way to view this is to define the *Krylov space* as

$$
K_m \equiv \{x \mid x = c_0 r^0 + c_1 A r^0 + \cdots + c_{m-1} A^{m-1} r^0, c_i \in \mathbb{R}\}.
$$

The Krylov spaces have these very useful properties:

$$
\begin{aligned}
K_m &\subset K_{m+1} & (9.1.1) \\
A K_m &\subset K_{m+1} \text{ and} & (9.1.2) \\
K_m &\equiv \{x \mid x = a_0 r^0 + a_1 r^1 + \cdots + a_{m-1} r^{m-1}, a_i \in \mathbb{R}\}. & (9.1.3)
\end{aligned}
$$

So the *alternate definition of the conjugate gradient method* is to choose $x^{m+1} \in x^0 + K_{m+1}$ so that

$$J(x^{m+1}) = \min_{y \in x^0 + K_{m+1}} J(y). \tag{9.1.4}$$

This approach can be shown to be equivalent to the original version, see C. T. Kelley [11, Chapter 2].

In order to gain some insight to this let $x^m \in x^0 + K_m$ be such that it minimizes $J(y)$ where $y \in x^0 + K_m$. Let $y = x^m + tz$ where $z \in K_m$ and use the identity in (8.4.3)

$$J(x^m + tz) = J(x^m) - r(x^m)^T tz + \frac{1}{2}(tz)^T A(tz).$$

Then the derivative of $J(x^m + tz)$ with respect to the parameter t evaluated at $t = 0$ must be zero so that

$$r(x^m)^T z = 0 \text{ for all } z \in K_m. \tag{9.1.5}$$

Since K_m contains the previous residuals,

$$r(x^m)^T r(x^l) = 0 \text{ for all } l < m.$$

Next consider the difference between two iterates given by (9.1.4), $x^{m+1} = x^m + w$ where $w \in K_{m+1}$. Use (9.1.5) with m replaced by $m + 1$

$$
\begin{aligned}
r(x^{m+1})^T z &= 0 \text{ for all } z \in K_{m+1} \\
(d - A(x^m + w))^T z &= 0 \\
(d - Ax^m)^T z - w^T A^T z &= 0 \\
r(x^m)^T z - w^T Az &= 0.
\end{aligned}
$$

Since $K_m \subset K_{m+1}$, we may let $z \in K_m$ and use (9.1.5) to get

$$w^T Az = (x^{m+1} - x^m)^T Az = 0. \tag{9.1.6}$$

This implies that $w = x^{m+1} - x^m \in K_{m+1}$ can be expressed as a multiple of $r(x^m) + \hat{z}$ for some $\hat{z} \in K_m$. Additional inductive arguments show for suitable α and β

$$x^{m+1} - x^m = \alpha(r(x^m) + \beta p^{m-1}). \tag{9.1.7}$$

This is a noteworthy property because one does not need to store all the conjugate search directions $p^0, \cdots, p^m$.

The proof of (9.1.7) is done by mathematical induction. The term $x^{0+1} \in x^0 + K_{0+1}$ is such that

$$
\begin{aligned}
J(x^{0+1}) &= \min_{y \in x^0 + K_{0+1}} J(y) \\
&= \min_{c_0} J(x^0 + c_0 r^0).
\end{aligned}
$$

Thus, $x^1 = x^0 + \alpha_0 r^0$ for an appropriate α_0 as given by the identity in (8.4.3). Note

$$
\begin{aligned}
r^1 &= d - Ax^1 \\
&= d - A(x^0 + \alpha_0 r^0) \\
&= (d - Ax^0) - \alpha_0 Ar^0 \\
&= r^0 - \alpha_0 Ap^0.
\end{aligned}
$$

The next step is a little more interesting. By definition $x^{1+1} \in x^0 + K_{1+1}$ is such that

$$
J(x^{1+1}) = \min_{y \in x^0 + K_{1+1}} J(y).
$$

Properties (9.1.5) and (9.1.6) imply

$$
\begin{aligned}
(r^1)^T r^0 &= 0 \\
(x^2 - x^1)^T Ar^0 &= 0.
\end{aligned}
$$

Since $x^2 - x^1 \in K_2, x^2 - x^1 = \alpha(r^1 + \beta p^0)$ and

$$
\begin{aligned}
(x^2 - x^1)^T Ar^0 &= 0 \\
(r^1 + \beta p^0)^T Ar^0 &= 0 \\
(r^1)^T Ar^0 + (\beta p^0)^T Ar^0 &= 0.
\end{aligned}
$$

Since $r^1 = r^0 - \alpha_0 Ap^0$ and $(r^1)^T r^0 = 0$,

$$
\begin{aligned}
(r^1)^T r^0 &= (r^0 - \alpha_0 Ap^0)^T r^0 \\
&= (r^0)^T r^0 - \alpha_0 (Ap^0)^T r^0 \\
&= 0.
\end{aligned}
$$

So, if $r^0 \neq 0$, $(Ap^0)^T r^0 \neq 0$ and we may choose

$$
\beta = \frac{-(r^1)^T Ar^0}{(Ap^0)^T r^0}.
$$

The inductive step of the formal proof is similar.

Theorem 9.1.1 *(Alternate CG Method) Let A be SPD and let x^{m+1} be the alternate definition of the conjugate gradient method in (9.1.4). If the residuals are not zero vectors, then equations (9.1.5-9.1.7) are true.*

The utility of the Krylov approach to both the conjugate gradient and the generalized residual methods is a very nice analysis of convergence properties. These are based on the following algebraic identities. Let x be the solution of $Ax = d$ so that $r(x) = d - Ax = 0$, and use the identity in line (8.4.3) for symmetric matrices

$$
\begin{aligned}
J(x^{m+1}) - J(x) &= \frac{1}{2}(x^{m+1} - x)^T A(x^{m+1} - x) \\
&= \frac{1}{2} \left\| x^{m+1} - x \right\|_A^2.
\end{aligned}
$$

Now write the next iterate in terms of the Krylov vectors

$$
\begin{aligned}
x - x^{m+1} &= x - (x^0 + c_0 r^0 + c_1 A r^0 + \cdots + c_m A^m r^0) \\
&= x - x^0 - (c_0 r^0 + c_1 A r^0 + \cdots + c_m A^m r^0) \\
&= x - x^0 - (c_0 I + c_1 A + \cdots + c_m A^m) r^0.
\end{aligned}
$$

Note $r^0 = d - Ax^0 = Ax - Ax^0 = A(x - x^0)$ so that

$$
\begin{aligned}
x - x^{m+1} &= x - x^0 - (c_0 I + c_1 A + \cdots + c_m A^m) A(x - x^0) \\
&= (I - (c_0 I + c_1 A + \cdots + c_m A^m)) A(x - x^0) \\
&= (I - (c_0 A + c_1 A + \cdots + c_m A^{m+1}))(x - x^0).
\end{aligned}
$$

Thus

$$
2(J(x^{m+1}) - J(x)) = \left\| x^{m+1} - x \right\|_A^2 \leq \left\| q_{m+1}(A)(x - x^0) \right\|_A^2 \tag{9.1.8}
$$

where $q_{m+1}(z) = 1 - (c_0 z + c_1 z^2 + \cdots + c_m z^{m+1})$. One can make appropriate choices of the polynomial $q_{m+1}(z)$ and use some properties of eigenvalues and matrix algebra to prove the following theorem, see [11, Chapter 2].

Theorem 9.1.2 *(CG Convergence Properties) Let A be an $n \times n$ SPD matrix, and consider $Ax = d$.*

1. *The conjugate gradient method will obtain the solution within n iterations.*

2. *If d is a linear combination of k eigenvectors of A, then the conjugate gradient method will obtain the solution within k iterations.*

3. *If the set of all eigenvalues of A has at most k distinct eigenvalues, then the conjugate gradient method will obtain the solution within k iterations.*

4. *Let $\kappa_2 \equiv \lambda_{\max}/\lambda_{\min}$ be the condition number of A, which is the ratio of the largest and smallest eigenvalues of A. Then the following error estimate holds so that a condition number closer to one is desirable*

$$
\left\| x^m - x \right\|_A \leq 2 \left\| x^0 - x \right\|_A \left(\frac{\sqrt{\kappa_2} - 1}{\sqrt{\kappa_2} + 1} \right)^m. \tag{9.1.9}
$$

9.1.1   Exercises

1. Show the three properties if Krylov spaces in lines (9.1.1-9.1.3) are true.
2. Show equation (9.1.7) is true for $m = 2$.
3. Give a mathematical induction proof of Theorem 9.1.1.

9.2   Preconditioners

Consider the SPD matrix A and the problem $Ax = d$. For the preconditioned conjugate gradient method a *preconditioner* is another SPD matrix M. The matrix M is chosen so that the equivalent problem $M^{-1}Ax = M^{-1}d$ can be more easily solved. Note $M^{-1}A$ may not be SPD. However M^{-1} is also SPD and must have a Cholesky factorization, which we write as

$$M^{-1} = S^T S.$$

The system $M^{-1}Ax = M^{-1}d$ can be rewritten in terms of a SPD matrix $\widehat{A}\widehat{x} = \widehat{d}$

$$
\begin{aligned}
S^T S A x &= S^T S d \\
(SAS^T)(S^{-T}x) &= Sd \text{ where}
\end{aligned}
$$

$\widehat{A} = SAS^T$, $\widehat{x} = S^{-T}x$ and $\widehat{d} = Sd$. $\widehat{A}$ is SPD. It is similar to $M^{-1}A$, that is, $S^{-T}(M^{-1}A)S^T = \widehat{A}$ so that the eigenvalues of $M^{-1}A$ and $\widehat{A}$ are the same. The idea is use the conjugate gradient method on $\widehat{A}\widehat{x} = \widehat{d}$ and to choose M so that the properties in Theorem 9.1.1 favor rapid convergence of the method.

At first glance it may appear to be very costly to do this. However the following identities show that the application of the conjugate gradient method to $\widehat{A}\widehat{x} = \widehat{d}$ is relatively simple:

$$\widehat{A} = SAS^T, \widehat{x} = S^{-T}x, \widehat{r} = S(d - Ax) \text{ and } \widehat{p} = S^{-T}p.$$

Then

$$\widehat{\alpha} = \frac{\widehat{r}^T \widehat{r}}{\widehat{p}^T \widehat{A}\widehat{p}} = \frac{(Sr)^T (Sr)}{(S^{-T}p)^T (SAS^T) (S^{-T}p)} = \frac{r^T(S^T Sr)}{p^T Ap}.$$

Similar calculations hold for $\widehat{\beta}$ and $\widehat{p}$. This eventually gives the following preconditioned conjugate gradient algorithm, which does require an "easy" solution of $Mz = r$ for each iterate.

Preconditioned Conjugate Gradient Method.

> Let x^0 be an initial guess
> $r^0 = d - Ax^0$
> solve $Mz^0 = r^0$ and set $p^0 = z^0$
> for m = 0, maxm
> $\alpha = (z^m)^T r^m / (p^m)^T Ap^m$
> $x^{m+1} = x^m + \alpha p^m$
> $r^{m+1} = r^m - \alpha Ap^m$
>   test for convergence
>   solve $Mz^{m+1} = r^{m+1}$
> $\beta = (z^{m+1})^T r^{m+1} / (z^m)^T r^m$
> $p^{m+1} = z^{m+1} + \beta p^m$
> endloop.

Examples of Preconditioners.

1.   Block diagonal part of A. For the Poisson problem in two space variables

$$M = \begin{bmatrix} B & 0 & 0 & 0 \\ 0 & B & 0 & 0 \\ 0 & 0 & \ddots & \vdots \\ 0 & 0 & \cdots & B \end{bmatrix} \text{ where}$$

$$A = \begin{bmatrix} B & -I & 0 & 0 \\ -I & B & -I & 0 \\ 0 & -I & \ddots & \vdots \\ 0 & 0 & \cdots & B \end{bmatrix} \text{ and } B = \begin{bmatrix} 4 & -1 & 0 & 0 \\ -1 & 4 & -1 & 0 \\ 0 & -1 & \ddots & \vdots \\ 0 & 0 & \cdots & 4 \end{bmatrix}.$$

2.   Incomplete Cholesky factorization of A. Let $A = M + E = \widehat{G}\widehat{G}^T + E$ where E is chosen so that either the smaller components in A are neglected or some desirable structure of M is attained. This can be done by defining a subset $S \subset \{1, \ldots, n\}$ and overwriting the a_{ij} when $i, j \in S$

$$a_{ij} = a_{ij} - a_{ik}\frac{1}{a_{kk}}a_{kj}.$$

This common preconditioner and some variations are described in [6].

3.   Incomplete domain decomposition. Let $A = M + E$ where M has the form of an "arrow" associated with domain decompositions. For example, consider a problem with two large blocks separated by a smaller block, which is listed last.

$$M = \begin{bmatrix} A_{11} & 0 & A_{13} \\ 0 & A_{22} & A_{23} \\ A_{31} & A_{32} & A_{33} \end{bmatrix} \text{ where}$$

$$A = \begin{bmatrix} A_{11} & A_{12} & A_{13} \\ A_{21} & A_{22} & A_{23} \\ A_{31} & A_{32} & A_{33} \end{bmatrix}.$$

Related preconditioners are described in the paper by E. Chow and Y. Saad [5].

4.   ADI sweeps. Form M by doing one or more ADI sweeps. For example, if $A = H + V$ and one sweep is done in each direction, then for $A = M - N$ as defined by

$$\begin{aligned} (\alpha I + H)x^{m+\frac{1}{2}} &= d + (\alpha I - V)x^m \\ (\alpha I + V)x^{m+1} &= d + (\alpha I - H)x^{m+\frac{1}{2}} \\ &= d + (\alpha I - H)((\alpha I + H)^{-1}(d + (\alpha I - V)x^m)). \end{aligned}$$

Solve for $x^{m+1} = M^{-1}d + M^{-1}Nx^m$ where

$$
\begin{aligned}
M^{-1} &= (\alpha I + V)^{-1}(I + (\alpha I - H)(\alpha I + H)^{-1}) \\
&= (\alpha I + V)^{-1}((\alpha I + H) + (\alpha I - H))(\alpha I + H)^{-1} \\
&= (\alpha I + V)^{-1}2\alpha(\alpha I + H)^{-1} \text{ so that}
\end{aligned}
$$

$$
M = (\alpha I + H)\frac{1}{2\alpha}(\alpha I + V).
$$

The parallelism in this approach comes from independent tridiagonal solves in each direction. Another approach is to partition each tridiagonal solve as is indicated in [15].

5.   SSOR. Form M by doing one forward SOR and then one backward SOR sweep. Let $A = D - L - U$ where $L^T = U$ and $0 < \omega < 2$.

$$
\begin{aligned}
\frac{1}{\omega}(D - \omega L)x^{m+\frac{1}{2}} &= d + \frac{1}{\omega}((1-\omega)D + \omega U)x^m \\
\frac{1}{\omega}(D - \omega U)x^{m+1} &= d + \frac{1}{\omega}((1-\omega)D + \omega L)x^{m+\frac{1}{2}} \\
&= d + \frac{1}{\omega}((1-\omega)D + \omega L)((\frac{1}{\omega}(D - \omega L))^{-1} \\
&\quad (d + \frac{1}{\omega}((1-\omega)D + \omega U)x^m).
\end{aligned}
$$

Solve for $x^{m+1} = M^{-1}d + M^{-1}Nx^m$ where

$$
\begin{aligned}
M^{-1} &= (\frac{1}{\omega}(D - \omega U))^{-1}(I + \frac{1}{\omega}((1-\omega)D + \omega L)(\frac{1}{\omega}(D - \omega L))^{-1} \\
&= (\frac{1}{\omega}(D - \omega U))^{-1}((\frac{1}{\omega}(D - \omega L)) + \frac{1}{\omega}((1-\omega)D + \omega L)(\frac{1}{\omega}(D - \omega L))^{-1} \\
&= (\frac{1}{\omega}(D - \omega U))^{-1}\frac{2-\omega}{\omega}D(\frac{1}{\omega}(D - \omega L))^{-1} \text{ so that}
\end{aligned}
$$

$$
M = \frac{1}{\omega}(D - \omega L)(\frac{2-\omega}{\omega}D)^{-1}\frac{1}{\omega}(D - \omega U).
$$

6.   Additive Schwarz. As motivation consider the Poisson problem and divide the space grid into subsets of n_i unknowns so that $\Sigma n_i \geq n$. Let B_i be associated with a splitting of a restriction of the coefficient matrix to subset i of unknowns, and let R_i be the restriction operator so that

$$
\begin{aligned}
A &: \mathbb{R}^n \longrightarrow \mathbb{R}^n \\
B_i &: \mathbb{R}^{n_i} \longrightarrow \mathbb{R}^{n_i} \\
R_i &: \mathbb{R}^n \longrightarrow \mathbb{R}^{n_i} \\
R_i^T &: \mathbb{R}^{n_i} \longrightarrow \mathbb{R}^n.
\end{aligned}
$$

Define $\widehat{M_i} = R_i^T B_i^{-1} R_i : \mathbb{R}^n \longrightarrow \mathbb{R}^n$. Although these matrices are not invertible, one may be able to associate a splitting with the summation $\Sigma \widehat{M_i}$. Often a coarse mesh is associated with the problem. In this case let A_0 be $n_0 \times n_0$ and let $R_0^T : \mathbb{R}^{n_0} \longrightarrow \mathbb{R}^n$ be an extension operator from the coarse to fine mesh. The following may accelerate the convergence

$$M^{-1} \equiv R_0^T A_0^{-1} R_0 + \Sigma \widehat{M}_i = R_0^T A_0^{-1} R_0 + \Sigma R_i^T B_i^{-1} R_i.$$

A common example is to apply SSOR to the subsets of unknowns with zero boundary conditions. A nice survey of Schwarz methods can be found in the paper written by Xiao-Chuan Cai in the first chapter of [12].

7.   Least squares approximations for $A\widehat{M} = I$. This is equivalent to n least squares problems

$$A\widehat{m}_j = e_j.$$

For example, if the column vectors $\widehat{m}_j$ were to have nonzero components in rows $i = j - 1, j$ and $j + 1$, then this becomes a least squares problem with n equations and three unknowns $\widetilde{m}_j = [m_{j-1}, m_j, m_{j+1}]^T$ where A is restricted to columns $j - 1, j$ and $j + 1$

$$A(1 : n, j - 1 : j + 1)\widetilde{m}_j = e_j.$$

The preconditioner is formed by collecting the column vectors $\widehat{m}_j$

$$M^{-1} = [\; \widehat{m}_1 \quad \widehat{m}_2 \quad \cdots \quad \widehat{m}_n \;].$$

Additional information on this approach can be found in the paper by M. J. Grote and T. Huckle [9].

    The following MATLAB code is a slight variation of precg.m that was described in Section 3.6. Here the SSOR and the block diagonal preconditioners are used. The choice of preconditioners is made in lines 28-33. The number of iterates required for convergence was 19, 55 and 73 for SSOR, block diagonal and no preconditioning, respectively. In the ssorpc.m preconditioner function the forward solve is done in lines 3-7, and the backward solve is done in lines 9-13. In the bdiagpc.m preconditioner function the diagonal blocks are all the same and are defined in lines 3-12 every time the function is evaluated. The solves for each block are done in lines 13-15.

MATLAB Code pccg.m with ssorpc.m and bdiagpc.m

```
1.    % Solves -u_xx - u_yy = 200+200sin(pi x)sin(pi y).
2.    % Uses PCG with SSOR or block diagonal preconditioner.
3.    % Uses 2D arrays for the column vectors.
4.    % Does not explicity store the matrix.
5.    clear;
6.    w = 1.6;
7.    n = 65;
8.    h = 1./n;
9.    u(1:n+1,1:n+1)= 0.0;
```

```
10.     r(1:n+1,1:n+1)= 0.0;
11.     rhat(1:n+1,1:n+1) = 0.0;
12.     % Define right side of PDE
13.     for j= 2:n
14.           for i = 2:n
15.                 r(i,j)= h*h*(200+200*sin(pi*(i-1)*h)*sin(pi*(j-1)*h));
16.           end
17.     end
18.     errtol = .0001*sum(sum(r(2:n,2:n).*r(2:n,2:n)))^.5;
19.     p(1:n+1,1:n+1)= 0.0;
20.     q(1:n+1,1:n+1)= 0.0;
21.     err = 1.0;
22.     m = 0;
23.     rho = 0.0;
24.     % Begin PCG iterations
25.     while ((err > errtol)&(m < 200))
26.           m = m+1;
27.           oldrho = rho;
28.     % Execute SSOR preconditioner
29.           rhat = ssorpc(n,n,1,1,1,1,4,.25,w,r,rhat);
30.     % Execute block diagonal preconditioner
31.     % rhat = bdiagpc(n,n,1,1,1,1,4,.25,w,r,rhat);
32.     % Use the following line for no preconditioner
33.     % rhat = r;
34.     % Find conjugate direction
35.           rho = sum(sum(r(2:n,2:n).*rhat(2:n,2:n)));
36.           if (m==1)
37.                 p = rhat;
38.           else
39.                 p = rhat + (rho/oldrho)*p;
40.           end
41.     % Use the following line for steepest descent method
42.     % p=r;
43.     % Executes the matrix product q = Ap without storage of A
44.           for j= 2:n
45.                 for i = 2:n
46.                       q(i,j)=4.*p(i,j)-p(i-1,j)-p(i,j-1)-p(i+1,j)-p(i,j+1);
47.                 end
48.           end
49.     % Executes the steepest descent segment
50.           alpha = rho/sum(sum(p.*q));
51.           u = u + alpha*p;
52.           r = r - alpha*q;
53.     % Test for convergence via the infinity norm of the residual
54.           err = max(max(abs(r(2:n,2:n))));
```

```
55.              reserr(m) = err;
56.      end
57.      m
58.      semilogy(reserr)
```

```
1.      function r = ssorpc(nx,ny,ae,aw,as,an,ac,rac,w,d,r)
2.      % This preconditioner is SSOR.
3.      for j= 2:ny
4.          for i = 2:nx
5.              r(i,j) = w*(d(i,j) + aw*r(i-1,j) + as*r(i,j-1))*rac;
6.          end
7.      end
8.      r(2:nx,2:ny) = ((2.-w)/w)*ac*r(2:nx,2:ny);
9.      for j= ny:-1:2
10.         for i = nx:-1:2
11.             r(i,j) = w*(r(i,j)+ae*r(i+1,j)+an*r(i,j+1))*rac;
12.         end
13.     end
```

```
1.      function r = bdiagpc(nx,ny,ae,aw,as,an,ac,rac,w,d,r)
2.      % This preconditioner is block diagonal.
3.      Adiag = zeros(nx-1);
4.      for i = 1:nx-1
5.          Adiag(i,i) = ac;
6.          if i>1
7.              Adiag(i,i-1) = -aw;
8.          end
9.          if i<nx-1
10.             Adiag(i,i+1) = -ae;
11.         end
12.     end
13.     for j = 2:ny
14.         r(2:nx,j) = Adiag\d(2:nx,j);
15.     end
```

9.2.1   Exercises

1.   In the derivation of the preconditioned conjugate gradient method do the calculations for $\widehat{\beta}$ and $\widehat{p}$. Complete the derivation of the PCG algorithm.

2.   Verify the calculations for the MATLAB code pccg.m. Experiment with some variations of the SSOR and block preconditioners.

3.   Experiment with the incomplete Cholesky preconditioner.

4.   Experiment with the incomplete domain decomposition preconditioner.

5.   Experiment with the ADI preconditioner.

6.   Experiment with the additive Schwarz preconditioner.

7.   Experiment with the least squares preconditioner. You may want to review least squares as in Section 9.4 and see the interesting paper by Grote and Huckle [9].

9.3   PCG and MPI

This section contains the MPI Fortran code of the preconditioned conjugate gradient algorithm for the solution of a Poisson problem. The preconditioner is an implementation of the additive Schwarz preconditioner with no coarse mesh acceleration. It uses SSOR on large blocks of unknowns by partitioning the second index with zero boundary conditions on the grid boundaries. So, this could be viewed as a block diagonal preconditioner where the diagonal blocks of the coefficient matrix are split by the SSOR splitting. Since the number of blocks are associated with the number of processors, the preconditioner really does change with the number of processors.

The global initialization is done in lines 1-36, and an initial guess is the zero vector. In lines 37-42 MPI is started and the second index is partitioned according to the number of processors. The conjugate gradient loop is executed in lines 48-120, and the partial outputs are given by lines 121-138.

The conjugate gradient loop has substages, which are done in parallel. The preconditioner is done on each block in lines 50-62. The local dot products for β are computed, then *mpi_allreduce()* is used to total the local dot products and to broadcast the result to all the processors. The local parts of the updated search direction are computed in lines 67-71. In order to do the sparse matrix product Ap, the top and bottom grid rows of p are communicated in lines 72-107 to the adjacent processors. This communication scheme is similar to that used in Section 6.6 and is illustrated in Figures 6.6.1 and 6.6.2. Lines 108-109 contain the local computation of Ap. In lines 111-114 the local dot products for the computation of α in the steepest descent direction computation are computed, and then *mpi_allreduce()* is used to total the local dot products and to broadcast the result to all the processors. Lines 114 and 115 contain the updated local parts of the approximated solution and residual. Lines 117-118 contain the local computation of the residual error, and then *mpi_allreduce()* is used to total the errors and to broadcast the result to all the processors. Once the error criteria have been satisfied for all processors, the conjugate gradient loop will be exited.

MPI/Fortran Code cgssormpi.f

```
1.     program cgssor
2.!     This code approximates the solution of
3.!         -u_xx - u_yy = f
4.!     PCG is used with a SSOR verson of the
5.!     Schwarz additive preconditioner.
6.!     The sparse matrix product, dot products and updates
7.!     are also done in parallel.
```

```
8.     implicit none
9.     include 'mpif.h'
10.    real,dimension(0:1025,0:1025):: u,p,q,r,rhat
11.    real,dimension (0:1025) :: x,y
12.    real :: oldrho,ap, rho,alpha,error,dx2,w,t0,timef,tend
13.    real :: loc_rho,loc_ap,loc_error
14.    integer :: i,j,n,m
15.    integer :: my_rank,proc,source,dest,tag,ierr,loc_n
16.    integer :: status(mpi_status_size),bn,en
17.    integer :: maxit,sbn
18.    w = 1.8
19.    u = 0.0
20.    n = 1025
21.    maxit = 200
22.    dx2 = 1./(n*n)
23.    do i=0,n
24.        x(i) = float(i)/n
25.        y(i) = x(i)
26.    end do
27.    r = 0.0
28.    rhat = 0.0
29.    q = 0.0
30.    p = 0.0
31.    do j = 1,n-1
32.        r(1:n-1,j)=200.0*dx2*(1+sin(3.14*x(1:n-1))*sin(3.14*y(j)))
33.    end do
34.    error = 1.
35.    m = 0
36.    rho = 0.0
37.    call mpi_init(ierr)
38.    call mpi_comm_rank(mpi_comm_world,my_rank,ierr)
39.    call mpi_comm_size(mpi_comm_world,proc,ierr)
40.    loc_n = (n-1)/proc
41.    bn = 1+(my_rank)*loc_n
42.    en = bn + loc_n -1
43.    call mpi_barrier(mpi_comm_world,ierr)
44.    if (my_rank.eq.0) then
45.        t0 = timef()
46.    end if
47.    do while ((error>.0001).and.(m<maxit))
48.        m = m+1
49.        oldrho = rho
50.    ! Execute Schwarz additive SSOR preconditioner.
51.    ! This preconditioner changes with the number of processors!
52         do j= bn,en
```

```
53.                 do i = 1,n-1
54.                     rhat(i,j) = w*(r(i,j)+rhat(i-1,j)+rhat(i,j-1))*.25
55.                 end do
56.             end do
57.             rhat(1:n-1,bn:en) = ((2.-w)/w)*4.*rhat(1:n-1,bn:en)
58.             do j= en,bn,-1
59.                 do i = n-1,1,-1
60.                     rhat(i,j) = w*(rhat(i,j)+rhat(i+1,j)+rhat(i,j+1))*.25
61.                 end do
62.             end do
63.     ! rhat = r
64.     ! Find conjugate direction.
65.             loc_rho = sum(r(1:n-1,bn:en)*rhat(1:n-1,bn:en))
66.             call mpi_allreduce(loc_rho,rho,1,mpi_real,mpi_sum,&
                                    mpi_comm_world,ierr)
67.             if (m.eq.1) then
68.                 p(1:n-1,bn:en) = rhat(1:n-1,bn:en)
69.             else
70.                 p(1:n-1,bn:en) = rhat(1:n-1,bn:en)&
                                    + (rho/oldrho)*p(1:n-1,bn:en)
71.             endif
72.     ! Execute matrix product q = Ap.
73.     ! First, exchange information between processors.
74.             if (my_rank.eq.0) then
75.                 call mpi_recv(p(0,en+1),(n+1),mpi_real,my_rank+1,50,&
76.                                     mpi_comm_world,status,ierr)
77.                 call mpi_send(p(0,en),(n+1),mpi_real,my_rank+1,50,&
78.                                     mpi_comm_world,ierr)
79.             end if
80.             if ((my_rank.gt.0).and.(my_rank.lt.proc-1)&
81.                             .and.(mod(my_rank,2).eq.1)) then
82.                 call mpi_send(p(0,en),(n+1),mpi_real,my_rank+1,50,&
83.                                     mpi_comm_world,ierr)
84.                 call mpi_recv(p(0,en+1),(n+1),mpi_real,my_rank+1,50,&
85.                                     mpi_comm_world,status,ierr)
86.                 call mpi_send(p(0,bn),(n+1),mpi_real,my_rank-1,50,&
87.                                     mpi_comm_world,ierr)
88.                 call mpi_recv(p(0,bn-1),(n+1),mpi_real,my_rank-1,50,&
89.                                     mpi_comm_world,status,ierr)
90.             end if
91.             if ((my_rank.gt.0).and.(my_rank.lt.proc-1)&
92.                             .and.(mod(my_rank,2).eq.0)) then
93.                 call mpi_recv(p(0,bn-1),(n+1),mpi_real,my_rank-1,50,&
94.                                     mpi_comm_world,status,ierr)
95.                 call mpi_send(p(0,bn),(n+1),mpi_real,my_rank-1,50,&
```

```
96.                                        mpi_comm_world,ierr)
97.            call mpi_recv(p(0,en+1),(n+1),mpi_real,my_rank+1,50,&
98.                                        mpi_comm_world,status,ierr)
99.            call mpi_send(p(0,en),(n+1),mpi_real,my_rank+1,50,&
100.                                       mpi_comm_world,ierr)
101.       end if
102.       if (my_rank.eq.proc-1) then
103.           call mpi_send(p(0,bn),(n+1),mpi_real,my_rank-1,50,&
104.                                       mpi_comm_world,ierr)
105.           call mpi_recv(p(0,bn-1),(n+1),mpi_real,my_rank-1,50,&
106.                                       mpi_comm_world,status,ierr)
107.       end if
108.       q(1:n-1,bn:en)=4.0*p(1:n-1,bn:en)-p(0:n-2,bn:en)-p(2:n,bn:en)&
109.                       - p(1:n-1,bn-1:en-1) - p(1:n-1,bn+1:en+1)
110.    ! Find steepest descent.
111.       loc_ap = sum(p(1:n-1,bn:en)*q(1:n-1,bn:en))
112.       call mpi_allreduce(loc_ap,ap,1,mpi_real,mpi_sum,&
113.                                       mpi_comm_world,ierr)
114.       alpha = rho/ap
115.       u(1:n-1,bn:en) = u(1:n-1,bn:en) + alpha*p(1:n-1,bn:en)
116.       r(1:n-1,bn:en) = r(1:n-1,bn:en) - alpha*q(1:n-1,bn:en)
117.       loc_error = maxval(abs(r(1:n-1,bn:en)))
118.       call mpi_allreduce(loc_error,error,1,mpi_real, mpi_sum,&
119.                                       mpi_comm_world,ierr)
120.    end do
121.    ! Send local solutions to processor zero.
122.    if (my_rank.eq.0) then
123.        do source = 1,proc-1
124.            sbn = 1+(source)*loc_n
125.            call mpi_recv(u(0,sbn),(n+1)*loc_n,mpi_real,source,50,&
126.                                       mpi_comm_world,status,ierr)
127.        end do
128.    else
129.        call mpi_send(u(0,bn),(n+1)*loc_n,mpi_real,0,50,&
130.                                       mpi_comm_world,ierr)
131.    end if
132.    if (my_rank.eq.0) then
133.        tend = timef()
134.        print*, 'time =', tend
135.        print*, 'time per iteration = ', tend/m
136.        print*, m,error, u(512 ,512)
137.        print*, 'w = ',w
138.    end if
139.    call mpi_finalize(ierr)
140.    end program
```

Table 9.3.1: MPI Times for cgssormpi.f

p	time	iteration
2	35.8	247
4	16.8	260
8	07.9	248
16	03.7	213
32	03.0	287

The Table 9.3.1 contains computations for $n = 1025$ and using $w = 1.8$. The computation times are in seconds, and note the number of iterations vary with the number of processors.

9.3.1   Exercises

1.   Verify the computations in Table 9.3.1. Experiment with the convergence criteria.

2.   Experiment with variations on the SSOR preconditioner and include different n and ω.

3.   Experiment with variations of the SSOR preconditioner to include the use of a coarse mesh in the additive Schwarz preconditioner.

4.   Use an ADI preconditioner in place of the SSOR preconditioner.

9.4   Least Squares

Consider an algebraic system where there are more equations than unknowns. This will be a subproblem in the next two sections where the unknowns will be the coefficients of the Krylov vectors. Let A be $n \times m$ where $n > m$. In this case it may not be possible to find x such that $Ax = d$, that is, the residual vector $r(x) = d - Ax$ may never be the zero vector. The next best alternative is to find x so that in some way the residual vector is as small as possible.

Definition. Let $R(x) \equiv r(x)^T r(x)$ where A is $n \times m$, $r(x) = d - Ax$ and x is $m \times 1$. The least squares solution of $Ax = d$ is

$$R(x) = \min_y R(y).$$

The following identity is important in finding a least squares solution

$$
\begin{aligned}
R(y) &= (d - Ay)^T (d - Ay) \\
&= d^T d - 2(Ay)^T d + (Ay)^T Ay \\
&= d^T d + 2[1/2\, y^T (A^T A)y - y^T (A^T d)].
\end{aligned}
\tag{9.4.1}
$$

If $A^T A$ is SPD, then by Theorem 8.4.1 the second term in (9.4.1) will be a minimum if and only if

$$A^T Ax = A^T d. \tag{9.4.2}$$

This system is called the *normal equations*.

Theorem 9.4.1 *(Normal Equations) If A has full column rank (Ax = 0 implies x = 0), then the least squares solution is characterized by the solution of the normal equations (9.4.2).*

Proof. Clearly $A^T A$ is symmetric. Note $x^T(A^T A)x = (Ax)^T(Ax) = 0$ if and only if $Ax = 0$. The full column rank assumption implies $x = 0$ so that $x^T(A^T A)x > 0$ if $x \neq 0$. Thus $A^T A$ is SPD. Apply the first part of Theorem 8.4.1 to the second term in (9.4.1). Since the first term in (9.4.1) is constant with respect to y, $R(y)$ will be minimized if and only if the normal equations (9.4.2) are satisfied. ∎

Example 1. Consider the 3×2 algebraic system

$$\begin{bmatrix} 1 & 1 \\ 1 & 2 \\ 1 & 3 \end{bmatrix} \begin{bmatrix} x_1 \\ x_2 \end{bmatrix} = \begin{bmatrix} 4 \\ 7 \\ 8 \end{bmatrix}.$$

This could have evolved from the linear curve $y = mt + c$ fit to the data $(t_i, y_i) = (1, 4), (2, 7)$ and $(3, 8)$ where $x_1 = c$ and $x_2 = m$. The matrix has full column rank and the normal equations are

$$\begin{bmatrix} 3 & 6 \\ 6 & 14 \end{bmatrix} \begin{bmatrix} x_1 \\ x_2 \end{bmatrix} = \begin{bmatrix} 19 \\ 42 \end{bmatrix}.$$

The solution is $x_1 = c = 7/3$ and $x_2 = m = 2$.

The normal equations are often ill-conditioned and prone to significant accumulation of roundoff errors. A good alternative is to use a QR factorization of A.

Definition. *Let A be $n \times m$. Factor $A = QR$ where Q is $n \times m$ such that $Q^T Q = I$, and R is $m \times m$ is upper triangular. This is called a QR factorization of A.*

Theorem 9.4.2 *(QR Factorization) If $A = QR$ and has full column rank, then the solution of the normal equations is given by the solution of $Rx = Q^T d$.*

Proof. The normal equations become

$$\begin{aligned} (QR)^T(QR)x &= (QR)^T d \\ R^T(Q^T Q)Rx &= R^T Q^T d \\ R^T Rx &= R^T Q^T d. \end{aligned}$$

Because A is assumed to have full column rank, R must have an inverse. Thus we only need to solve $Rx = Q^T d$. ∎

There are a number of ways to find the QR factorization of the matrix. The modified Gram-Schmidt method is often used when the matrix has mostly nonzero components. If the matrix has a small number of nonzero components, then one can use a small sequence of Givens transformations to find the QR factorization. Other methods for finding the QR factorization are the row version of Gram-Schmidt, which generates more numerical errors, and the Householder transformation, see [16, Section 5.5].

In order to formulate the *modified (also called the column version) Gram-Schmidt* method, write the $A = QR$ in columns

$$
\begin{bmatrix} a_1 & a_2 & \cdots & a_m \end{bmatrix} = \begin{bmatrix} q_1 & q_2 & \cdots & q_m \end{bmatrix}
\begin{bmatrix}
r_{11} & r_{12} & \cdots & r_{1m} \\
 & r_{22} & \cdots & r_{2m} \\
 & & \ddots & \vdots \\
 & & & r_{nm}
\end{bmatrix}
$$

$$
\begin{aligned}
a_1 &= q_1 r_{11} \\
a_2 &= q_1 r_{12} + q_2 r_{22} \\
&\vdots \\
a_m &= q_1 r_{1m} + q_2 r_{2m} + \cdots + q_m r_{mm}.
\end{aligned}
$$

First, choose $q_1 = a_1/r_{11}$ where $r_{11} = (a_1^T a_1)^{\frac{1}{2}}$. Second, since $q_1^T q_k = 0$ for all $k > 1$, compute $q_1^T a_k = 1 r_{1k} + 0$. Third, for $k > 1$ move the columns $q_1 r_{1k}$ to the left side, that is, update column vectors $k = 2, ..., m$

$$
\begin{aligned}
a_2 - q_1 r_{12} &= q_2 r_{22} \\
&\vdots \\
a_m - q_1 r_{1m} &= q_2 r_{2m} + \cdots + q_m r_{mm}.
\end{aligned}
$$

This is a reduction in dimension so that the above three steps can be repeated on the $n \times (m-1)$ reduced problem.

Example 2. Consider the 4×3 matrix

$$
A = \begin{bmatrix}
1 & 1 & 1 \\
1 & 1 & 0 \\
1 & 0 & 2 \\
1 & 0 & 0
\end{bmatrix}.
$$

$r_{11} = (a_1^T a_1)^{\frac{1}{2}} = (\begin{bmatrix} 1 & 1 & 1 & 1 \end{bmatrix}^T \begin{bmatrix} 1 & 1 & 1 & 1 \end{bmatrix})^{\frac{1}{2}} = 2.$
$q_1 = \begin{bmatrix} 1/2 & 1/2 & 1/2 & 1/2 \end{bmatrix}^T.$
$q_1^T a_2 = r_{12} = 1$ and $a_2 - q_1 r_{12} = \begin{bmatrix} 1/2 & 1/2 & -1/2 & -1/2 \end{bmatrix}^T.$
$q_1^T a_3 = r_{13} = 3/2$ and $a_3 - q_1 r_{13} = \begin{bmatrix} 1/4 & -3/4 & 5/4 & -3/4 \end{bmatrix}^T.$
This reduces to a 4×2 matrix QR factorization. Eventually, the QR factorization is obtained

$$
A = \begin{bmatrix}
1/2 & 1/2 & 1/\sqrt{10} \\
1/2 & 1/2 & -1/\sqrt{10} \\
1/2 & -1/2 & 2/\sqrt{10} \\
1/2 & -1/2 & -2/\sqrt{10}
\end{bmatrix}
\begin{bmatrix}
2 & 1 & 3/2 \\
0 & 1 & -1/2 \\
0 & 0 & \sqrt{10}/2
\end{bmatrix}.
$$

The modified Gram-Schmidt method allows one to find QR factorizations where the column dimension of the coefficient matrix is increasing, which is the case for the application to the GMRES methods. Suppose A, initially $n \times (m-1)$, is a matrix, whose QR factorization has already been computed. Augment this matrix by another column vector. We must find q_m so that

$$a_m = q_1 r_{1,m} + \cdots + q_{m-1} r_{m-1,m} + q_m r_{m,m}.$$

If the previous modified Gram-Schmidt method is to be used for the $n \times (m-1)$ matrix, then none of the updates for the new column vector have been done. The first update for column m is $a_m - q_1 r_{1,m}$ where $r_{1,m} = q_1^T a_m$. By overwriting the new column vector one can obtain all of the needed vector updates. The following loop completes the modified Gram-Schmidt QR factorization when an additional column is augmented to the matrix, *augmented modified Gram-Schmidt*,

$$q_m = a_m$$
for $i = 1, m - 1$
$\qquad r_{i,m} = q_i^T q_m$
$\qquad q_m = q_m - q_i r_{i,m}$
endloop
$r_{m,m} = (q_m^T q_m)^{\frac{1}{2}}$
if $r_{m,m} = 0$ then
$\qquad$ stop
else
$\qquad q_m = q_m / r_{m,m}$
endif.

When the above loop is used with $a_m = A q_{m-1}$ and within a loop with respect to m, this gives the *Arnoldi algorithm*, which will be used in the next section.

In order to formulate the *Givens transformation* for a matrix with a small number of nonzero components, consider the 2×1 matrix

$$A = \begin{bmatrix} a \\ b \end{bmatrix}.$$

The QR factorization has a simple form

$$Q^T A = Q^T (QR) = (Q^T Q) R = R$$
$$Q^T \begin{bmatrix} a \\ b \end{bmatrix} = \begin{bmatrix} r_{11} \\ 0 \end{bmatrix}.$$

By inspection one can determine the components of a 2×2 matrix that does this

$$Q^T = G^T = \begin{bmatrix} c & -s \\ s & c \end{bmatrix}$$

where $s = -b/r_{11}, c = a/r_{11}$ and $r_{11} = \sqrt{a^2 + b^2}$. G is often called the Givens rotation because one can view s and c as the sine and cosine of an angle.

Example 3. Consider the 3×2 matrix

$$\begin{bmatrix} 1 & 1 \\ 1 & 2 \\ 1 & 3 \end{bmatrix}.$$

Apply three Givens transformations so as to zero out the lower triangular part of the matrix:

$$G_{21}^T A = \begin{bmatrix} 1/\sqrt{2} & 1/\sqrt{2} & 0 \\ -1/\sqrt{2} & 1/\sqrt{2} & 0 \\ 0 & 0 & 1 \end{bmatrix} \begin{bmatrix} 1 & 1 \\ 1 & 2 \\ 1 & 3 \end{bmatrix}$$

$$= \begin{bmatrix} \sqrt{2} & 3/\sqrt{2} \\ 0 & 1/\sqrt{2} \\ 1 & 3 \end{bmatrix},$$

$$G_{31}^T G_{21}^T A = \begin{bmatrix} \sqrt{2}/\sqrt{3} & 0 & 1/\sqrt{3} \\ 0 & 1 & 0 \\ -1/\sqrt{3} & 0 & \sqrt{2}/\sqrt{3} \end{bmatrix} \begin{bmatrix} \sqrt{2} & 3/\sqrt{2} \\ 0 & 1/\sqrt{2} \\ 1 & 3 \end{bmatrix}$$

$$= \begin{bmatrix} \sqrt{3} & 2\sqrt{3} \\ 0 & 1/\sqrt{2} \\ 0 & \sqrt{3}/\sqrt{2} \end{bmatrix} \quad \text{and}$$

$$G_{32}^T G_{31}^T G_{21}^T A = \begin{bmatrix} 1 & 0 & 0 \\ 0 & 1/2 & \sqrt{3}/2 \\ 0 & -\sqrt{3}/2 & 1/2 \end{bmatrix} \begin{bmatrix} \sqrt{3} & 2/\sqrt{3} \\ 0 & 1/\sqrt{2} \\ 0 & \sqrt{3}/\sqrt{2} \end{bmatrix}$$

$$= \begin{bmatrix} \sqrt{3} & 2\sqrt{3} \\ 0 & \sqrt{2} \\ 0 & 0 \end{bmatrix}.$$

This gives the "big" or "fat" version of the QR factorization where $\widehat{Q}$ is square with a third column and $\widehat{R}$ has a third row of zero components

$$A = G_{21}G_{31}G_{32}\widehat{R} = \widehat{Q}\widehat{R}$$

$$= \begin{bmatrix} .5774 & -.7071 & .4082 \\ .5774 & 0 & -.8165 \\ .5774 & .7071 & .4082 \end{bmatrix} \begin{bmatrix} 1.7321 & 3.4641 \\ 0 & 1.4142 \\ 0 & 0 \end{bmatrix}.$$

The solution to the least squares problem in the first example can be found by solving $Rx = Q^T d$

$$\begin{bmatrix} 1.7321 & 3.4641 \\ 0 & 1.4142 \end{bmatrix} \begin{bmatrix} x_1 \\ x_2 \end{bmatrix} = \begin{bmatrix} .5774 & .5774 & .5774 \\ -.7071 & 0 & .7071 \end{bmatrix} \begin{bmatrix} 4 \\ 7 \\ 8 \end{bmatrix}$$

$$= \begin{bmatrix} 10.9697 \\ 2.8284 \end{bmatrix}.$$

The solution is $x_2 = 2.0000$ and $x_1 = 2.3333$, which is the same as in the first example. A very easy computation is in MATLAB where the single command A\d will produce the least squares solution of $Ax = d$! Also, the MATLAB command [q r] = qr(A) will generate the QR factorization of A.

9.4.1   Exercises

1.   Verify by hand and by MATLAB the calculations in Example 1.

2.   Verify by hand and by MATLAB the calculations in Example 2 for the modified Gram-Schmidt method.

3.   Consider Example 2 where the first two columns in the Q matrix have been computed. Verify by hand that the loop for the *augmented modified Gram-Schmidt* will give the third column in Q.

4.   Show that if the matrix A has full column rank, then the matrix R in the QR factorization must have an inverse.

5.   Verify by hand and by MATLAB the calculations in Example 3 for the sequence of Givens transformations.

6.   Show $Q^T Q = I$ where Q is a product of Givens transformations.

9.5   GMRES

If A is not a SPD matrix, then the conjugate gradient method cannot be directly used. One alternative is to replace $Ax = d$ by the normal equations $A^T Ax = A^T d$, which may be ill-conditioned and subject to significant roundoff errors. Another approach is to try to minimize the residual $R(x) = r(x)^T r(x)$ in place of $J(x) = \frac{1}{2}x^T Ax - x^T d$ for the SPD case. As in the conjugate gradient method, this will be done on the Krylov space.

Definition. *The generalized minimum residual method (GMRES) is*

$$x^{m+1} = x^0 + \sum_{i=0}^{m} \alpha_i A^i r^0$$

where $r^0 = d - Ax^0$ and $\alpha_i \in \mathbb{R}$ are chosen so that

$$R(x^{m+1}) = \min_{y} R(y)$$

$$y \in x^0 + K_{m+1} \text{ and}$$

$$K_{m+1} = \{z \mid z = \sum_{i=0}^{m} c_i A^i r^0, c_i \in \mathbb{R}\}.$$

Like the conjugate gradient method the Krylov vectors are very useful for the analysis of convergence. Consider the residual after $m + 1$ iterations

$$
\begin{aligned}
d - Ax^{m+1} &= d - A(x^0 + \alpha_0 r^0 + \alpha_1 Ar^0 + \cdots + \alpha_m A^m r^0) \\
&= r^0 - A(\alpha_0 r^0 + \alpha_1 Ar^0 + \cdots + \alpha_m A^m r^0) \\
&= (I - A(\alpha_0 I + \alpha_1 A + \cdots + \alpha_m A^m))r^0.
\end{aligned}
$$

Thus

$$
\left\| r^{m+1} \right\|_2^2 \leq \left\| q_{m+1}(A)r^0 \right\|_2^2 \tag{9.5.1}
$$

where $q_{m+1}(z) = 1 - (\alpha_0 z + \alpha_1 z^2 + \cdots + \alpha_m z^{m+1})$. Next one can make appropriate choices of the polynomial $q_{m+1}(z)$ and use some properties of eigenvalues and matrix algebra to prove the following theorem, see C. T. Kelley [11, Chapter 3].

Theorem 9.5.1 (*GMRES Convergence Properties*) *Let A be an $n \times n$ invertible matrix and consider $Ax = d$.*

1. *GMRES will obtain the solution within n iterations.*

2. *If d is a linear combination of k of the eigenvectors of A and $A = V\Lambda V^H$ where $VV^H = I$ and Λ is a diagonal matrix, then the GMRES will obtain the solution within k iterations.*

3. *If the set of all eigenvalues of A has at most k distinct eigenvalues and if $A = V\Lambda V^{-1}$ where Λ is a diagonal matrix, then GMRES will obtain the solution within k iterations.*

The Krylov space of vectors has the nice property that $AK_m \subset K_{m+1}$. This allows one to reformulate the problem of finding the α_i

$$
A\left(x^0 + \sum_{i=0}^{m-1} \alpha_i A^i r^0\right) = d
$$

$$
Ax^0 + \sum_{i=0}^{m-1} \alpha_i A^{i+1} r^0 = d
$$

$$
\sum_{i=0}^{m-1} \alpha_i A^{i+1} r^0 = r^0. \tag{9.5.2}
$$

Let bold $\mathbf{K}_m$ be the $n \times m$ matrix of Krylov vectors

$$
\mathbf{K}_m = \begin{bmatrix} r^0 & Ar^0 & \cdots & A^{m-1}r^0 \end{bmatrix}.
$$

The equation in (9.5.2) has the form

$$
A\mathbf{K}_m \alpha = r^0 \text{ where} \tag{9.5.3}
$$

$$
\begin{aligned}
A\mathbf{K}_m &= A\begin{bmatrix} r^0 & Ar^0 & \cdots & A^{m-1}r^0 \end{bmatrix} \text{ and} \\
\alpha &= \begin{bmatrix} \alpha_0 & \alpha_1 & \cdots & \alpha_{m-1} \end{bmatrix}^T.
\end{aligned}
$$

The equation in (9.5.3) is a least squares problem for $\alpha \in \mathbb{R}^m$ where $A\mathbf{K}_m$ is an $n \times m$ matrix.

In order to efficiently solve this sequence of least squares problems, we construct an orthonormal basis of K_m one column vector per iteration. Let $V_m = \{v_1, v_2, \dots, v_m\}$ be this basis, and let bold $\mathbf{V}_m$ be the $n \times m$ matrix whose columns are the basis vectors

$$\mathbf{V}_m = \begin{bmatrix} v_1 & v_2 & \cdots & v_m \end{bmatrix}.$$

Since $AK_m \subset K_{m+1}$, each column in $A\mathbf{V}_m$ should be a linear combination of columns in $\mathbf{V}_{m+1}$. This allows one to construct $\mathbf{V}_m$ one column per iteration by using the modified Gram-Schmidt process.

Let the first column of $\mathbf{V}_m$ be the normalized initial residual

$$r^0 = bv_1$$

where $b = ((r^0)^T r^0)^{\frac{1}{2}}$ is chosen so that $v_1^T v_1 = 1$. Since $AK_0 \subset K_1$, A times the first column should be a linear combination of v_1 and v_2

$$Av_1 = v_1 h_{11} + v_2 h_{21}.$$

Find h_{11} and h_{21} by requiring $v_1^T v_1 = v_2^T v_2 = 1$ and $v_1^T v_2 = 0$ and assuming $Av_1 - v_1 h_{11}$ is not the zero vector

$$
\begin{aligned}
h_{11} &= v_1^T A v_1, \\
z &= Av_1 - v_1 h_{11}, \\
h_{21} &= (z^T z)^{\frac{1}{2}} \text{ and} \\
v_2 &= z/h_{21}.
\end{aligned}
$$

For the next column

$$Av_2 = v_1 h_{12} + v_2 h_{22} + v_3 h_{32}.$$

Again require the three vectors to be orthonormal and $Av_2 - v_1 h_{12} - v_2 h_{22}$ is not zero to get

$$
\begin{aligned}
h_{12} &= v_1^T A v_2 \text{ and } h_{22} = v_2^T A v_2, \\
z &= Av_2 - v_1 h_{12} - v_2 h_{22}, \\
h_{32} &= (z^T z)^{\frac{1}{2}} \text{ and} \\
v_3 &= z/h_{32}.
\end{aligned}
$$

Continue this and represent the results in matrix form

$$AV_m \;=\; \mathbf{V}_{m+1}H \quad \text{where} \tag{9.5.4}$$

$$AV_m \;=\; \begin{bmatrix} Av_1 & Av_2 & \cdots & Av_m \end{bmatrix},$$
$$\mathbf{V}_{m+1} \;=\; \begin{bmatrix} v_1 & v_2 & \cdots & v_{m+1} \end{bmatrix},$$

$$H \;=\; \begin{bmatrix} h_{11} & h_{12} & \cdots & & h_{1m} \\ h_{21} & h_{22} & \cdots & & h_{2m} \\ 0 & h_{32} & \cdots & & h_{3m} \\ 0 & 0 & \ddots & & \vdots \\ 0 & 0 & 0 & & h_{m+1,m} \end{bmatrix},$$

$$h_{i,m} \;=\; v_i^T A v_m \text{ for } i \le m, \tag{9.5.5}$$
$$z \;=\; Av_m - v_1 h_{1,m} \cdots - v_m h_{m,m} \ne 0,$$
$$h_{m+1,m} \;=\; (z^T z)^{\frac{1}{2}} \text{ and} \tag{9.5.6}$$
$$v_{m+1} \;=\; z/h_{m+1,m}. \tag{9.5.7}$$

Here A is $n \times n$, $\mathbf{V}_m$ is $n \times m$ and H is an $(m+1) \times m$ *upper Hessenberg matrix* ($h_{ij} = 0$ when $i > j + 1$). This allows for the easy solution of the least squares problem (9.5.3).

Theorem 9.5.2 *(GMRES Reduction)   The solution of the least squares problem (9.5.3) is given by the solution of the least squares problem*

$$H\beta = e_1 b \tag{9.5.8}$$

where e_1 is the first unit vector, $b = ((r^0)^T r^0)^{\frac{1}{2}}$ and $AV_m = \mathbf{V}_{m+1}H$.

Proof.   Since $r^0 = b v_1$, $r^0 = \mathbf{V}_{m+1} e_1 b$. The least squares problem in (9.5.3) can be written in terms of the orthonormal basis

$$AV_m \beta = \mathbf{V}_{m+1} e_1 b.$$

Use the orthonormal property in the expression for

$$\hat{r}(\beta) = \mathbf{V}_{m+1} e_1 b - AV_m \beta = \mathbf{V}_{m+1} e_1 b - \mathbf{V}_{m+1} H \beta$$

$$
\begin{aligned}
(\hat{r}(\beta))^T \hat{r}(\beta) &= (\mathbf{V}_{m+1} e_1 b - \mathbf{V}_{m+1} H\beta)^T (\mathbf{V}_{m+1} e_1 b - \mathbf{V}_{m+1} H\beta) \\
&= (e_1 b - H\beta)^T \mathbf{V}_{m+1}^T \mathbf{V}_{m+1} (e_1 b - H\beta) \\
&= (e_1 b - H\beta)^T (e_1 b - H\beta).
\end{aligned}
$$

Thus the least squares solution of (9.5.8) will give the least squares solution of (9.5.3) where $\mathbf{K}_m \alpha = \mathbf{V}_m \beta$. ∎

If $z = Av_m - v_1 h_{1,m} \cdots - v_m h_{m,m} = 0$, then the next column vector v_{m+1} cannot be found and
$$AV_m = \mathbf{V}_m H(1:m.1:m).$$

Now $H = H(1:m.1:m)$ must have an inverse and $H\beta = e_1 b$ has a solution. This means
$$
\begin{aligned}
0 &= r^0 - A\mathbf{V}_m \beta \\
&= d - Ax^0 - A\mathbf{V}_m \beta \\
&= d - A(x^0 + \mathbf{V}_m \beta).
\end{aligned}
$$

If $z = Av_m - v_1 h_{1,m} \cdots - v_m h_{m,m} \neq 0$, then $h_{m+1,m} = (z^T z)^{\frac{1}{2}} \neq 0$ and $AV_m = \mathbf{V}_{m+1} H$. Now H is an upper Hessenberg matrix with nonzero components on the subdiagonal. This means H has full column rank so that the least squares problem in (9.5.8) can be solved by the QR factorization of $H = QR$. The normal equation for (9.5.8) gives
$$
\begin{aligned}
H^T H \beta &= H^T e_1 b \text{ and} \\
R\beta &= Q^T e_1 b. \qquad\qquad (9.5.9)
\end{aligned}
$$

The QR factorization of the Hessenberg matrix can easily be done by Givens rotations. An implementation of the GMRES method can be summarized by the following algorithm.

GMRES Method.

      let x^0 be an initial guess for the solution
      $r^0 = d - Ax^0$ and $V(:,1) = r^0/((r^0)^T r^0)^{\frac{1}{2}}$
      for k = 1, m
          $V(:,k+1) = AV(:,k)$
          compute columns $k+1$ of V_{k+1} and H in (9.5.4)-(9.5.7)
              (use modified Gram-Schmidt)
          compute the QR factorization of H
              (use Givens rotations)
          test for convergence
          solve (9.5.8) for β
          $x^{k+1} = x^0 + V_{k+1}\beta$
      endloop.

The following MATLAB code is for a two variable partial differential equation with both first and second order derivatives. The discrete problem is obtained by using centered differences and upwind differences for the first order derivatives. The sparse matrix implementation of GMRES is used along with the SSOR preconditioner, and this is a variation of the code in [11, chapter 3].

The code is initialized in lines 1-42, the GMRES loop is done in lines 43-87, and the output is generated in lines 88-98. The GMRES loop has the sparse matrix product in lines 47-49, SSOR preconditioning in lines 51-52, the modified Gram-Schmidt orthogonalization in lines 54-61, and Givens rotations are done in lines 63-83. Upon exiting the GMRES loop the upper triangular solve in (9.5.8) is done in line 89, and the approximate solution $x^0 + \mathbf{V}_{k+1}\beta$ is generated in the loop 91-93.

MATLAB Code pcgmres.m

```
1.      % This code solves the partial differential equation
2.      % -u_xx - u_yy + a1 u_x + a2 u_y + a3 u = f.
3.      % It uses gmres with the SSOR preconditioner.
4.      clear;
5.      % Input data.
6.      nx = 65;
7.      ny = nx;
8.      hh = 1./nx;
9.      errtol=.0001;
10.     kmax = 200;
11.     a1 = 1.;
12.     a2 = 10.;
13.     a3 = 1.;
14.     ac = 4.+a1*hh+a2*hh+a3*hh*hh;
15.     rac = 1./ac;
16.     aw = 1.+a1*hh;
17.     ae = 1.;
18.     as = 1.+a2*hh;
19.     an = 1.;
20.     % Initial guess.
21.     x0(1:nx+1,1:ny+1) = 0.0;
22.     x = x0;
23.     h = zeros(kmax);
24.     v = zeros(nx+1,ny+1,kmax);
25.     c = zeros(kmax+1,1);
26.     s = zeros(kmax+1,1);
27.     for j= 1:ny+1
28.         for i = 1:nx+1
29.             b(i,j) = hh*hh*200.*(1.+sin(pi*(i-1)*hh)*sin(pi*(j-1)*hh));
30.         end
31.     end
32.     rhat(1:nx+1,1:ny+1) = 0.;
33.     w = 1.60;
34.     r = b;
35.     errtol = errtol*sum(sum(b(2:nx,2:ny).*b(2:nx,2:ny)))^.5;
36.     % This preconditioner is SSOR.
37.     rhat = ssorpc(nx,ny,ae,aw,as,an,ac,rac,w,r,rhat);
38.     r(2:nx,2:ny) = rhat(2:nx,2:ny);
39.     rho = sum(sum(r(2:nx,2:ny).*r(2:nx,2:ny)))^.5;
```

```
40.     g = rho*eye(kmax+1,1);
41.     v(2:nx,2:ny,1) = r(2:nx,2:ny)/rho;
42.     k = 0;
43.     % Begin gmres loop.
44.     while((rho > errtol) & (k < kmax))
45.         k = k+1;
46.     % Matrix vector product.
47.         v(2:nx,2:ny,k+1) = -aw*v(1:nx-1,2:ny,k)-ae*v(3:nx+1,2:ny,k)-...
48.                         as*v(2:nx,1:ny-1,k)-an*v(2:nx,3:ny+1,k)+...
49.                         ac*v(2:nx,2:ny,k);
50.     % This preconditioner is SSOR.
51.         rhat = ssorpc(nx,ny,ae,aw,as,an,ac,rac,w,v(:,:,k+1),rhat);
52.         v(2:nx,2:ny,k+1) = rhat(2:nx,2:ny);
53.     % Begin modified GS. May need to reorthogonalize.
54.         for j=1:k
55.             h(j,k) = sum(sum(v(2:nx,2:ny,j).*v(2:nx,2:ny,k+1)));
56.             v(2:nx,2:ny,k+1) = v(2:nx,2:ny,k+1)-h(j,k)*v(2:nx,2:ny,j);
57.         end
58.         h(k+1,k) = sum(sum(v(2:nx,2:ny,k+1).*v(2:nx,2:ny,k+1)))^.5;
59.         if(h(k+1,k) ~= 0)
60.             v(2:nx,2:ny,k+1) = v(2:nx,2:ny,k+1)/h(k+1,k);
61.         end
62.     % Apply old Givens rotations to h(1:k,k).
63.         if k>1
64.             for i=1:k-1
65.                 hik       = c(i)*h(i,k)-s(i)*h(i+1,k);
66.                 hipk      = s(i)*h(i,k)+c(i)*h(i+1,k);
67.                 h(i,k)    = hik;
68.                 h(i+1,k)  = hipk;
69.             end
70.         end
71.         nu = norm(h(k:k+1,k));
72.     % May need better Givens implementation.
73.     % Define and Apply new Givens rotations to h(k:k+1,k).
74.         if nu~=0
75.             c(k)       = h(k,k)/nu;
76.             s(k)       = -h(k+1,k)/nu;
77.             h(k,k)     = c(k)*h(k,k)-s(k)*h(k+1,k);
78.             h(k+1,k)   = 0;
79.             gk         = c(k)*g(k) -s(k)*g(k+1);
80.             gkp        = s(k)*g(k) +c(k)*g(k+1);
81.             g(k)       = gk;
82.             g(k+1)     = gkp;
83.         end
84.         rho=abs(g(k+1));
```

```
85.                mag(k) = rho;
86.        end
87.        % End of gmres loop.
88.        % h(1:k,1:k) is upper triangular matrix in QR.
89.        y = h(1:k,1:k)\g(1:k);
90.        % Form linear combination.
91.        for i=1:k
92.                x(2:nx,2:ny) = x(2:nx,2:ny) + v(2:nx,2:ny,i)*y(i);
93.        end
94.        k
95.        semilogy(mag)
96.        x((nx+1)/2,(nx+1)/2)
97.        % mesh(x)
98.        % eig(h(1:k,1:k))
```

With the SSOR preconditioner convergence of the above code is attained in 25 iterations, and 127 iterations are required with no preconditioner. Larger numbers of iterations require more storage for the increasing number of basis vectors. One alternative is to restart the iteration and to use the last iterate as an initial guess for the restarted GMRES. This is examined in the next section.

9.5.1   Exercises

1.   Experiment with the parameters nx, errtol and w in the code pcgmres.m.
2.   Experiment with the parameters a1, a2 and a3 in the code pcgmres.m.
3.   Verify the calculations with and without the SSOR preconditioner. Compare the SSOR preconditioner with others such as block diagonal or ADI preconditioning.

9.6   GMRES(m) and MPI

In order to avoid storage of the basis vectors that are constructed in the GMRES method, after doing a number of iterates one can restart the GMRES iteration using the last GMRES iterate as the initial iterate of the new GMRES iteration.

GMRES(m) Method.

```
        let x⁰ be an initial guess for the solution
        for i = 1, imax
            for k = 1, m
                find xᵏ via GMRES
                test for convergence
            endloop
            x⁰ = xᵐ
        endloop.
```

The following is a partial listing of an MPI implementation of GMRES(m). It solves the same partial differential equation as in the previous section where the MATLAB code pcgmres.m used GMRES. Lines 1-66 are the initialization of the code. The outer loop of GMRES(m) is executed in the while loop in lines 66-256. The inner loop is expected in lines 135-230, and here the restart m is given by kmax. The new initial guess is defined in lines 112-114 where the new initial residual is computed. The GMRES implementation is similar to that used in the MATLAB code pcgmres.m. The additive Schwarz SSOR preconditioner is also used, but here it changes with the number of processors. Concurrent calculations used to do the matrix products, dot products and vector updates are similar to the MPI code cgssormpi.f.

MPI/Fortran Code gmresmmpi.f

```
1.     program gmres
2.!      This code approximates the solution of
3.!          -u_xx - u_yy + a1 u_x + a2 u_y + a3 u = f
4.!      GMRES(m) is used with a SSOR verson of the
5.!      Schwarz additive preconditioner.
6.!      The sparse matrix product, dot products and updates
7.!      are also done in parallel.
8.     implicit none
9.     include 'mpif.h'
10.    real, dimension(0:1025,0:1025,1:51):: v
11.    real, dimension(0:1025,0:1025):: r,b,x,rhat
12.    real, dimension(0:1025):: xx,yy
13.    real, dimension(1:51,1:51):: h
14.    real, dimension(1:51):: g,c,s,y,mag
15.    real:: errtol,rho,hik,hipk,nu,gk,gkp,w,t0,timef,tend
16.    real :: loc_rho,loc_ap,loc_error,temp
17.    real :: hh,a1,a2,a3,ac,ae,aw,an,as,rac
18.    integer :: nx,ny,n,kmax,k,i,j,mmax,m,sbn
19.    integer :: my_rank,proc,source,dest,tag,ierr,loc_n
20.    integer :: status(mpi_status_size),bn,en
Lines 21-56 initialize arrays and are not listed
57.    call mpi_init(ierr)
58.    call mpi_comm_rank(mpi_comm_world,my_rank,ierr)
59.    call mpi_comm_size(mpi_comm_world,proc,ierr)
60.    loc_n = (n-1)/proc
61.    bn = 1+(my_rank)*loc_n
62.    en = bn + loc_n -1
63.    call mpi_barrier(mpi_comm_world,ierr)
64.    if (my_rank.eq.0) then
65.        t0 = timef()
66.    end if
67.! Begin restart loop.
```

```
68.        do while ((rho>errtol).and.(m<mmax))
69.            m = m+1
70.            h = 0.0
71.            v= 0.0
72.            c= 0.0
73.            s= 0.0
74.            g = 0.0
75.            y = 0.0
76.! Matrix vector product for the initial residual.
77.! First, exchange information between processors.
Lines 78-111 are not listed.
112.          r(1:nx-1,bn:en) = b(1:nx-1,bn:en)+aw*x(0:nx-2,bn:en)+&
113.                     ae*x(2:nx,bn:en)+as*x(1:nx-1,bn-1:en-1)+&
114.                     an*x(1:nx-1,bn+1:en+1)-ac*x(1:nx-1,bn:en)
115.! This preconditioner changes with the number of processors!
Lines 116-126 are not listed.
127.          r(1:n-1,bn:en) = rhat(1:n-1,bn:en)
128.          loc_rho = (sum(r(1:nx-1,bn:en)*r(1:nx-1,bn:en)))
129.          call mpi_allreduce(loc_rho,rho,1,mpi_real,mpi_sum,&
130.                                  mpi_comm_world,ierr)
131.          rho = sqrt(rho)
132.          g(1) =rho
133.          v(1:nx-1,bn:en,1)=r(1:nx-1,bn:en)/rho
134.          k=0
135.! Begin gmres loop.
136.          do while((rho > errtol).and.(k < kmax))
137.              k=k+1
138.! Matrix vector product.
139.! First, exchange information between processors.
Lines 140-173 are not listed.
174.              v(1:nx-1,bn:en,k+1) = -aw*v(0:nx-2,bn:en,k)&
175.                  -ae*v(2:nx,bn:en,k)-as*v(1:nx-1,bn-1:en-1,k)&
176.                  -an*v(1:nx-1,bn+1:en+1,k)+ac*v(1:nx-1,bn:en,k)
177.! This preconditioner changes with the number of processors!
Lines 178-188 are not listed.
189.              v(1:n-1,bn:en,k+1) = rhat(1:n-1,bn:en)
190.! Begin modified GS. May need to reorthogonalize.
191.              do j=1,k
192.                  temp = sum(v(1:nx-1,bn:en,j)*v(1:nx-1,bn:en,k+1))
193.                  call mpi_allreduce(temp,h(j,k),1,mpi_real,&
194.                              mpi_sum,mpi_comm_world,ierr)
195.                  v(1:nx-1,bn:en,k+1) = v(1:nx-1,bn:en,k+1)-&
196.                              h(j,k)*v(1:nx-1,bn:en,j)
197.              end do
198.              temp = (sum(v(1:nx-1,bn:en,k+1)*v(1:nx-1,bn:en,k+1)))
```

```
199.            call mpi_allreduce(temp,h(k+1,k),1,mpi_real,&
200.                            mpi_sum,mpi_comm_world,ierr)
201.         h(k+1,k) = sqrt(h(k+1,k))
202.         if (h(k+1,k).gt.0.0.or.h(k+1,k).lt.0.0) then
203.             v(1:nx-1,bn:en,k+1)=v(1:nx-1,bn:en,k+1)/h(k+1,k)
204.         end if
205.         if (k>1) then
206.! Apply old Givens rotations to h(1:k,k).
207.            do i=1,k-1
208.                hik     = c(i)*h(i,k)-s(i)*h(i+1,k)
209.                hipk    = s(i)*h(i,k)+c(i)*h(i+1,k)
210.                h(i,k)    = hik
211.                h(i+1,k) = hipk
212.            end do
213.         end if
214.         nu = sqrt(h(k,k)**2 + h(k+1,k)**2)
215.! May need better Givens implementation.
216.! Define and Apply new Givens rotations to h(k:k+1,k).
217.         if (nu.gt.0.0) then
218.            c(k)      =h(k,k)/nu
219.            s(k)      =-h(k+1,k)/nu
220.            h(k,k)    =c(k)*h(k,k)-s(k)*h(k+1,k)
221.            h(k+1,k)  =0
222.            gk        =c(k)*g(k) -s(k)*g(k+1)
223.            gkp       =s(k)*g(k) +c(k)*g(k+1)
224.            g(k)      = gk
225.            g(k+1)    = gkp
226.         end if
227.         rho = abs(g(k+1))
228.         mag(k) = rho
229.! End of gmres loop.
230.         end do
231.! h(1:k,1:k) is upper triangular matrix in QR.
232.      y(k) = g(k)/h(k,k)
233.      do i = k-1,1,-1
234.         y(i) = g(i)
235.         do j = i+1,k
236.            y(i) = y(i) -h(i,j)*y(j)
237.         end do
238.         y(i) = y(i)/h(i,i)
239.      end do
240.! Form linear combination.
241.      do i = 1,k
242.         x(1:nx-1,bn:en) = x(1:nx-1,bn:en) + v(1:nx-1,bn:en,i)*y(i)
243.      end do
```

Table 9.6.1: MPI Times for gmresmmpi.f

p	time	iteration
2	358.7	10,9
4	141.6	9,8
8	096.6	10,42
16	052.3	10,41
32	049.0	12,16

```
244.! Send the local solutions to processor zero.
245.        if (my_rank.eq.0) then
246.            do source = 1,proc-1
247.                sbn = 1+(source)*loc_n
248.                call mpi_recv(x(0,sbn),(n+1)*loc_n,mpi_real,&
249.                            source,50,mpi_comm_world,status,ierr)
250.            end do
251.        else
252.            call mpi_send(x(0,bn),(n+1)*loc_n,mpi_real,0,50,&
253.                        mpi_comm_world,ierr)
254.        end if
255.    ! End restart loop.
256.    end do
257.    if (my_rank.eq.0) then
258.        tend = timef()
259.        print*, m, mag(k)
260.        print*, m,k,x(513,513)
261.        print*, 'time =', tend
262.    end if
263.    call mpi_finalize(ierr)
264.    end program
```

The Table 9.6.1 contains computations for $n = 1025$ using $w = 1.8$. The computation times are in seconds, and note the number of iterations changes with the number of processors. The restarts are after 50 inner iterations, and the iterations in the third column are (outer, inner) so that the total is outer * 50 + inner.

9.6.1 Exercises

1.   Examine the full code gmresmmpi.f and identify the concurrent computations. Also study the communications that are required to do the matrix-vector product, which are similar to those used in Section 6.6 and illustrated in Figures 6.6.1 and 6.6.2.

2.   Verify the computations in Table 9.6.1. Experiment with different number of iterations used before restarting GMRES.

3.   Experiment with variations on the SSOR preconditioner and include different n and ω.

4.   Experiment with variations of the SSOR preconditioner to include the use of a coarse mesh in the additive Schwarz preconditioner.

5.   Use an ADI preconditioner in place of the SSOR preconditioner.

Bibliography

[1] E. Anderson, Z. Bai, C. Bischof, J. Demmel, J. Dongarra, J. Du Croz, A. Greenbaum, S. Hammarling, A. McKenney, S. Ostrouchov and D. Sorensen, *LAPACK Users' Guide*, SIAM, 2nd ed., 1995.

[2] Edward Beltrami, *Mathematical Models for Society and Biology*, Academic Press, 2002.

[3] M. Bertero and P. Boccacci, *Introduction to Inverse Problems in Imaging*, IOP Publishing, Bristol, UK, 1998.

[4] Richard J. Burden and Douglas J. Faires, *Numerical Analysis*, Brooks Cole, 7th ed., 2000.

[5] Edmond Chow and Yousef Saad, *Approximate inverse techniques for block-partitioned matrices*, SIAM J. Sci. Comp., vol. 18, no. 6, pp. 1657-1675, Nov. 1997.

[6] Jack J. Dongarra, Iain S. Duff, Danny C. Sorensen and Henk A. van der Vorst, *Numerical Linear Algebra for High-Performance Computers*, SIAM, 1998.

[7] Loyd D. Fosdick, Elizabeth J. Jessup, Carolyn J. C. Schauble and Gitta Domik, *Introduction to High-Performance Scientific Computing*, MIT Press, 1996.

[8] William Gropp, Ewing Lusk, Anthony Skjellum and Rajeev Thahur, *Using MPI 2nd Edition: Portable Parallel Programming with Message Passing Interface*, MIT Press, 2nd ed., 1999.

[9] Marcus J. Grote and Thomas Huckle, *Parallel preconditioning with sparse approximate inverses*, SIAM J. Sci. Comp., vol. 18, no. 3, pp. 838-853, May 1997.

[10] Michael T. Heath, *Scientific Computing, Second Edition*, McGraw-Hill, 2001.

[11] C. T. Kelley, *Iterative Methods for Linear and Nonlinear Equations*, SIAM, 1995.

[12] David E. Keyes, Yousef Saad and Donald G. Truhlar (editors), *Domain-Based Parallelism and Problem Decomposition Methods in Computional Science and Engineering*, SIAM, 1995.

[13] Rubin H. Landau and M. J. Paez, *Computational Physics, Problem Solving with Computers*, John Wiley, 1997.

[14] The MathWorks Inc., http://www.mathworks.com.

[15] Nathan Mattor, Timothy J. Williams and Dennis W. Hewett, *Algorithm for solving tridiagonal matrix problems in parallel*, Parallel Computing, vol. 21, pp. 1769-1782, 1995.

[16] Carl D. Meyer, *Matrix Analysis and Applied Linear Algebra*, SIAM, 2000.

[17] NPACI (National Parternship for Advanced Computational Infrastructure), http://www.npaci.edu.

[18] NCSC (North Carolina Computing Center), *NCSC User Guide*, http://www.ncsc.org/usersupport/USERGUIDE/toc.html.

[19] Akira Okubo and Simon A. Levin, *Diffusion and Ecological Problems: Modern Perspectives*, 2nd ed., Springer-Verlag, 2001.

[20] James J. Ortega, *Introduction to Parallel and Vector Solution of Linear Systems*, Plenum Press, 1988.

[21] P. S. Pacheco, *Parallel Programming with MPI*, Morgan-Kaufmann, 1996.

[22] Shodor Education Foundation, Inc., http://www.shodor.org.

[23] G. D. Smith, *Numerical Solution of Partial Differential Equations*, Oxford, 3rd ed., 1985.

[24] J. Stoer and R. Bulirsch, *Introduction to Numerical Analysis*, Springer-Verlag, 1992.

[25] *UMAP Journal*, http://www.comap.com.

[26] C. R. Vogel, *Computation Methods for Inverse Problems*, SIAM, 2002.

[27] R. E. White, *An Introduction to the Finite Element Method with Applications to Nonlinear Problems*, Wiley, 1985.

[28] Paul Wilmott, Sam Howison and Jeff Dewynne, *The Mathematics of Financial Derivatives*, Cambridge, 1995.

[29] Joseph L. Zachary, *Introduction to Scientific Programming: Computational Problem Solving Using Maple and C,* Springer-Verlag, 1996.

Index